Undergraduate Texts in Mathematics

Editors

S. Axler

F. W. Gehring

K. A. Ribet

Springer
New York
Berlin
Heidelberg
Barcelona
Hong Kong
London
Milan
Paris
Singapore
Tokyo

Undergraduate Texts in Mathematics

(continued after index)

Richard S. Millman
George D. Parker

Geometry

A Metric Approach with Models

Second Edition

With 261 Figures

 Springer

Richard S. Millman
Vice President for Academic
    Affairs
California State University
San Marcos, CA 92096 USA

George D. Parker
Department of Mathematics
Southern Illinois University
Carbondale, IL 62901 USA

Mathematics Subject Classification (2000): 51-01, 51Mxx

Library of Congress Cataloging-in-Publication Data Available.
Printed on acid-free paper.

Printed and bound by R. R. Donnelley & Sons, Harrisonburg, Virginia.
Printed in the United States of America.

9 8 7 6 5 4

ISBN 0-387-97412-1
ISBN 3-540-97412-1   SPIN 10830910

Springer-Verlag   New York Berlin Heidelberg
A member of BertelsmannSpringer Science+Business Media GmbH

To Sherry and Margie
for their love and support

Preface

This book is intended as a first rigorous course in geometry. As the title indicates, we have adopted Birkhoff's metric approach (i.e., through use of real numbers) rather than Hilbert's synthetic approach to the subject. Throughout the text we illustrate the various axioms, definitions, and theorems with models ranging from the familiar Cartesian Plane to the Poincaré Upper Half Plane, the Taxicab Plane, and the Moulton Plane. We hope that through an intimate acquaintance with examples (and a model is just an example), the reader will obtain a real feeling and intuition for non-Euclidean (and in particular, hyperbolic) geometry. From a pedagogical viewpoint this approach has the advantage of reducing the reader's tendency to reason from a picture. In addition, our students have found the strange new world of the non-Euclidean geometries both interesting and exciting.

Our basic approach is to introduce and develop the various axioms slowly, and then, in a departure from other texts, illustrate major definitions and axioms with two or three models. This has the twin advantages of showing the richness of the concept being discussed and of enabling the reader to picture the idea more clearly. Furthermore, encountering models which do not satisfy the axiom being introduced or the hypothesis of the theorem being proved often sheds more light on the relevant concept than a myriad of cases which do.

The fundamentals of neutral (i.e., absolute) geometry are covered in the first six chapters. In addition to developing the general theory, these chapters include a rigorous demonstration of the existence of angle measures in our two major models, the Euclidean Plane and the Poincaré Plane. Chapter Seven begins the theory of parallels, which continues with an introduction to hyperbolic geometry in Chapter Eight and some classical Euclidean geometry in Chapter Nine. The existence of an area function in any neutral geometry is proved in Chapter Ten along with the beautiful cut and reassemble theory of Bolyai. The last (and most sophisticated) chapter studies the classification of isometries of a neutral geometry and computes the isometry groups for our two primary models.

The basic prerequisite for a course built on this book is mathematical maturity. Certain basic concepts from calculus are used in the development of some of the models. In particular, the intermediate value theorem as it is presented in calculus is needed at the end of Chapter Six. The latter part of the last chapter of the book requires an elementary course in group theory.

Courses of various lengths can be based on this book. The first six chapters (with the omission of Sections 5.2 and 5.4) would be ideal for a one quarter course. A semester course could consist of the first seven chapters, culminating in the All or None Theorem and the Euclidean/hyperbolic dichotomy. Alternatively, a Euclidean oriented course could include Section 7.1 and parts of Chapter 9. (The dependence of Chapter 9 on Chapter 8 is discussed at the beginning of Section 9.1.) A third alternative would include the first six chapters and the first three sections of Chapter 11. This gives the student a thorough background in classical geometry and adds the flavor of transformation geometry. A two quarter course allows a wider variety of topics from the later chapters, including area theory and Bolyai's Theorem in Chapter 10. The entire book can be covered in a year.

Mathematics is learned by doing, not by reading. Therefore, we have included more than 750 problems in the exercise sets. These range from routine applications of the definitions to challenging proofs. They may involve filling in the details of a proof, supplying proofs for major parts of the theory, developing areas of secondary interest, or calculations in a model. The reader should be aware that an asterisk on a problem does not indicate that it is difficult, but rather that its result will be needed later in the book. Most sections include a second set of problems which ask the reader to supply a proof or, if the statement is false, a counterexample. Part of the challenge in these latter problems is determining whether the stated result is true or not. The most difficult problems have also been included in these Part B problems.

In this second edition we have added a selection of expository exercises. There is renewed emphasis on writing in colleges and universities which extends throughout the four years of the undergraduate experience. Going under the name of "writing across the curriculum", this effort involves writing in all disciplines, not just in the traditional areas of the humanities and social sciences. Writing in a geometry course is discussed in more detail in Millman [1990]. We feel that the expository exercises add another dimension to the course and encourage the instructor to assign some of them both as writing exercises and as enrichment devices. We have found that a multiple draft format is very effective for writing assignments. In this approach, there is no finished product for grading until the student has handed in a number of drafts. Each version is examined carefully by the instructor and returned with copious notes for a rewrite. The final product should show that the student has learned quite a bit about a geometric topic and has improved his or her writing skills. The students get a chance to investigate either a topic of interest to them at the present or one that will be used later in their careers. This approach is especially useful and effective when many of the students are pre-service teachers.

A few words about the book's format are in order. We have adopted the standard triple numbering system (Theorem 7.4.9) for our results. Within each section one consecutive numbering system has been used for all theorems, lemmas, propositions, and examples for ease in locating references. The term *proposition* has been reserved for results regarding particular models. Reference citations are made in the form Birkhoff [1932] where the year refers to the date of publication as given in the bibliography.

We would like to thank our students at Southern Illinois University, Michigan Technological University, and Wright State University whose feedback over the past ten years has led to the changes and (we hope) the improvements we have made in this new edition. Our sincere thanks go to Sharon Champion and Shelley Castellano, who typed the original manuscript, and to Linda Macak, who typed the changes for this edition. Finally, we would like to thank our wives for putting up without us while we closeted ourselves, preparing this new edition.

Computers and Hyperbolic Geometry

After teaching a course out of the first edition of this book for several years, it became clear to the second author that there were all sorts of interesting computational problems in the Poincaré Plane that were a bit beyond the range of the average student. In addition, graphical aids could be very useful in developing intuition in hyperbolic geometry.

Out of this realization grew a computer program POINCARE. The program was written in Pascal over a three year period and runs on MS-DOS computers. It allows graphical explorations in the Poincaré Plane as well as various calculations such as finding the midpoint of a segment, finding an angle bisector, finding the common perpendicular of two lines (if it exists), carrying out geometric constructions, testing quadrilaterals for convexity and the Saccheri property, solving triangles, finding the cycle through three points, and finding the pencil determined by two lines.

All of the code is based on the theory presented in this book, with the exception of the hyperbolic trigonometry. Except for transformation geometry, all topics in the book are represented. Perhaps in the near future a module on isometries will be added to the program. POINCARE is currently in its third major version and is ready for distribution.

Readers of this book who are interested in more information, who wish to obtain a personal copy, or who wish to obtain it for use in their school computer lab should contact the author:

George D. Parker
1702 West Taylor
Carbondale, IL 62901

Contents

Contents xiii

CHAPTER 1
Preliminary Notions

1.1 Axioms and Models

Our study of geometry begins with two basic concepts. One is the notion of points, and the other is the notion of lines. These are then related to each other by a collection of *axioms*, or first principles. For example, when we discuss incidence geometry below, we shall assume as a first principle that if A and B are distinct points then there is a unique line that contains both A and B.

In the early development of geometry the point of view was that an axiom was a statement that described the true state of the universe. Axioms were thought of as "basic truths." Such axioms should be "self-evident." Of the basic axioms stated by Euclid in his *Elements*, all but one was accepted by the mathematical community as "true" and self-evident. However, his fifth axiom, which dealt with parallel lines, was not as well received. While everyone agreed it was true (whatever that meant) it was by no means obvious. For over two thousand years mathematicians tried to show that the fifth axiom was a theorem which could be proved on the basis of the remaining axioms. As we shall see, such efforts were doomed to fail. With great foresight Euclid chose an axiom whose value was justified not only by its intuitive content but also by the rich theory it implied.

The modern view is that an axiom is a statement of a useful property. When we assume an axiom holds, usually in a definition, we are saying that we want to discuss only those objects which possess this special property. We are making no statement as to whether the axiom is a statement about the real world. Rather, we are saying "accept the following as a hypothesis." (For a nice discussion of the modern axiomatic method see Kennedy [1972].)

Although we may use any consistent axioms we wish, the choice of axioms is really guided by three underlying principles. First, the axioms must be "reasonable" or "appealing" because they correspond to some intuitive picture which we have of some geometric property. Second, the axioms should be useful and lead to a rich variety of theorems and hence a rich mathematical structure. Third, the axioms must be consistent—there must not be any internal inconsistency or contradiction. As we shall see, Euclid's choice of axioms (or rather the modern version which we shall present), does satisfy these conditions and lead to a rich subject, Euclidean geometry. BUT, there are other different choices of axioms that also lead to rich theories which are significantly different from the Euclidean ones. In particular there is an alternative to Euclid's fifth axiom which develops into a particularly beautiful and interesting subject, hyperbolic geometry. One of the goals of this book is to investigate this alternative structure.

The moment we mentioned points and lines above you probably started to visualize a picture, namely the plane with straight lines from high school geometry. This is proper and also very useful in helping you understand all the definitions that will follow. However, it does have its drawbacks. It is very tempting to try to prove a theorem or proposition or to do a homework problem by looking at a picture. If this is done, the picture may be confused with the geometry itself. A picture may be misleading, either by not covering all possibilities, or, even worse, by reflecting our unconscious bias as to what is "correct." This often leads to an incorrect "proof by picture." *It is crucially important in a proof to use only the axioms and the theorems which have been derived from them, and not depend on any preconceived idea or picture.* (Of course, we may use a picture as an aid to our intuition. The point here is that when the "final" proof of the result is written it cannot depend on the picture.)

The discussion of pictures leads us to the idea of a *model* for a geometry. A model is nothing more than an example. Each model of a geometry is determined by giving a set whose elements will be called "points" and a collection of subsets of this set which will be called "lines." For instance, if we are given the definition of an incidence geometry, we may write down as an example the standard Euclidean geometry we met in high school. We then must check to see if this example satisfies all the axioms that are listed in the definition of an incidence geometry. When we are done we will have one example. But there are many other examples of an incidence geometry and hence many models, as we shall see.

Throughout this book there will be several models, but we will concentrate on two particular ones: the Euclidean Plane and the Poincaré Plane. This will mean that we will have at our fingertips two strikingly different examples. Two main purposes will be served. The first is to insure that we do not reason by pictures (pictures in the Poincaré Plane are very different from those in the familiar Euclidean Plane). The second purpose is to give valuable insight by allowing us to work with several examples while attacking some of the problems in the text. We can also benefit from different examples by examining a newly defined concept in light of the definition. For example, a circle in the "taxicab metric" of Chapter 2 is an amusing phenomenon. This examination will add to our understanding as to what the definition is saying (and what is is not saying).

At this point we cannot emphasize enough that the *only* proof that can be given of a theorem or proposition in a geometry is one based just on the axioms of that geometry. We must *not* go to a model and show that the theorem holds there. All we would have shown in that case is that the theorem is true *in that particular model*. It might be false in another model (and hence false in general). There are many statements that are true in the Euclidean Plane but false in the Poincaré Plane, and vice versa.

Whenever we introduce a new axiom system or add axioms to an old system, we are changing the requirements for the system. For instance, consider the statement that "a *slurb* is a set which has only letters in it." The set $S = \{X, Y, z\}$ satisfies this statement and so is an example or model of a slurb. If we further define that "a *big slurb* is a slurb that contains only capital letters" then the set S may or may not be a model for the new (enlarged) axiom system. (Of course, S is not a model for this system; that is, S is not a big slurb.) This means that if we continue to use certain models we must prove that they obey the new axioms that are cited. Do not confuse this with proving an axiom. Axioms are statements of desirable properties to be studied and cannot be proved. Verifying that something is a model of a certain axiom system is very much like making a general statement and then showing that it is applicable in the particular instance you want.

What then is a geometry? Formally, a geometry consists of two sets—a set of points and a set of lines—together with a collection of relationships, called axioms, between those two sets. On the other hand, a model for the geometry is just an example. That is, a model for a geometry is a mathematical entity which satisfies all of the axioms for the geometry. It is important not to confuse a model with the geometry of which it is a model.

1.2 Sets and Equivalence Relations

Intuitively a set, S, is a collection of objects which are called elements. It must be described by a very specific rule which lets us determine if any particular object belongs to the collection. (The collection of tall people is not a set because the terminology is not precise—it is subjective. The collection of all living people at least 2 meters tall is a set—the characteristic for tall is made precise.) We write $a \in S$ to mean that the object a belongs to S, and read this as "a is an element of S." Similarly we write $a \notin S$ to mean that a does not belong to the set S.

Definition. The set T is a **subset** of the set S (written $T \subset S$) if every element of T is also an element of S.

The set T **equals** the set S (written $T = S$) if every element of T is in S, and every element of S is in T. (Hence $T = S$ if and only if $T \subset S$ and $S \subset T$.)

The **empty set** is the set with no members, and is denoted $\varnothing$. Note $\varnothing \subset S$ for every set S.

As usual, the notation $T = \{x \in S \mid \cdots\}$ means that the elements of T are precisely those elements of S which satisfy the property listed after the bar, $\mid$.

Definition. The **union** of two sets A and B is the set $A \cup B = \{x \mid x \in A$ or $x \in B\}$.

The **intersection** of two sets A and B is the set $A \cap B = \{x \mid x \in A$ and $x \in B\}$. If $A \cap B = \varnothing$ then A and B are **disjoint**.

The **difference** of two sets A and B is the set $A - B = \{x \mid x \in A$ and $x \notin B\}$.

The following example illustrates several of the above ideas. Note in particular the basic way we show two sets S and T are equal: we show that $S \subset T$ and that $T \subset S$.

Example 1.2.1. Show that $A \cap (B \cup C) = (A \cap B) \cup (A \cap C)$.

SOLUTION. We first show that $A \cap (B \cup C) \subset (A \cap B) \cup (A \cap C)$. Let $x \in A \cap (B \cup C)$. Then $x \in A$ and $x \in B \cup C$. Since $x \in B \cup C$ either $x \in B$ or $x \in C$ (or both!). If $x \in B$ then $x \in A \cap B$. If $x \in C$ then $x \in A \cap C$. Either way $x \in (A \cap B) \cup (A \cap C)$. Thus $A \cap (B \cup C) \subset (A \cap B) \cup (A \cap C)$.

Next we show that $(A \cap B) \cup (A \cap C) \subset A \cap (B \cup C)$. Let $x \in (A \cap B) \cup (A \cap C)$. If $x \in A \cap B$ then $x \in A$ and $x \in B$. Hence $x \in B \cup C$ and $x \in A \cap (B \cup C)$. Similarly, if $x \in A \cap C$ then $x \in A$ and $x \in C$. Hence, $x \in B \cup C$ and $x \in A \cap (B \cup C)$. In either case, $x \in A \cap (B \cup C)$. Thus $(A \cap B) \cup (A \cap C) \subset A \cap (B \cup C)$.

Since $A \cap (B \cup C) \subset (A \cap B) \cup (A \cap C)$ and $(A \cap B) \cup (A \cap C) \subset A \cap (B \cup C)$, we have $A \cap (B \cup C) = (A \cap B) \cup (A \cap C)$. $\qquad\square$

Definition. Let A and B be sets. An **ordered pair** is a symbol (a, b) where $a \in A$ and $b \in B$. Two ordered pairs (a, b) and (c, d) are **equal** if $a = c$ and $b = d$. The **Cartesian product** of A and B is the set

$$A \times B = \{(a, b) \mid a \in A \text{ and } b \in B\}.$$

Because of the use of the word "symbol", the definition above is somewhat informal. The basic idea that the entries are "ordered" comes from the definition of equality: (a, b) and (b, a) are not equal unless $a = b$. Thus, changing the order of the entries leads to a different object. It is possible to give a purely set-theoretic definition of (a, b). This is done in problem B16, where the reader is asked to prove that $(a, b) = (c, d)$ if and only if $a = c$ and $b = d$ using the formal definition given in that problem.

Note that the notation $\mathbb{R}^2$ to denote the set of ordered pairs of real numbers is an adaptation of exponential notation to represent $\mathbb{R} \times \mathbb{R}$.

As a first use of the concept of ordered pairs we present a way to say that two elements in a set are related in some particular way. A motivating example is given by the idea "less than". The graph of the inequality $x < y$ in $\mathbb{R}^2$ consists of all ordered pairs $(a, b) \in \mathbb{R}^2$ such that $a < b$. (See Figure 1-1.) When we say that $2 < 3$, we are saying that $(2, 3)$ is part of the graph. Conversely, since $(-3, 2)$ is part of the graph, $-3 < 2$. Thus the graph carries all the information of the "less than" relation. A binary relation is a generalization of "less than" that is described in terms of a graph.

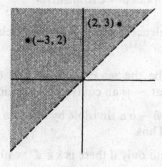

Figure 1-1

Definition. A **binary relation**, R, on a set S is a subset of $S \times S$. If $(s, t) \in R$ then we say that s is **related** to t.

Example 1.2.2. Each of the following is a binary relation on the set $\mathbb{R}$ of real numbers.

$$A = \{(s, t) \in \mathbb{R}^2 \mid s = t + 2\}.$$
$$B = \{(s, t) \in \mathbb{R}^2 \mid st \text{ is an integer}\}.$$
$$C = \{(s, t) \in \mathbb{R}^2 \mid s \leq t\}.$$
$$D = \{(s, t) \in \mathbb{R}^2 \mid s^2 + t^2 = 1\}$$

$\square$

We frequently name relations using symbols such as $\leq$ (for relation C above), $\equiv$, $\simeq$, $\parallel$, or $\sim$ instead of letters. We then indicate that two elements are related by placing the name of the relation between the elements: $(3, 5) \in C$ becomes $(3, 5) \in \leq$, which becomes $3 \leq 5$. Thus we may make statements about "the relation $\sim$" and write statements such as "$a \sim b$". Note that if two elements a, b are not related by the relation $\sim$ we write $a \nsim b$.

Because the idea of a relation depends on ordered pairs, the order that we write the symbols is important: $2 < 5$ but $5 \nless 2$. For some special relations, like those below, the order is not important—the relation is symmetric. Note that if $\sim$ is a relation on S and $a \in S$, then it is possible that there is no element b with $a \sim b$. For example, if S is the set of positive integers, and if the relation is $>$ (greater than) then there is no $b \in S$ with $1 > b$. In this case 1 is not related to anything.

Definition. A binary relation, $\sim$, on S is an **equivalence relation** if for every a, b, and $c \in S$

(i) $a \sim a$ (reflexive)
(ii) $a \sim b$ implies $b \sim a$ (symmetric)
(iii) $a \sim b$ and $b \sim c$ implies $a \sim c$ (transitive).

Note that an equivalence relation is a binary relation that satisfies three axioms.

Example 1.2.3. Let $\mathbb{Z}$ be the set of integers and define $a \sim b$ if $a - b$ is divisible by 2. Show that $\sim$ is an equivalence relation.

SOLUTION. To say that $a - b$ is divisible by 2 means that there is an integer k such that $a - b = 2k$. Thus

$$a \sim b \quad \text{if and only if there is } k \in \mathbb{Z} \quad \text{with } a - b = 2k.$$

(i) Let $a \in \mathbb{Z}$. Then $a - a = 0 = 2 \cdot 0$ so that $a \sim a$ and $\sim$ is reflexive.
(ii) Suppose that $a, b \in \mathbb{Z}$ and $a \sim b$. Then there is a $k \in \mathbb{Z}$ with $a - b = 2k$. This means that $b - a = 2(-k)$. Since $-k \in \mathbb{Z}$, we have $b \sim a$. Thus $\sim$ is symmetric.
(iii) If $a \sim b$ and $b \sim c$ then there are numbers $k_1 \in \mathbb{Z}$ and $k_2 \in \mathbb{Z}$ with

$$a - b = 2k_1 \quad \text{and} \quad b - c = 2k_2.$$

Adding these equations we obtain $a - c = 2(k_1 + k_2)$ and so $a \sim c$. Thus $\sim$ is transitive. $\sim$ is therefore an equivalence relation. $\square$

Definition. If a and b are integers then a is **equivalent to b modulo n** if $a - b = kn$ for some integer k. This is written $a \equiv b(n)$ and means that $a - b$ is divisible by n.

The above example shows that $\equiv(2)$ is an equivalence relation. In Problem A7 you will show that $\equiv(n)$ is an equivalence relation for any n.

Example 1.2.4. Show that none of the binary relations in Example 1.2.2. is an equivalence relation.

SOLUTION. A is not reflexive. It is certainly not true that $a = a + 2$ for all a. (Neither is it symmetric or transitive.)

B is not reflexive or transitive.
C is not symmetric.
D is not reflexive or transitive.   □

In an equivalence relation we view several elements of S as being alike (or equivalent) if they have similar properties. In Example 1.2.3 all the odd numbers are related to each other and thus are equivalent. It is convenient to have a name for the set of all elements related (or equivalent) to a given element.

Definition. If $\sim$ is an equivalence relation on the set S and $s \in S$, then the **equivalence class** of s is the subset of S given by

$$[s] = \{x \in S | x \sim s\} = \{x \in S | s \sim x\}.$$

Example 1.2.5. In Example 1.2.3 the equivalence class of 3 is the set of odd integers, and the equivalence class of 2 is the set of even integers. Note in this case that if $x, y \in \mathbb{Z}$ then either $[x] = [y]$ or $[x] \cap [y] = \varnothing$.   □

Example 1.2.6. Let $S = \{1, 2, \ldots, 100\}$ and define $x \sim y$ if x and y have the same number of digits in their base 10 representation. Then $\sim$ is an equivalence relation and, for example,

$$[5] = \{1, 2, 3, 4, 5, 6, 7, 8, 9\} = [7] = [9] = [8], \quad \text{etc.}$$
$$[11] = \{s \in S | 10 \le s \le 99\} = [63] = [43], \quad \text{etc.}$$
$$[100] = \{100\}.$$

Note that, although different equivalence classes may have a different number of elements, we still have the result that two equivalence classes are either equal or disjoint. This is true in general as we now see.   □

Theorem 1.2.7. *If $\sim$ is an equivalence relation on S and if $s, t \in S$ then either*

$$[s] \cap [t] = \varnothing \quad \text{or} \quad [s] = [t].$$

PROOF. We will show that if the first case is not true (i.e., $[s] \cap [t] \neq \varnothing$) then the second holds. This is the standard way we show either... or... results.

Assume that $[s] \cap [t] \neq \varnothing$. Then there is an $x \in [s] \cap [t]$. Hence $x \in [s]$ and $x \in [t]$. Thus $x \sim s$ and $x \sim t$. By symmetry $s \sim x$, and then by transitivity, $s \sim x$ and $x \sim t$ imply that $s \sim t$. We use this to show $[s] \subset [t]$.

Let $y \in [s]$. Then $y \sim s$ and, since $s \sim t$, we also have $y \sim t$ by transitivity. Thus $y \in [t]$. Hence $[s] \subset [t]$. Similarly, since $t \sim s$, we can show $[t] \subset [s]$. Hence $[s] = [t]$.   □

PROBLEM SET 1.2   Throughout this problem set, A, B, and C are sets.

Part A.

1. If $A \subset B$ prove that $A \cap C \subset B \cap C$.

2. Prove that $C \cap (B—A) = (C \cap B)—A$.

3. Prove that $A \cup (B \cap C) = (A \cup B) \cap (A \cup C)$.

4. Suppose that $A \subset C$ and $B \subset C$. Show that $A \cap B = \varnothing$ implies that $B \subset C—A$.

5. If $B \subset C—A$ show that $A \cap B = \varnothing$.

6. a. If $x \sim y$ means that $x - y$ is divisible by 3, show that $\sim$ is an equivalence relation on the set of integers.
   b. What is $[3]$? $[6]$? $[9]$? $[1]$? $[5]$?

7. Show that $\equiv (n)$ is an equivalence relation on the integers for any n. What are the equivalence classes?

8. Let $\mathbb{R}^2 = \{P = (x, y) \mid x$ and y are real numbers$\}$. We say that $P_1 = (x_1, y_1)$ and $P_2 = (x_2, y_2)$ are equivalent if $x_1^2 + y_1^2 = x_2^2 + y_2^2$. Prove that this gives an equivalence relation on $\mathbb{R}^2$. What is $[(1,0)]$? $[(0,1)]$? $[(2,2)]$? $[(0,0)]$? What does an equivalence class "look like?"

9. The height, h, of a rectangle is by definition the length of the longer of the sides. The width, w, is the length of the shorter of the sides (thus $h \geq w > 0$). If the rectangle R_1 has height h_1 and width w_1 and the rectangle R_2 has height h_2 and width w_2, we say that $R_1 \sim R_2$ if $h_1/w_1 = h_2/w_2$. Prove that this defines an equivalence relation on the set of all rectangles.

10. Let the triangle T_1 have height h_1 and base b_1 and the triangle T_2 have height h_2 and base b_2. We say that $T_1 \sim T_2$ if $b_1 h_1 = b_2 h_2$. Show that this is an equivalence relation on the set of all triangles. Show that $T_1 \in [T]$ if and only if T_1 has the same area as T.

Part B.   "Prove" may mean "find a counterexample".

11. Prove that $(A \cup B)—C = (A—C) \cup (B—C)$.

12. Prove that $(A—C) \cap (B—C) = (A \cap B)—C$.

13. Let S be the set of all real numbers. Let $s_1 \sim s_2$ if $s_1^2 = s_2^2$. Prove that $\sim$ is an equivalence relation on S. What are the equivalence classes?

14. Let X be the set of all people. We say that $p_1 \sim p_2$ if p_1 and p_2 have the same father. Prove that this is an equivalence relation. What are the equivalence classes?

15. Let X be the set of all people. We say that $p_1 \sim p_2$ if p_1 lives within 100 kilometers of p_2. Show that $\sim$ is an equivalence relation. What are the equivalence classes?

16. A careful definition of ordered pair would be the following: If $a \in A$ and $b \in B$, then the **ordered pair** (a, b) is the set $(a, b) = \{\{a, 1\}, \{b, 2\}\}$. Use this definition to

prove that $(a, b) = (c, d)$ if and only if $a = c$ and $b = d$. (Be careful: $\{\{a, 1\}, \{b, 2\}\} = \{\{c, 1\}, \{d, 2\}\}$ does not immediately imply that $\{a, 1\} = \{c, 1\}$.)

17. Give a careful definition of an ordered triple (a, b, c) and then prove that $(a, b, c) = (d, e, f)$ if and only if $a = d$, $b = e$, and $c = f$.

18. Let $\mathbb{R}^3 = \{\vec{v} = (x, y, z) | x, y, \text{ and } z \in \mathbb{R}\}$. Let $\vec{v}, \vec{w} \in \mathbb{R}^3 - \{\vec{0}\}$. We say that $\vec{v} \sim \vec{w}$ if there is a non-zero real number, λ, with $\vec{v} = \lambda\vec{w}$. Show that $\sim$ is an equivalence relation on $\mathbb{R}^3 - \{\vec{0}\}$. What are the equivalence classes? The set of all equivalence classes will be called the (real) **Projective Plane.**

19. Let $X = \{(x, y) | x \in \mathbb{Z} \text{ and } y \in \mathbb{Z} \text{ and } y \neq 0\}$. We define a binary relation on X by

$$(x_1, y_1) \sim (x_2, y_2) \quad \text{if and only if } x_1 y_2 = x_2 y_1.$$

Prove that $\sim$ is an equivalence relation. What are the equivalence classes?

20. Let $\sim$ be a relation on S that is both symmetric and transitive. What is wrong with the following "proof" that $\sim$ is also reflexive? "Suppose $a \sim b$. Then by symmetry $b \sim a$. Finally by transitivity $a \sim b$ and $b \sim a$ imply $a \sim a$. Thus $\sim$ is reflexive."

Part C. Expository exercises.

21. How would you explain to a high school audience the notion of an equivalence relation? In that context, what would an equivalence mean? Where would it occur in their daily lives?

22. Think of a binary relation as a graph. What is the geometric (graphical) meaning of the three axioms for an equivalence relation?

1.3 Functions

In this section we review the standard material about functions and bijections. The latter notion is an important part of the Ruler Postulate which will appear in the next chapter. We will continue to use $\mathbb{R}$ to denote the set of real numbers and $\mathbb{Z}$ to denote the set of integers.

Definition. If S and T are sets, then a **function** $f : S \to T$ is a subset $f \subset S \times T$ such that for each $s \in S$ there is exactly one $t \in T$ with $(s, t) \in f$. This unique element t is usually denoted $f(s)$. S is called the **domain** of f and T is called the **range** of f.

In a very intuitive manner we may view a function as an archer who takes arrows (elements) from her quiver S and shoots them at a target T. In this analogy we say that the element $s \in S$ "hits" the element $f(s) \in T$.

Following standard conventions, we frequently describe a function f by giving a formula (or rule) for computing $f(s)$ from s. Note that the function consists of this rule *together with the two sets S and T.*

Example 1.3.1. Let $f:\mathbb{R} \to \mathbb{R}$ by the rule $f(x) = x^2$. Let $g:\mathbb{Z} \to \mathbb{R}$ by the rule $g(x) = x^2$. Note that f is not equal to g—they have different domains. Now let $\mathbb{R}^+ = \{x \in \mathbb{R} \mid x \geq 0\}$ and let $h:\mathbb{R} \to \mathbb{R}^+$ by the rule $h(x) = x^2$. Note that f and h are not equal—they have different ranges.   □

Definition. If $f:S \to T$ is a function then the **image** of f is

$$\text{Im}(f) = \{t \in T \mid t = f(s) \text{ for some } s \in S\}.$$

Thus, $\text{Im}(f)$ consists of the elements of T that are actually "hit" by f. Of course, $\text{Im}(f) \subset \text{Range}(f)$, but these sets need not be equal. In Example 1.3.1, $\text{Im}(f) = \mathbb{R}^+$ but $\text{Range}(f) = \mathbb{R}$. (Some mathematicians use the word "range" to mean "image". They then use "codomain" to mean what we call the "range".)

Definition. $f:S \to T$ is **surjective** if for every $t \in T$ there is an $s \in S$ with $f(s) = t$.

·In keeping with the target analogy above, this means that all elements of the target T are "hit." Of course, an element may be hit more than once. That is, there may be several $s \in S$ such that $f(s) = t$. It is common usage to say that a function is "onto" instead of "surjective". In this text we shall use the more correct terminology "surjective".

Example 1.3.2. Show that $f:\mathbb{R} \to \mathbb{R}$ by $f(x) = x^3 - 1$ is surjective while $g:\mathbb{R} \to \mathbb{R}$ by $g(x) = x^2 - 1$ is not surjective.

SOLUTION. To show that f is surjective we must show that for every $t \in \text{Range}(f) = \mathbb{R}$ there is an $s \in \text{Domain}(f)$ with $f(s) = t$. That is, we must show that the equation

$$s^3 - 1 = t \tag{3-1}$$

has a solution for every value of t. Since every real number has a cube root, we may set $s = \sqrt[3]{t + 1}$. Then

$$f(s) = (\sqrt[3]{t + 1})^3 - 1 = t + 1 - 1 = t.$$

Hence f is surjective.

To show that g is not surjective we need only produce one value of t such that the equation

$$s^2 - 1 = t \tag{3-2}$$

does not have a solution. Let $t = -2$. Then a solution to Equation (3-2) must satisfy

$$s^2 - 1 = -2 \quad \text{or} \quad s^2 = -1.$$

This clearly cannot occur for any real number s. Hence g is not surjective.   □

Example 1.3.2 illustrates how we actually attempt to prove that a function is surjective. We set up the equation $f(s) = t$ and try to solve for s (given t). At least one solution for every value of t must be found.

Example 1.3.3. Let $T = \{t \in \mathbb{R} | t \geq e^{-1/4}\}$ and $h: \mathbb{R} \to T$ be given by $h(s) = e^{s^2 - s}$. Prove that h is surjective.

SOLUTION. Given $t \in T$, note that $\ln t \geq -\frac{1}{4}$. Thus $1 + 4 \ln t \geq 0$. Let s be the "obvious choice":

$$s = \frac{1 + \sqrt{1 + 4 \ln t}}{2}. \tag{3-3}$$

It is easy enough to show that Equation (3-3) defines an $s \in \mathbb{R}$ with $h(s) = t$. $\square$

The above represents an absolutely correct (and totally unmotivated) solution to the problem. The way that we came up with the "obvious choice" (and the way to attack the problem) is to attempt to solve the equation $h(s) = t$. Thus on scrap paper you might write: For fixed t, solve for s:

$$e^{s^2 - s} = t.$$

First take the natural logarithm of both sides to get

$$s^2 - s = \ln t \quad \text{or} \quad s^2 - s - \ln t = 0.$$

An application of the quadratic formula gives

$$s = \frac{1 \pm \sqrt{1 + 4 \ln t}}{2}.$$

Because we only need to exhibit one solution, we take the "+" sign: $s = \frac{1}{2}(1 + \sqrt{1 + 4 \ln t})$. Of course, taking $s = \frac{1}{2}(1 - \sqrt{1 + 4 \ln t})$ would also show that f is surjective. Once we have found what we believe is the solution, we must verify that it is correct as in Example 1.3.3. This is similar to solving an equation in algebra and then checking the solution to be sure that we have not found an extraneous solution. Note that it was crucial that we verified that $1 + 4 \ln t \geq 0$. If the range had included numbers $t < e^{-1/4}$ we would not always have $1 + 4 \ln t \geq 0$ and the function would not have been surjective.

The concept of a surjective function deals with whether or not an equation can be solved. Another important idea is the notion of an injective function, which deals with the number of solutions to an equation. Thus "surjective" deals with existence of a solution while "injective" deals with uniqueness.

Definition. $f : S \to T$ is **injective** if $f(s_1) = f(s_2)$ implies $s_1 = s_2$.

In terms of the target analogy, $f : S \to T$ is injective if no two arrows hit the same place on the target. (Note that this says nothing about whether all of the target is hit.) It is common practice to use the term "one-to-one" to mean injective. In keeping with our practice regarding the word "surjective" we shall only use the term "injective" and not "one-to-one."

An alternative way of defining injective would be: f is injective if $s_1 \neq s_2$ implies $f(s_1) \neq f(s_2)$. While this is not as convenient to use, there are times (e.g., Theorem 1.3.8) when it is helpful.

Example 1.3.4. Let $f : \mathbb{R} \to \mathbb{R}$ be given by $f(s) = e^{s^3 + 1}$. Show that f is injective.

SOLUTION. We assume that there are real numbers s_1 and s_2 such that

$$e^{s_1^3 + 1} = e^{s_2^3 + 1}. \tag{3-4}$$

We must show that the only way that this occurs is if $s_1 = s_2$. If we take the natural logarithm of both sides of Equation (3-4) we obtain

$$s_1^3 + 1 = s_2^3 + 1 \quad \text{or} \quad s_1^3 = s_2^3.$$

Since every real number has a *unique* cube root we must have $s_1 = s_2$. Thus f is injective.   □

Example 1.3.5. Show that $h : \mathbb{R}^+ \to \mathbb{R}^+$ by $h(x) = x^2$ is injective.

SOLUTION. Assume that $h(s_1) = h(s_2)$ so that $s_1^2 = s_2^2$. Then by taking square roots we have $s_1 = \pm s_2$. However, since the elements of $\mathbb{R}^+$ are not negative, both s_1 and s_2 must be greater than or equal to zero. Hence $s_1 \neq -s_2$ (unless both are 0) and so $s_1 = s_2$. Thus h is injective.   □

The words "injective" and "surjective" are adjectives. If we wish to have a noun it is common to say "injection" for "injective function" and "surjection" for "surjective function." Note that in Example 1.3.4 the function was an injection but not a surjection, whereas in Example 1.3.5 the function was both a surjection and an injection. The function h of Example 1.3.3 gives an example of a function which is a surjection but not an injection (since $h(0) = h(1) = 1$). There are many examples of functions which are neither injective or surjective. However, a function which is both has a special name.

Definition. $f : S \to T$ is a **bijection** if f is both an injection and a surjection.

Example 1.3.5 is a bijection. The term "one-to-one correspondence" is also common, but we shall not use it here because readers have a tendency to confuse the term with the idea of "one-to-one."

Recall the definition of the composition of functions.

Definition. If $f : S \to T$, $g : U \to V$, and $\text{Im}(f) \subset U$, then the **composition** of f and g is the function $g \circ f : S \to V$ given by $(g \circ f)(s) = g(f(s))$.

Notice that the domain of g must contain the image of f in order for the composition of f and g to be defined.

Example 1.3.6. If $f: \mathbb{R} \to \mathbb{R} - \{0\}$ is given by $f(s) = 2 + \sin(s)$ and $g: \mathbb{R} - \{0\} \to \mathbb{R}$ is given by $g(t) = 1/t$ find $g \circ f$.

SOLUTION. $g \circ f: \mathbb{R} \to \mathbb{R}$ is given by

$$(g \circ f)(s) = g(f(s)) = \frac{1}{f(s)}$$

$$= \frac{1}{2 + \sin(s)}. \qquad \square$$

Theorem 1.3.7. If $f: S \to T$ and $g: T \to V$ are both surjections then $g \circ f$ is also a surjection.

PROOF. Let $v \in V$. We must show that there is an $s \in S$ such that $(g \circ f)(s) = v$. Since g is surjective, there is a $t \in T$ with $g(t) = v$. Then since f is surjective there is an $s \in S$ with $f(s) = t$. Now

$$(g \circ f)(s) = g(f(s)) = g(t) = v$$

so that $g \circ f$ is a surjection. $\qquad \square$

The next two results are left as Problems A6 and A7.

Theorem 1.3.8. If $f: S \to T$ and $g: T \to V$ are both injections then $g \circ f: S \to V$ is an injection.

Theorem 1.3.9. If $f: S \to T$ and $g: T \to V$ are both bijections then $g \circ f: S \to V$ is also a bijection.

If $f: S \to T$ is a bijection then for each $t \in T$ there is a unique $s \in S$ with $f(s) = t$. This allows us to assign to each $t \in T$ a corresponding element $s \in S$. Thus we have manufactured a new function (called the inverse) which goes backwards. More formally we have

Definition. If $f: S \to T$ is a bijection, then the **inverse** of f is the function $g: T \to S$ which is defined by

$$g(t) = s, \qquad \text{where } s \text{ is the unique element of } S \text{ with } f(s) = t \qquad (3\text{-}5)$$

The function g is frequently denoted f^{-1}.

If f is the natural logarithm function given by $f(s) = \ln(s)$, then the inverse of f is the exponential function g given by $g(t) = e^t$ since $e^{\ln(s)} = s$.

Definition. If S is a set, then the **identity function** $\text{id}_S: S \to S$ is given by

$$\text{id}_S(s) = s.$$

Theorem 1.3.10. *If* $f: S \to T$, *then* f *is a bijection if and only if* $g \circ f = \text{id}_S$ *and* $f \circ g = \text{id}_T$ *for some function* $g: T \to S$. *Furthermore, the inverse of* f *is* g *in this case.*

PROOF. First we shall prove that if there is a function $g: T \to S$ with $f \circ g = \text{id}_T$ and $g \circ f = \text{id}_S$ then f is a bijection and g is its inverse.

Assume there is a function $g: T \to S$ with $f \circ g = \text{id}_T$ and $g \circ f = \text{id}_S$. If $t \in T$ then $g(t) \in S$ and $f(g(t)) = \text{id}_T(t) = t$. Hence $t \in \text{Im}(f)$ and f is surjective. If $f(s_1) = f(s_2)$ for $s_1, s_2 \in S$, then $g(f(s_1)) = g(f(s_2))$ or $\text{id}_S(s_1) = \text{id}_S(s_2)$ or $s_1 = s_2$. Thus f is injective and hence is a bijection. Finally if $t = f(s)$ then $g(t) = g(f(s)) = \text{id}_S(s) = s$, so that g satisfies Equation (3-5). Thus g is the inverse of f.

Next we shall show that if f is a bijection then there is a function $g: T \to S$ with $f \circ g = \text{id}_T$ and $g \circ f = \text{id}_S$. Since f is a bijection it has an inverse. Call this inverse $g: T \to S$. Then $g(t) = s$ whenever $f(s) = t$.

In particular, if $t \in T$ then

$$f(g(t)) = f(s) = t \quad \text{for all } t \in T$$

so that $f \circ g = \text{id}_T$. Also if $s \in T$ let $t = f(s)$. Then by Equation (3-5), $g(t) = s$ so that

$$g(f(s)) = s \quad \text{for all } s \in S.$$

Thus $g \circ f = \text{id}_S$.   □

Example 1.3.11. Let $\mathbb{P}^+ = \{t \in \mathbb{R} | t > 0\}$ and set $f: \mathbb{R} \to \mathbb{P}^+$ by $f(s) = e^s$. What is $f^{-1}: \mathbb{P}^+ \to \mathbb{R}$?

SOLUTION. Equation (3-5) says that we must find a function $g = f^{-1}: \mathbb{P}^+ \to \mathbb{R}$ with the property that $g(t) = s$ whenever $e^s = t$. This function is $g(t) = \ln t$.

Since

$$e^{\ln t} = t \quad \text{and} \quad \ln e^s = s,$$

Theorem 1.3.10 gives a formal proof that our solution is correct.   □

Theorem 1.3.10 may be used to prove the next result.

Theorem 1.3.12. *If* $f: S \to T$ *and* $h: T \to V$ *are bijections then* $(h \circ f)^{-1} = f^{-1} \circ h^{-1}$.

PROBLEM SET 1.3

Part A.

1. Prove that each of the following functions is surjective.
   a. $f: \mathbb{R} - \{0\} \to \mathbb{R} - \{2\}$; $f(s) = 2 + 1/s$
   b. $g: \mathbb{R} \to \mathbb{R}$; $g(s) = s^3 - 6s$

   c. $h:\mathbb{R}^2 \to \mathbb{R}$; $h(x, y) = xy$

   d. $l:\mathbb{R} \to \{x \mid -2 \le x \le 2\}$; $l(t) = 2 \cos t$

2. Prove that each of the following functions is injective.
   a. $f:\mathbb{R} \to \mathbb{R}$ by $f(s) = \sqrt[3]{s^3 + 1}$
   b. $g:\mathbb{R}-\{0\} \to \mathbb{R}$ by $g(x) = 3 - 1/(2x)$
   c. $h:\mathbb{R}^2 \to \mathbb{R}^2$ by $h(x, y) = (y^3, xy^2 + x + 1)$
   d. $l:\mathbb{P}^+ \to \mathbb{R}^2$ by $l(t) = (t, \ln t)$, where $\mathbb{P}^+$ is the set of positive real numbers

3. Which of the functions in Problem 1 are bijections? In these cases, find the inverse.

4. Which of the functions in Problem 2 are bijections? In these cases, find the inverse.

5. Give an example of a function $f:\mathbb{R} \to \mathbb{R}$ which is neither injective nor surjective.

6. Prove Theorem 1.3.8.

7. Prove Theorem 1.3.9.

8. Prove Theorem 1.3.12.

*9. If $f:S \to T$ is a bijection prove that $f^{-1}:T \to S$ is also a bijection.

Part B. "Prove" may mean "find a counterexample".

10. If $g: T \to V$ and $f:S \to T$ and if $g \circ f$ is surjective prove that both g and f are surjective.

11. If $g: T \to V$ and $f:S \to T$ and if $g \circ f$ is injective, prove that both g and f are injective.

12. If $f:S \to T$ we define a binary relation on S by $s_1 \sim s_2$ if $f(s_1) = f(s_2)$. Prove that $\sim$ is an equivalence relation on S.

13. If $h: X \to Y$ and $g: Y \to Z$ and $g \circ h$ is a bijection, prove that g is surjective and h is injective.

14. If $f: X \to Y$ and $A \subset B \subset X$, prove that $f(B-A) \subset f(B)-f(A)$.

15. If $f:S \to \mathbb{R}$ and $g:S \to \mathbb{R}$ are both injective and $h:S \to \mathbb{R}$ is defined by $h(s) = f(s)+g(s)$, prove that h is injective.

16. If $f:S \to \mathbb{R}$ and $g:S \to \mathbb{R}$ are both surjective and $h:S \to \mathbb{R}$ is defined by $h(s) = f(s) + g(s)$, prove that h is surjective.

17. Prove that the set of all bijections from S to S forms a group. (Hint: Use composition for the multiplication.)

18. Let S be the set of all polynomials. Let $f:S \to S$ by $f(p(x)) = d/dx(p(x))$ and $g:S \to S$ by $g(p(x)) = \int_0^x p(t)\, dt$. Show that $f \circ g = \text{id}_S$ but that f is not a bijection. Does this contradict Theorem 1.3.10?

Part C. Expository exercises.

19. Discuss at an intuitive level what injective and surjective functions are. How would you explain them to an engineering student? to your parents or spouse? to high school students? Note how your answer changes depending on your audience!

CHAPTER 2
Incidence and Metric Geometry

2.1 Definition and Models of Incidence Geometry

In this section we shall define the notions of an abstract geometry and an incidence geometry. These are given by listing a set of axioms that must be satisfied. After the definitions are made, we will give a number of examples which will serve as models for these geometries. Two of these models, the Cartesian Plane and the Poincaré Plane, will be used throughout the rest of the book.

As we discussed in Section 1.1, a geometry is a set $\mathscr{S}$ of points and a set $\mathscr{L}$ of lines together with relationships between the points and lines. What relationship shall we insist on first? We would certainly want every two points to be on some line, and we would want to avoid the pathology of a line with only one point. We will add more relationships between points and lines later and so change the nature of the geometry.

Definition. An **abstract geometry** $\mathscr{A}$ consists of a set $\mathscr{S}$, whose elements are called **points**, together with a collection $\mathscr{L}$ of non-empty subsets of $\mathscr{S}$, called **lines**, such that:

(i) For every two points $A, B \in \mathscr{S}$ there is a line $l \in \mathscr{L}$ with $A \in l$ and $B \in l$.
(ii) Every line has at least two points.

If $\mathscr{A} = \{\mathscr{S}, \mathscr{L}\}$ is an abstract geometry with $P \in \mathscr{S}$, $l \in \mathscr{L}$ and $P \in l$, we say that P *lies on the line* l, or that l *passes through* P. In this language the first axiom of an abstract geometry reads: "every pair of points lies on some line." A word of warning is necessary, however. Just because we use the word "line," you should not think "straight line." "Straight" is a biased term that comes from your previous exposure to geometry, and particularly Euclidean geometry. To us a "line" is just an element of $\mathscr{L}$. See Proposition 2.1.2 below, where the "lines" do not "look straight."

Proposition 2.1.1. *Let* $\mathscr{S} = \mathbb{R}^2 = \{(x, y)\,|\,x, y \in \mathbb{R}\}$. *We define a set of "lines" as follows. A* **vertical line** *is any subset of* $\mathbb{R}^2$ *of the form*

$$L_a = \{(x, y) \in \mathbb{R}^2\,|\,x = a\} \tag{1-1}$$

where a is a fixed real number. A **non-vertical line** *is any subset of* $\mathbb{R}^2$ *of the form*

$$L_{m,b} = \{(x, y) \in \mathbb{R}^2\,|\,y = mx + b\} \tag{1-2}$$

where m and b are fixed real numbers. (See Figure 2-1.) Let $\mathscr{L}_E$ *be the set of all vertical and non-vertical lines. Then* $\mathscr{C} = \{\mathbb{R}^2, \mathscr{L}_E\}$ *is an abstract geometry.*

PROOF. We must show that if $P = (x_1, y_1)$ and $Q = (x_2, y_2)$ are any two distinct points of $\mathbb{R}^2$ then there is an $l \in \mathscr{L}_E$ containing both. This is done by considering two cases.

Case 1. If $x_1 = x_2$ let $a = x_1 = x_2$. Then both P and Q belong to $l = L_a \in \mathscr{L}_E$.

Case 2. If $x_1 \neq x_2$ we show how to find m and b with $P, Q \in L_{m,b}$. Motivated by the idea of the "slope" of a line we define m and b by the equations:

$$m = \frac{y_2 - y_1}{x_2 - x_1} \quad \text{and} \quad b = y_2 - mx_2.$$

It is easy to show that $y_2 = mx_2 + b$ and that $y_1 = mx_1 + b$, so that both P and Q belong to $l = L_{m,b} \in \mathscr{L}_E$.

It is easy to see that each line has at least two points so that $\mathscr{C}$ is an abstract geometry. $\qquad\square$

Definition. The model $\mathscr{C} = \{\mathbb{R}^2, \mathscr{L}_E\}$ is called the **Cartesian Plane** (The notation L_a and $L_{m,b}$ will be reserved for the lines of the Cartesian Plane and certain other models that are developed later using the same set of points and lines.)

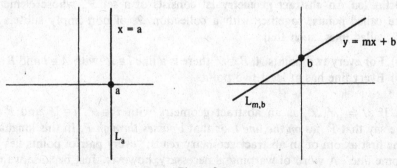

Figure 2-1

We use the letter E in the name of the set of Cartesian lines $(\mathscr{L}_E)$ to remind us of Euclid, the author (c. 300 B.C.E.) of the first axiomatic treatment of geometry. Later we shall add the additional structures of distance and angle measurement to the Cartesian model to obtain the familiar Euclidean model of geometry that is studied in high school. The name Cartesian is used to honor the French mathematician and philosopher René Descartes (1596–1650), who had the revolutionary idea of putting coordinates on the plane. Our verification that $\mathscr{C}$ satisfied the axioms depended heavily on the use of coordinates. Descartes is also responsible for many of our conventions in algebra, such as using x, y, z for unknown quantities and a, b, c for known quantities, and for introducing the exponential notation x^n.

Recall from your elementary courses that there are other ways to describe straight lines in $\mathbb{R}^2$. The way chosen above (that is, through L_a and $L_{m,b}$) is the best suited for this chapter. In Chapter 3, the results are proved most easily if the vector form of the equation of a line is used, and so we will start to use that approach there. Another approach is included in Problem A14.

Proposition 2.1.2. Let $\mathscr{S} = \mathbb{H} = \{(x, y) \in \mathbb{R}^2 \mid y > 0\}$. As in the case of the Cartesian plane, we shall describe two types of lines. A **type I** line is any subset of $\mathbb{H}$ of the form

$$_aL = \{(x, y) \in \mathbb{H} \mid x = a\} \qquad (1\text{-}3)$$

where a is a fixed real number. A **type II** line is any subset of $\mathbb{H}$ of the form

$$_cL_r = \{(x, y) \in \mathbb{H} \mid (x - c)^2 + y^2 = r^2\} \qquad (1\text{-}4)$$

where c and r are fixed real numbers with $r > 0$. (See Figure 2-2.) Let $\mathscr{L}_{\mathbb{H}}$ be the set of all type I and type II lines. Then $\mathscr{H} = \{\mathbb{H}, \mathscr{L}_{\mathbb{H}}\}$ is an abstract geometry.

PROOF. Let $P = (x_1, y_1)$ and $Q = (x_2, y_2)$ be distinct points in $\mathbb{H}$ so that $y_1 > 0$ and $y_2 > 0$.

Case 1. If $x_1 = x_2$ then P and Q both belong to $l = {}_aL \in \mathscr{L}_{\mathbb{H}}$ where $a = x_1 = x_2$.

Case 2. If $x_1 \neq x_2$ define c and r by

$$c = \frac{y_2^2 - y_1^2 + x_2^2 - x_1^2}{2(x_2 - x_1)} \qquad (1\text{-}5)$$

$$r = \sqrt{(x_1 - c)^2 + y_1^2}. \qquad (1\text{-}6)$$

(In Proposition 2.1.5 below we will see what led to this choice of c and r.) In Problem A6 you will show that P and Q both belong to $l = {}_cL_r \in \mathscr{L}_{\mathbb{H}}$. It is easy to see that each line has at least two points so that $\mathscr{H}$ is an abstract geometry.   □

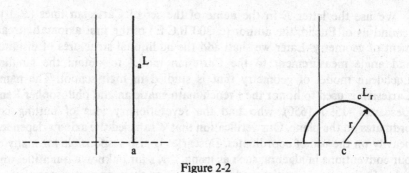

Figure 2-2

Definition. The model $\mathcal{H} = \{H, \mathcal{L}_H\}$ will be called the **Poincaré Plane**. (The notation $_aL$ and $_cL_r$ will be used only to refer to lines in $\mathcal{H}$.)

$\mathcal{H}$ is called the Poincaré Plane in honor of the French mathematican Henri Poincaré (1854–1912) who first used it. Poincaré was a prolific researcher in many areas of pure and applied mathematics. He is particularly remembered for his work in mechanics, for his study of elliptic functions which tied analysis and group theory together, and for his work in geometry which led to the development of modern topology. The letters $\mathcal{H}$, H, and H are used to remind us of the word "hyperbolic". We shall see later in this chapter that the hyperbolic functions are important in this model, just as the trigonometric functions are important in Euclidean geometry. Once we have added more structure to $\mathcal{H}$ it will be a model of what we call a hyperbolic geometry.

In the models given in Propositions 2.1.1 and 2.1.2 it seems clear that through any two points there is a *unique* line. This need not be the case in all abstract geometries as we see in the next example. This example will have a particular subset of $\mathbb{R}^3 = \{(x, y, z)|x, y, z \in \mathbb{R}\}$ as its set of points, $\mathcal{S}$.

Definition. The **unit sphere** in $\mathbb{R}^3$ is

$$S^2 = \{(x, y, z) \in \mathbb{R}^3 \,|\, x^2 + y^2 + z^2 = 1\}.$$

A **plane** in $\mathbb{R}^3$ is a set of the form

$$\{(x, y, z) \in \mathbb{R}^3 \,|\, ax + by + cz = d\}$$

where a, b, c, d are fixed real numbers, and not all of a, b, c are zero.

Note that in the definition of a plane if the constant $d = 0$, then the plane goes through the origin $(0, 0, 0)$.

Definition. A **great circle**, $\mathcal{G}$, of the sphere S^2 is the intersection of S^2 with a plane through the origin. Thus $\mathcal{G}$ is a great circle if there are a, b, $c \in \mathbb{R}$, not all zero, with

$$\mathcal{G} = \{(x, y, z) \in S^2 \,|\, ax + by + cz = 0\}.$$

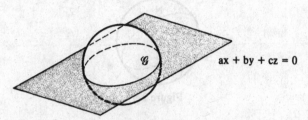

$$ax + by + cz = 0$$

Figure 2-3

Proposition 2.1.3. *Let $\mathscr{S} = S^2$ and let $\mathscr{L}_R$ be the set of great circles on S^2. $\{S^2, \mathscr{L}_R\}$ is an abstract geometry.*

PROOF. We must show that if $P = (x_1, y_1, z_1) \in S^2$ and $Q = (x_2, y_2, z_2) \in S^2$ then there is a great circle $\mathscr{G}$ with $P \in \mathscr{G}$ and $Q \in \mathscr{G}$. Thus we must find a, b, c real numbers (not all zero) such that

$$ax_1 + by_1 + cz_1 = 0 \quad \text{and} \quad ax_2 + by_2 + cz_2 = 0. \qquad (1\text{-}7)$$

View Equations (1-7) as two equations in the three unknowns a, b, c. Since two homogeneous linear equations in three unknowns always have a non-zero solution (in fact, infinitely many solutions), we may always find a, b, and c solving Equations (1-7). Thus, there is a great circle $\mathscr{G}$ with $P \in \mathscr{G}$ and $Q \in \mathscr{G}$. Finally each great circle has at least two points. $\square$

Definition. The **Riemann Sphere** is the abstract geometry $\mathscr{R} = \{S^2, \mathscr{L}_R\}$.

The Riemann Sphere is named after G. B. F. Riemann (1826–1866) who wrote foundational papers in geometry, topology and analysis. His paper on geometry, *Uber die Hypothesen, welche der Geometrie zu Grunde liegen* (On the Hypotheses which lie at the Foundation of Geometry), which was written in 1854 (see Spivak [1970, vol. II] or Smith [1929]), provided geometry with a great unifying idea, that of a Riemannian metric. This concept, which is quite advanced, is the basis for modern differential geometry (see Millman and Parker [1977]) and the mathematics of Einstein's theory of general relativity. The name Riemann Sphere comes from Riemann's work in functions of a complex variable and not from his work in geometry.

Note that it is "geometrically obvious" and was proven above that any two points on S^2 lie on a great circle. However, unlike the first two examples, two points on S^2 may have more than one great circle joining them. Consider the north and south poles N and S as in Figure 2-4. There are infinitely many great circles joining N to S. The uniqueness of lines joining two points is such an important concept that it is singled out in the definition of incidence geometry.

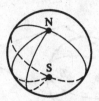

Figure 2-4

Definition. An abstract geometry $\{\mathcal{S}, \mathcal{L}\}$ is an **incidence geometry** if

(i) Every two distinct points in $\mathcal{S}$ lie on a unique line.
(ii) There exist three points A, B, $C \in \mathcal{S}$ which do not lie all on one line.

> **Notation.** If $\{\mathcal{S}, \mathcal{L}\}$ is an incidence geometry and P, $Q \in \mathcal{S}$, then the unique line l on which both P and Q lie will be written as $l = \overleftrightarrow{PQ}$.

It is useful to restate the second axiom of an incidence geometry in terms of the concept of collinearity.

Definition. A set of points $\mathcal{P}$ is **collinear** if there is a line l such that $\mathcal{P} \subset l$. $\mathcal{P}$ is **non-collinear** if $\mathcal{P}$ is not a collinear set.

Sometimes we will say that "A, B, and C are collinear" instead of saying "$\{A, B, C\}$ is a collinear set." This abuse of notation and language makes it easier to state some results. Axiom (ii) of the definition above can be restated as

(ii)′ There exists a set of three non-collinear points.

Although the Riemann Sphere is not an incidence geometry both the Cartesian Plane and the Poincaré Plane are, as we shall now see.

Proposition 2.1.4. *The Cartesian Plane $\mathscr{C}$ is an incidence geometry.*

PROOF. We must show that two distinct points uniquely determine a Cartesian line. Let $P = (x_1, y_1)$ and $Q = (x_2, y_2)$ with $P \neq Q$. We shall assume that P, Q belong to two distinct lines and reach a contradiction.

Case 1. Suppose P, Q belong to both L_a and $L_{a'}$ with $a \neq a'$. Then $a = x_1 = x_2$ and $a' = x_1 = x_2$ so that $a = a'$, which is a contradiction.

Case 2. If P, Q belong to both L_a and $L_{m,b}$, then $P = (a, y_1)$ and $Q = (a, y_2)$. Since both belong to $L_{m,b}$ we also have

$$y_1 = mx_1 + b = ma + b \quad \text{and} \quad y_2 = mx_2 + b = ma + b.$$

Thus $y_1 = y_2$, which contradicts $(a, y_1) = P \neq Q = (a, y_2)$.

Case 3. Suppose that P, Q belong to both $L_{m,b}$ and $L_{n,c}$ and that $L_{m,b} \neq L_{n,c}$. Then

$$y_1 = mx_1 + b, \qquad y_2 = mx_2 + b. \tag{1-8}$$

By Case 2, P, Q cannot both belong to a vertical line so $x_1 \neq x_2$. Hence we may solve Equation (1-8) for m:

$$m = \frac{y_2 - y_1}{x_2 - x_1}. \tag{1-9}$$

From this value of m we obtain b:

$$b = y_1 - mx_1 \tag{1-10}$$

A similar calculation for the line $L_{n,c}$ yields

$$n = \frac{y_2 - y_1}{x_2 - x_1}, \qquad c = y_1 - nx_1.$$

But this implies $m = n$ and $b = c$, which contradicts $L_{m,b} \neq L_{n,c}$.

Thus in all cases, the assumption that P, Q belong to two different lines leads to a contradiction so that P, Q belong to a unique line. In Problem A5 you will show there exists a set of three non-collinear points. Hence $\mathscr{C}$ is an incidence geometry. $\qquad\square$

Note that in the above proof we did not depend on any pictures or "facts" we already know about "straight lines." Instead we were careful to use only the definition of the model and results from elementary algebra.

Proposition 2.1.5. *The Poincaré Plane $\mathscr{H}$ is an incidence geometry.*

PROOF. Let P, $Q \in \mathbb{H}$ with $P \neq Q$. If P and Q lie on two type I lines $_aL$ and $_{a'}L$ then we can show that $a = a'$ just as in Proposition 2.1.4. Thus P and Q cannot lie on two different type I lines. In Problem A7 you will show that P and Q cannot lie on both a type I line and a type II line.

We are left with proving that if $P = (x_1, y_1)$ and $Q = (x_2, y_2)$ are on both $_cL_r$ and $_dL_s$ then $_cL_r = {_dL_s}$. We will show that $c = d$ and $r = s$. This will be done by deriving Equations (1-5) and (1-6) and so will motivate the choice of c and r in Proposition 2.1.2. Since P and Q are on $_cL_r$,

$$(x_1 - c)^2 + y_1^2 = r^2 \quad \text{and} \quad (x_2 - c)^2 + y_2^2 = r^2.$$

Subtracting, we obtain $(x_1 - c)^2 - (x_2 - c)^2 = y_2^2 - y_1^2$ or

$$x_1^2 - 2cx_1 - x_2^2 + 2cx_2 = y_2^2 - y_1^2.$$

We then solve for c:

$$c = \frac{y_2^2 - y_1^2 + x_2^2 - x_1^2}{2(x_2 - x_1)}$$

which is Equation (1-5). An identical computation using the fact that P and Q are on $_dL_s$ will yield

$$d = \frac{y_2^2 - y_1^2 + x_2^2 - x_1^2}{2(x_2 - x_1)}$$

so that $c = d$. Since

$$r = \sqrt{(x_1 - c)^2 + y_1^2} = \sqrt{(x_1 - d)^2 + y_1^2} = s$$

we see that $r = s$ and so $_cL_r = {_d}L_s$.

In Problem A8 you will show there is a set of three non-collinear points. $\square$

Theorem 2.1.6. *Let l_1 and l_2 be lines in an incidence geometry. If $l_1 \cap l_2$ has two or more points then $l_1 = l_2$.*

PROOF. Assume that $P \neq Q$, $P \in l_1 \cap l_2$, and $Q \in l_1 \cap l_2$. Then since both P and Q are on l_1, $\overleftrightarrow{PQ} = l_1$. However, P and Q are also on l_2 so that $\overleftrightarrow{PQ} = l_2$. Hence $l_1 = l_2$. $\square$

Definition. If l_1 and l_2 are lines in an abstract geometry then l_1 is **parallel** to l_2 (written $l_1 \| l_2$) if either $l_1 = l_2$ or $l_1 \cap l_2 = \varnothing$.

The study of parallel lines has a central place in the history of geometry. It and its history will be dealt with in detail later in this book. In Problems A10, A11, and A12 the different "parallel properties" of our models are highlighted. Theorem 2.1.6 can be restated in terms of parallelism as the next result shows.

Corollary 2.1.7. *In an incidence geometry, two lines are either parallel or they intersect in exactly one point.*

PROBLEM SET 2.1
Part A.

1. Find the Poincaré line through $(1, 2)$ and $(3, 4)$.

2. Find the Poincaré line through $(2, 1)$ and $(4, 3)$.

3. Find a spherical line (great circle) through $(\frac{1}{2}, \frac{1}{2}, \sqrt{\frac{1}{2}})$ and $(1, 0, 0)$.

4. Find a spherical line (great circle) through $(0, \frac{1}{2}, \frac{1}{2}\sqrt{3})$ and $(0, -1, 0)$.

5. Show by example that there are (at least) three non-collinear points in the Cartesian Plane.

6. Verify that $P = (x_1, y_1)$ and $Q = (x_2, y_2)$ do lie on $_cL_r$, where c and r are given by Equations (1-5) and (1-6).

7. Prove that if P and Q are distinct points in $\mathbb{H}$ then they cannot lie simultaneously on both $_aL$ and $_cL_r$.

8. Show by example that there are (at least) three non-collinear points in the Poincaré Plane.

9. Let P and Q be in $\mathbb{H}$ and $\overleftrightarrow{PQ} = {_c}L_r$. Use your knowledge of Euclidean geometry to prove that c is the x-coordinate of the intersection of the Euclidean perpendicular bisector of the Euclidean line segment from P to Q with the x-axis. (Hint: Use Equation (1-5).)

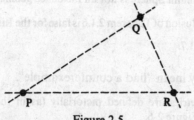

Figure 2-5

10. Find all lines through $(0, 1)$ which are parallel to the vertical line L_6 in the Cartesian Plane.

11. Find all lines in the Poincaré Plane through $(0, 1)$ which are parallel to the type I line $_6L$. (There will be infinitely many!)

12. Find all lines of $\mathcal{R}$ through $N = (0, 0, 1)$ which are parallel to the spherical line (great circle), $\mathscr{C}$, defined by the plane $z = 0$.

13. Let $\mathscr{S} = \{P, Q, R\}$ and $\mathscr{L} = \{\{P, Q\}, \{P, R\}, \{Q, R\}\}$. Show that $\{\mathscr{S}, \mathscr{L}\}$ is an incidence geometry. Note that this example has only finitely many (in fact, three) points. It may be pictured as in Figure 2-5. It is called the 3-point geometry. The dotted lines indicate which points lie on the same line.

14. Let $\mathscr{S} = \mathbb{R}^2$ and, for a given choice of a, b, and c, let

$$J_{a,b,c} = \{(x, y) \in \mathbb{R}^2 \mid ax + by = c\}.$$

Let $\mathscr{L}_J$ be the set of all $J_{a,b,c}$ with at least one of a and b nonzero. Prove that $\{\mathbb{R}^2, \mathscr{L}_J\}$ is an incidence geometry. (Note that this incidence geometry gives the same family of lines as the Cartesian Plane. The point here is that there are different ways to describe the set of lines of this geometry.)

15. Let $\mathscr{S} = \mathbb{R}^2 - \{(0, 0)\}$ and $\mathscr{L}$ be the set of all Cartesian lines which lie in $\mathscr{S}$. Show that $\{\mathscr{S}, \mathscr{L}\}$ is not an incidence geometry.

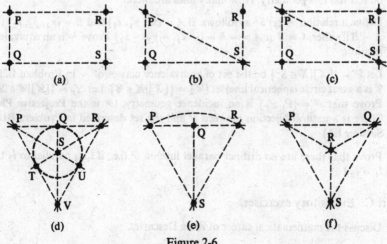

Figure 2-6

16. Prove that the Riemann Sphere is not an incidence geometry.

17. Show that the conclusion of Theorem 2.1.6 is false for the Riemann Sphere. Explain.

18. Prove Corollary 2.1.7.

Part B. "Prove" may mean "find a counterexample".

19. Some finite geometries are defined pictorially (as in the 3-point geometry of Problem A13) by Figure 2-6.

   i. In each example list the set of lines.
   ii. Which of these geometries are abstract geometries?
   iii. Which of these geometries are incidence geometries?

*20. Let $\{\mathscr{S}, \mathscr{L}\}$ be an abstract geometry and assume that $\mathscr{S}_1 \subset \mathscr{S}$. We define an $\mathscr{S}_1$-line to be any subset of $\mathscr{S}_1$ of the form $l \cap \mathscr{S}_1$ where l is a line of $\mathscr{S}$ and where $l \cap \mathscr{S}_1$ has at least two points. Let $\mathscr{L}_1$ be the collection of all $\mathscr{S}_1$-lines. Prove that $\{\mathscr{S}_1, \mathscr{L}_1\}$ is an abstract geometry. $\{\mathscr{S}_1, \mathscr{L}_1\}$ is called the **geometry induced** from $\{\mathscr{S}, \mathscr{L}\}$.

*21. If $\{\mathscr{S}_1, \mathscr{L}_1\}$ is the geometry induced from the incidence geometry $\{\mathscr{S}, \mathscr{L}\}$, prove that $\{\mathscr{S}_1, \mathscr{L}_1\}$ is an incidence geometry if $\mathscr{S}_1$ has a set of three non-collinear points.

22. Let $\{\mathscr{S}_1, \mathscr{L}_1\}$ and $\{\mathscr{S}_2, \mathscr{L}_2\}$ be abstract geometries. Let $\mathscr{S} = \mathscr{S}_1 \cup \mathscr{S}_2$ and $\mathscr{L} = \mathscr{L}_1 \cup \mathscr{L}_2$. Prove that $\{\mathscr{S}, \mathscr{L}\}$ is an abstract geometry.

23. Let $\{\mathscr{S}_1, \mathscr{L}_1\}$ and $\{\mathscr{S}_2, \mathscr{L}_2\}$ be abstract geometries. If $\mathscr{S} = \mathscr{S}_1 \cap \mathscr{S}_2$ and $\mathscr{L} = \mathscr{L}_1 \cap \mathscr{L}_2$ prove that $\{\mathscr{S}, \mathscr{L}\}$ is an abstract geometry.

24. Let $\{\mathscr{S}, \mathscr{L}\}$ be an abstract geometry. If l_1 and l_2 are lines in $\mathscr{L}$ we write $l_1 \sim l_2$ if l_1 is parallel to l_2. Prove that $\sim$ is an equivalence relation. If $\{\mathscr{S}, \mathscr{L}\}$ is the Cartesian Plane then each equivalence class can be characterized by a real number or infinity. What is this number?

25. There is a finite geometry with 7 points such that each line has exactly 3 points on it. Find this geometry. How many lines are there?

26. Define a relation $\sim$ on S^2 as follows. If $A = (x_1, y_1, z_1)$ and $B = (x_2, y_2, z_2)$ then $A \sim B$ if either $A = B$ or $A = -B = (-x_2, -y_2, -z_2)$. Prove $\sim$ is an equivalence relation.

27. Let $\mathbb{P} = \{[X] | X \in S^2\}$ be the set of equivalence classes of $\sim$ in Problem B26. If $\mathscr{C}$ is a great circle (spherical line) let $[\mathscr{C}] = \{[X] | X \in \mathscr{C}\}$. Let $\mathscr{L}_P = \{[\mathscr{C}] | \mathscr{C} \in \mathscr{L}_R\}$. Prove that $\mathscr{P} = \{\mathbb{P}, \mathscr{L}_P\}$ is an incidence geometry. ($\mathscr{P}$ is the **Projective Plane**. There is a natural bijection between $\mathbb{P}$ and the set described in Problem B18 of Section 1.2.)

28. Prove that there are no distinct parallel lines in $\mathscr{P}$ (i.e., if l_1 is parallel to l_2 then $l_1 = l_2$.)

Part C. Expository exercises.

29. Discuss the mathematical career of René Descartes.

30. Discuss the statement "parallel lines meet at infinity" in terms of the three models that are given in this section. Is there even a meaning to the phrase "at infinity" in the Poincaré Plane or the Riemann Sphere?

2.2 Metric Geometry

At this level there are two fundamental approaches to the type of geometry we are studying. The first, called the *synthetic approach*, involves deciding what are the important properties of the concepts you wish to study and then defining these concepts axiomatically by their properties. This approach was used by Euclid in his *Elements* (around 300 B.C.E.) and was made complete and precise by the German mathematician David Hilbert (1862–1943) in his book *Grundlagen der Geometrie* [1899; 8th Edition 1956; Second English Edition 1921]. Hilbert, as did Poincaré at the same time, worked in many areas of mathematics and profoundly affected the course of modern mathematics. He put several areas of mathematics on firm axiomatic footing. In an address to the International Congress of Mathematicians in 1900 he proposed a series of seventeen questions which he felt were the leading theoretical problems of his time. These questions (not all of which have been answered yet) directed mathematical research for years.

The second approach, called the *metric approach*, is due to the American mathematician, George David Birkhoff (1884–1944) in his paper "A Set of Postulates for Plane Geometry Based on Scale and Protractor" [1932]. In this approach, the concept of distance (or a *metric*) and angle measurement is added to that of an incidence geometry to obtain basic ideas of betweenness, line segments, congruence, etc. Such an approach brings some analytic tools (for example, continuity) into the subject and allows us to use fewer axioms. Birkhoff is also remembered for his work in relativity, differential equations, and dynamics.

A third approach, championed by Felix Klein (1849–1925), has a very different flavor—that of abstract algebra—and is more advanced because it uses group theory. Klein felt that geometry should be studied from the viewpoint of a group acting on a set. Concepts that are invariant under this action are the interesting geometric ideas. See Millman [1977] and Martin [1982]. In Chapter 11 we will study some of the ideas from this approach, which is called transformation geometry.

In this book we will follow the metric approach because the concept of distance is such a natural one. (Modern treatments of the synthetic approach can be found in Borsuk and Szmielew [1960] or Greenberg [1980]. We will briefly outline the synthetic approach in Section 6.7.) Intuitively, "distance" is a function which assigns a number $d(P, Q)$ to each pair of points P, Q. It should not matter whether we measure from P to Q or from Q to P (i.e.,

$d(P, Q) = d(Q, P)$). Furthermore, the only time the distance between two points is zero should be when the points are actually the same. More formally we have the following definition.

Definition. A **distance function** on a set $\mathscr{S}$ is a function $d: \mathscr{S} \times \mathscr{S} \to \mathbb{R}$ such that for all $P, Q \in \mathscr{S}$

(i) $d(P, Q) \geq 0$;
(ii) $d(P, Q) = 0$ if and only if $P = Q$; and
(iii) $d(P, Q) = d(Q, P)$.

The following definition gives a distance function for the Cartesian Plane. See Problem A1.

Definition. Let $\mathscr{S} = \mathbb{R}^2$, $P = (x_1, y_1)$ and $Q = (x_2, y_2)$. The **Euclidean distance** d_E is given by

$$d_E(P, Q) = \sqrt{(x_1 - x_2)^2 + (y_1 - y_2)^2}. \tag{2-1}$$

To give an example of a reasonable distance function in the Poincaré Plane requires more thought. Suppose that P and Q belong to a type I line. A reasonable guess for the distance between $P = (a, y_1)$ and $Q = (a, y_2)$ might be $|y_1 - y_2|$. However, this is somewhat displeasing because it means that as y_2 tends to zero (and thus Q goes toward the x-axis or "edge") the distance from P to Q tends to y_1, which is a finite number. It would be "nicer" if the "edge" were not a finite distance away. One way to avoid this is to use a logarithmic scale and say that the distance from (a, y_1) to (a, y_2) is $|\ln(y_1) - \ln(y_2)| = |\ln(y_1/y_2)|$. (Note that as $y_2 \to 0$, $\ln(y_1/y_2) \to \infty$.) This gives some justification for the following definition (which looks rather artificial.) The reasons for this definition will be clearer after we discuss the Ruler Postulate.

Definition. If $P = (x_1, y_1)$ and $Q = (x_2, y_2)$ are points in the Poincaré Plane $\mathscr{H}$, the **Poincaré distance** d_H is given by

$$d_H(P, Q) = \left| \ln\left(\frac{y_2}{y_1}\right) \right| \quad \text{if } x_1 = x_2 \tag{2-2}$$

$$d_H(P, Q) = \left| \ln\left(\frac{\dfrac{x_1 - c + r}{y_1}}{\dfrac{x_2 - c + r}{y_2}}\right) \right| \quad \text{if } P \text{ and } Q \text{ lie on } {}_cL_r. \tag{2-3}$$

The verification that d_H, as defined by Equations (2-2) and (2-3), actually satisfies axioms (i) and (iii) of a distance function is left to Problem A2. Axiom (ii) is more difficult, especially for points on a type II line. Essentially, we need to show that the function $f: {}_cL_r \to \mathbb{R}$ given by $f(x, y) = \ln\left(\dfrac{x - c + r}{y}\right)$ is injective. We will do this in the proof of Proposition 2.2.6.

We shall now present an example with a different twist. This example, called taxicab distance, comes from thinking of a taxi driving on the rectangular grid of a city's streets. The taxicab distance measures the distance the taxi would travel from point P to point Q if there were no one way streets. See Figure 2-7.

Definition. If $P = (x_1, y_1)$ and $Q = (x_2, y_2)$ are points in $\mathbb{R}^2$, the **taxicab distance** between them is given by

$$d_T(P, Q) = |x_1 - x_2| + |y_1 - y_2|. \tag{2-4}$$

Proposition 2.2.1. *The taxicab distance is a distance function on $\mathbb{R}^2$.*

PROOF. Note that $d_T(P, Q) \geq 0$ since it is a sum of absolute values, each of which is always nonnegative. Thus axiom (i) for a distance holds.

The second axiom states that $d_T(P, Q) = 0$ if and only if $P = Q$. Clearly if $P = Q$ then $d_T(P, Q) = 0$ by Equation (2-4). On the other hand, if $d_T(P, Q) = 0$ then $|x_1 - x_2| + |y_1 - y_2| = 0$. Since each of these two terms is at least zero, they must both be zero: $|x_1 - x_2| = 0$ and $|y_1 - y_2| = 0$. But this means $x_1 = x_2$ and $y_1 = y_2$. Therefore, if $d_T(P, Q) = 0$ then $P = Q$.

Finally axiom (iii), $d_T(P, Q) = d_T(Q, P)$, holds because $|a - b| = |b - a|$. $\square$

Note that d_T and d_E are both distance functions on the same underlying set $\mathbb{R}^2$. In general, a set may have many different distance functions on it (see, for example, Problem B16). Thus, when we want to talk about a property of distance on a set, we need to specify both the set $\mathscr{S}$ and the distance function d.

The concept of a ruler is central to the remainder of this book. This was the idea introduced by Birkhoff to move geometry away from the very

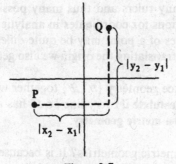

Figure 2-7

synthetic methods. Intuitively, a ruler is a line that has been marked so that it can be used to measure distances. We shall "mark" our lines by assuming that for every line there is a bijection between that line and $\mathbb{R}$ in such a way that the "markings" measure distance.

Definition. Let l be a line in an incidence geometry $\{\mathscr{S}, \mathscr{L}\}$. Assume that there is a distance function d on $\mathscr{S}$. A function $f: l \to \mathbb{R}$ is a **ruler** (or **coordinate system**) **for** l if

(i) f is a bijection;
(ii) for each pair of points P and Q on l

$$|f(P) - f(Q)| = d(P, Q). \tag{2-5}$$

Equation (2-5) is called the **Ruler Equation** and $f(P)$ is called the **coordinate of P with respect to f.**

Example 2.2.2. Let l be the nonvertical line $L_{2,3}$ in the Cartesian Plane $\mathscr{C}$ with the Euclidean distance. Show that if $Q = (x, y)$ then $f(Q) = \sqrt{5}x$ gives a ruler f for l and find the coordinate of $R = (1, 5)$ with respect to f.

SOLUTION. f is certainly a bijection so all we need verify is the Ruler Equation. Note that $(x, y) \in L_{2,3}$ if and only if $y = 2x + 3$ so that if $P = (x_1, y_1)$ then

$$d(P, Q) = \sqrt{(x_1 - x)^2 + (y_1 - y)^2} = \sqrt{(x_1 - x)^2 + 4(x_1 - x)^2}$$
$$= \sqrt{5}|x_1 - x| = |f(P) - f(Q)|.$$

Thus the Ruler Equation holds.
    The coordinate of $R = (1, 5)$ is $f(R) = \sqrt{5}$.   □

    Some comments are in order. The terms ruler and coordinate system are typically used interchangeably in the literature, and we will use both. Note also that since a point may lie on more than one line it may have different "coordinates" with respect to the various lines or rulers used. In particular, if P lies on the line l', and if l' has a ruler f', then there need not be any relation between the coordinate of P with respect to l and the coordinate of P with respect to l'. (See Problem A4.) In addition, we shall see that if a line has one ruler f, it has many rulers and thus many possible coordinates for P. (The analogous situations for coordinates in analytic geometry are that the rectangular coordinates of a point may be quite different from its polar coordinates and that by translating the origin we also get different coordinates.)

Definition. An incidence geometry $\{\mathscr{S}, \mathscr{L}\}$ together with a distance function d satisfies the **Ruler Postulate** if every line $l \in \mathscr{S}$ has a ruler. In this case we say $\mathscr{M} = \{\mathscr{S}, \mathscr{L}, d\}$ is a **metric geometry**.

    Why do we study metric geometries? It is because many of the concepts in the synthetic approach which must be added are already present in the metric geometry approach. This happens because we can transfer questions about a line l in $\mathscr{S}$ to the real numbers $\mathbb{R}$ by using a ruler f. In $\mathbb{R}$ we understand concepts like "between" and so can transfer them back (via f^{-1}) to l. This is the advantage of the metric approach alluded to in the beginning of the section. After we have more background (i.e., in Chapter 6), we will return to the question of a synthetic versus metric approach to geometry.

The definition states that in order to prove $\{\mathscr{S}, \mathscr{L}, d\}$ is a metric geometry, we need to find for each $l \in \mathscr{L}$ a function $f : l \to \mathbb{R}$ which is a bijection and which satisfies Equation (2-5). However, because of the next lemma, we do not really have to prove that f is an injection. This lemma will then prove useful in the problems at the end of the section as well as in Propositions 2.2.4 and 2.2.7.

Lemma 2.2.3. *Let $l \in \mathscr{L}$ and $f : l \to \mathbb{R}$ be surjective and satisfy Equation (2-5). Then f is a bijection and hence a ruler for l.*

PROOF. Since we assume that f is surjective we need only show that it is injective. Suppose that $f(P) = f(Q)$. Then by Equation (2-5) we have

$$d(P, Q) = |f(P) - f(Q)| = 0$$

so that $P = Q$ by the second axiom of distance. $\qquad\square$

Proposition 2.2.4. *The Cartesian Plane with the Euclidean distance, d_E, is a metric geometry.*

PROOF. Let l be a line. We need to find a ruler for l. This will be done in two cases.

Case 1. If $l = L_a$ is a vertical line then $P \in L_a$ means $P = (a, y)$ for some y. We define $f : l \to \mathbb{R}$ by

$$f(P) = f((a, y)) = y. \tag{2-6}$$

f is clearly surjective. If $P = (a, y_1)$ and $Q = (a, y_2)$, then

$$|f(P) - f(Q)| = |y_1 - y_2| = d(P, Q).$$

Therefore f is a ruler by Lemma 2.2.3.

Case 2. If $l = L_{m,b}$ then $P \in L_{m,b}$ means that $P = (x, y)$ where $y = mx + b$. Define $f : L_{m,b} \to \mathbb{R}$ by

$$f(P) = f((x, y)) = x\sqrt{1 + m^2}. \tag{2-7}$$

If $t \in \mathbb{R}$ let $x = t/\sqrt{1 + m^2}$, $y = (mt/\sqrt{1 + m^2}) + b$. Certainly, $P = (x, y) \in L_{m,b}$. Furthermore,

$$f(P) = \frac{t}{\sqrt{1 + m^2}} \cdot \sqrt{1 + m^2} = t$$

so that f is surjective.

Now suppose that $P = (x_1, y_1)$ and $Q = (x_2, y_2)$. Then

$$|f(P) - f(Q)| = |x_1\sqrt{1 + m^2} - x_2\sqrt{1 + m^2}|$$
$$= \sqrt{1 + m^2}|x_1 - x_2|.$$

On the other hand

$$d_E(P, Q) = \sqrt{(x_1 - x_2)^2 + (y_1 - y_2)^2}$$
$$= \sqrt{(x_1 - x_2)^2 + m^2(x_1 - x_2)^2}$$
$$= \sqrt{1 + m^2}\sqrt{(x_1 - x_2)^2}$$
$$= \sqrt{1 + m^2}|x_1 - x_2|.$$

Combining these two sets of equations we have $|f(P) - f(Q)| = d_E(P, Q)$. Hence by Lemma 2.2.3, f is a ruler.   □

Definition. The **Euclidean Plane** is the model

$$\mathscr{E} = \{\mathbb{R}^2, \mathscr{L}_E, d_E\}.$$

Our next step is to show that the Poincaré Plane with the Poincaré distance is a metric geometry. To do this it will be useful to use hyperbolic functions. Recall that the hyperbolic sine, hyperbolic cosine, hyperbolic tangent and hyperbolic secant are defined by

$$\sinh(t) = \frac{e^t - e^{-t}}{2}; \quad \cosh(t) = \frac{e^t + e^{-t}}{2};$$

$$\tanh(t) = \frac{\sinh(t)}{\cosh(t)} = \frac{e^t - e^{-t}}{e^t + e^{-t}}; \quad \text{sech}(t) = \frac{1}{\cosh(t)} = \frac{2}{e^t + e^{-t}}.$$

(2-8)

From the above it is easy to prove

Lemma 2.2.5. *For every value of* t:

(i) $[\cosh(t)]^2 - [\sinh(t)]^2 = 1$;
(ii) $[\tanh(t)]^2 + [\text{sech}(t)]^2 = 1$.

The first equation of Lemma 2.2.5 is particularly suggestive. Whereas the trigonometric (or circular) functions sine and cosine satisfy $\sin^2 t + \cos^2 t = 1$ and remind us of a circle: $x^2 + y^2 = 1$, the hyperbolic sine and cosine lead to an equation of a hyperbola: $x^2 - y^2 = 1$. We should also note that if $x = \tanh(t)$ and $y = \text{sech}(t)$ then (x, y) lies on the circle $x^2 + y^2 = 1$.

Proposition 2.2.6. d_H *is a distance function for the Poincaré Plane and* $\{\mathbb{H}, \mathscr{L}_H, d_H\}$ *is a metric geometry.*

PROOF. By Problem A2, d_H satisfies axioms (i) and (iii) of a distance function. We must verify axiom (ii) and find appropriate rulers. Clearly, if $P = Q$ then $d_H(P, Q) = 0$. We need to show that if $d_H(P, Q) = 0$ then $P = Q$. To do this we consider two cases depending on the type of line that P and Q belong to.

Suppose that P, Q belong to a type I line $_aL$ with $P = (a, y_1)$ and $Q = (a, y_2)$. If $d_H(P, Q) = 0$ then $|\ln(y_1/y_2)| = 0$ so that $y_1/y_2 = 1$ and $y_1 = y_2$. Thus if P, Q belong to a type I line and $d_H(P, Q) = 0$ then $P = Q$. In Problem A8 you will show that the function $g: _aL \to \mathbb{R}$ given by $g(a, y) = \ln(y)$ is a bijection and satisfies the Ruler Equation. Thus g is a ruler for $_aL$.

Now suppose that P, Q belong to a type II line $_cL_r$ and that $d_H(P, Q) = 0$.

Let $f: _cL_r \to \mathbb{R}$ be given by $f(x, y) = \ln\left(\dfrac{x - c + r}{y}\right)$. We will eventually show

that f is a ruler. First we must show it is a bijection. (Lemma 2.2.3 cannot be used because we do not yet know that d_H is a distance function.) To show that f is bijective we must show that for every $t \in \mathbb{R}$ there is one and only one pair (x, y) which satisfies

$$(x - c)^2 + y^2 = r^2, \quad y > 0, \quad \text{and} \quad f(x, y) = t. \tag{2-9}$$

We try to solve $f(x, y) = t$ for x and y.

If $f(x, y) = \ln\left(\dfrac{x - c + r}{y}\right) = t$ then $\dfrac{x - c + r}{y} = e^t$. Thus

$$e^{-t} = \frac{y}{x - c + r} = \frac{y(x - c - r)}{(x - c + r)(x - c - r)} = \frac{y(x - c - r)}{(x - c)^2 - r^2} = \frac{y(x - c - r)}{-y^2}$$

$$= -\frac{x - c - r}{y}$$

since $(x, y) \in {_cL_r}$. Hence

$$e^t + e^{-t} = \frac{x - c + r}{y} - \frac{x - c - r}{y} = \frac{2r}{y}$$

or

$$y = r \operatorname{sech}(t).$$

Also

$$\frac{e^t - e^{-t}}{e^t + e^{-t}} = \frac{\dfrac{x - c + r}{y} + \dfrac{x - c - r}{y}}{\dfrac{2r}{y}} = \frac{2(x - c)}{2r} = \frac{x - c}{r}$$

or

$$x - c = r \tanh(t).$$

Hence the *only* possible solution to Equation (2-9) is

$$x = c + r \tanh t, \qquad y = r \operatorname{sech} t. \tag{2-10}$$

A simple computation using Lemma 2.2.5 shows that x and y as given in Equations (2-10) satisfy $(x - c)^2 + y^2 = r^2$ and that $y > 0$. Thus Equations (2-10) define a point in $_cL_r$. Finally, a straightforward substitution verifies that for this x, y we have $f(x, y) = t$. Thus Equations (2-9) have one and only one solution for each $t \in \mathbb{R}$ and therefore $f : _cL_r \to \mathbb{R}$ is a bijection.

Next, if $P = (x_1, y_1)$ and $Q = (x_2, y_2)$ belong to $_cL_r$, then by Equation (2-3) and the properties of logarithms

$$d_H(P, Q) = |f(x_1, y_1) - f(x_2, y_2)|.$$

Hence, f satisfies the Ruler Equation. Finally, if $d_H(P, Q) = 0$, then $f(x_1, y_1) = f(x_2, y_2)$. Since f is bijective, this means $(x_1, y_1) = (x_2, y_2)$ and d_H satisfies axiom (ii) of a distance function.

Since we have proved that d_H is a distance and each line in $\mathcal{H}$ has a ruler (g and f above) $\{\mathbb{H}, \mathcal{L}_H, d_H\}$ is a metric geometry. $\qquad\square$

> **Convention.** From now on, the terminology Poincaré Plane and the symbol $\mathcal{H}$ will include the hyperbolic distance d_H:
>
> $$\mathcal{H} = \{\mathbb{H}, \mathcal{L}_H, d_H\}.$$

Note that with the given rulers in $\mathcal{H}$, if $P, Q \in l$ and if we let Q tend to the "edge" (i.e., the x-axis) along l, then $f(Q)$ tends to $\pm\infty$ so that

$$d(P, Q) = |f(P) - f(Q)| \to \infty.$$

That is, the "edge" of the Poincaré Plane is not a finite distance away from any point P. To a creature living in the geometry the edge is not reachable, hence not observable. The x-axis that we sketch in our pictures of the Poincaré Plane is the "horizon".

Proposition 2.2.7. *The Cartesian Plane with the taxicab distance is a metric geometry.*

PROOF. If l is a vertical line L_a we define $f : l \to \mathbb{R}$ by $f((a, y)) = y$. If l is a nonvertical line $L_{m,b}$ we define $f : l \to \mathbb{R}$ by $f((x, y)) = (1 + |m|)(x)$. We leave the proof that these really are coordinate systems to Problem A12. $\qquad\square$

Definition. The model $\mathcal{T} = \{\mathbb{R}^2, \mathcal{L}_E, d_T\}$ will be called the **Taxicab Plane.**

Note that we started with a single incidence geometry (the Cartesian Plane), put two different distances on it, and obtained two different metric geometries. Thus we have two metric geometries with the same underlying incidence geometry. In general, there are many metric geometries which have the same underlying incidence geometry.

How do we actually construct models of a metric geometry? In our three examples we started with an incidence geometry, defined a distance and hunted for rulers so that Equation (2-5) was satisfied. We can reverse this process in a certain sense. That is, we can start with a collection of bijections from the lines to $\mathbb{R}$ and use them to define a distance function which has these bijections as rulers. In fact, this method (which is described in Theorem 2.2.8 below) is really how we decided what the "right" definition was for a distance in $\mathbb{H}$.

Theorem 2.2.8. Let $\{\mathscr{S}, \mathscr{L}\}$ be an incidence geometry. Assume that for each line $l \in \mathscr{L}$ there exists a bijection $f_l: l \to \mathbb{R}$. Then there is a distance d such that $\{\mathscr{S}, \mathscr{L}, d\}$ is a metric geometry and each $f_l: l \to \mathbb{R}$ is a ruler.

PROOF. If $P, Q \in \mathscr{S}$ we must define $d(P, Q)$. If $P = Q$ let $d(P, Q) = 0$. If $P \neq Q$ let l be the unique line through P and Q, and $f_l: l \to \mathbb{R}$ be the bijection described in the hypothesis. Define $d(P, Q) = |f_l(P) - f_l(Q)|$. In Problem A13 you will verify that d satisfies the three properties of a distance. Finally each f_l is clearly a ruler for the line l. $\square$

The opposite problem in which we start with a distance and ask if there is metric geometry with that distance is more subtle. In Problem B15 we give an example of a distance on the incidence geometry, $\mathbb{R}^2$, which does not have rulers and hence does not give a metric geometry.

We close this section with a table which summarizes the rulers which we have discussed for the three major models of a metric geometry.

Model	Type of line	Standard Ruler or coordinate system for line
Euclidean Plane, $\mathscr{E}$	$L_a = \{(a, y) \mid y \in \mathbb{R}\}$	$f(a, y) = y$
	$L_{m,b} = \{(x, y) \in \mathbb{R}^2 \mid y = mx + b\}$	$f(x, y) = x\sqrt{1 + m^2}$
Poincaré Plane, $\mathscr{H}$	$_aL = \{(a, y) \in \mathbb{H} \mid y > 0\}$	$f(a, y) = \ln y$
	$_cL_r = \{(x, y) \in \mathbb{H} \mid (x - c)^2 + y^2 = r^2\}$	$f(x, y) = \ln\left(\dfrac{x - c + r}{y}\right)$
Taxicab Plane, $\mathscr{T}$	$L_a = \{(a, y) \mid y \in \mathbb{R}\}$	$f(a, y) = y$
	$L_{m,b} = \{(x, y) \in \mathbb{R}^2 \mid y = mx + b\}$	$f(x, y) = (1 + |m|)x$

Convention. In discussions about one of the three models above, the coordinate of a point with respect to a line l will always mean the coordinate with respect to the standard ruler for that line as given in the above table.

In the next section we will discuss some special rulers for a line. These should not be confused with the standard rulers defined above.

PROBLEM SET 2.2

Part A.

1. Prove that the Euclidean distance function as defined by Equation (2-1) is a distance function.

2. Verify that the function d_H defined by Equations (2-2) and (2-3) satisfies axioms (i) and (iii) of the definition of a distance function.

3. Prove Lemma 2.2.5.

4. In the Euclidean Plane, (i) find the coordinate of $(2, 3)$ with respect to the line $x = 2$; (ii) find the coordinate of $(2, 3)$ with respect to the line $y = -4x + 11$. (Note that your answers are different.)

5. Find the coordinate of $(2, 3)$ with respect to the line $y = -4x + 11$ for the Taxicab Plane. (Compare with Problem 4.)

6. Find the coordinates in $\mathbb{H}$ of $(2, 3)$ (i) with respect to the line $(x - 1)^2 + y^2 = 10$; (ii) with respect to the line $x = 2$.

7. Find the Poincaré distance between
   i. $(1, 2)$ and $(3, 4)$ (See Problem A1 of Section 2.1.)
   ii. $(2, 1)$ and $(4, 3)$ (See Problem A1 of Section 2.1.)

8. Show that the function $g: {}_aL \to \mathbb{R}$ given by $g(a, y) = \ln(y)$ is a bijection and that it satisfies the Ruler Equation. Show that the inverse of g is given by $g^{-1}(t) = (a, e^t)$.

9. Find a point P on the line $L_{2, -3}$ in the Euclidean Plane whose coordinate is -2.

10. Find a point P on the line $L_{2, -3}$ in the Taxicab Plane whose coordinate is -2.

11. Find a point P on the line ${}_{-3}L_{\sqrt{7}}$ in the Poincaré Plane whose coordinate is $\ln 2$.

12. Complete the proof of Proposition 2.2.7.

13. Complete the proof of Theorem 2.2.8.

Part B. "Prove" may mean "find a counterexample".

14. We shall define a new distance d^* on $\mathbb{R}^2$ by using d_E. Specifically:

$$d^*(P, Q) = \begin{cases} d_E(P, Q) & \text{if } d_E(P, Q) \leq 1 \\ 1 & \text{if } d_E(P, Q) > 1. \end{cases}$$

(i) Prove that d^* is a distance function. (ii) Find and sketch all points $P \in \mathbb{R}^2$ such that $d^*((0, 0), P) \leq 2$. (iii) Find all points $P \in \mathbb{R}^2$ such that $d^*((0, 0), P) = 2$.

15. Let d^* be the distance function of Problem B14. Prove that there is no incidence geometry on $\mathbb{R}^2$ such that $\{\mathbb{R}^2, \mathscr{L} \ d^*\}$ is a metric geometry. (Thus not every distance gives a metric geometry.) Hint: Suppose by way of contradiction that there is a ruler $f: l \to \mathbb{R}$ and that $P_0 \in l$ has coordinate zero. Consider the set of all points on l with coordinate ± 2.

16. If d_0 and d_1 are distance functions on $\mathscr{S}$, prove that if $s \geq 0$ and $t > 0$, then $sd_0 + td_1$ is also a distance function on $\mathscr{S}$.

17. If $\{\mathscr{S}, \mathscr{L}, d\}$ is a metric geometry and $P \in \mathscr{S}$, prove that for any $r > 0$ there is a point in $\mathscr{S}$ at distance r from P.

18. Define the **max distance** (or **supremum distance**), d_S, on $\mathbb{R}^2$ by

$$d_S(P, Q) = \max\{|x_1 - x_2|, |y_1 - y_2|\}$$

where $P = (x_1, y_1)$ and $Q = (x_2, y_2)$.
   i. Show that d_S is a distance function.
   ii. Show that $\{\mathbb{R}^2, \mathscr{L}_E, d_S\}$ is a metric geometry.

19. In a metric geometry $\{\mathscr{S}, \mathscr{L}, d\}$ if $P \in \mathscr{S}$ and $r > 0$, then the **circle with center** P and **radius** r is $\mathscr{C} = \{Q \in \mathscr{S} | d(P, Q) = r\}$. Draw a picture of the circle of radius 1 and center $(0, 0)$ in the $\mathbb{R}^2$ for each of the distances d_E, d_T, and d_S.

20. Let $\{\mathscr{S}, \mathscr{L}, d\}$ be a metric geometry, let $P \in \mathscr{S}$, let $l \in \mathscr{L}$ with $P \in l$, and let $\mathscr{C}$ be a circle with center P. Prove that $l \cap \mathscr{C}$ contains exactly two points.

21. Find the circle of radius 1 with center $(0, e)$ in the Poincaré Plane. Hint: As a set this circle "looks" like an ordinary circle. Carefully show this.

22. We may define a distance function for the Riemann Sphere as follows. On a great circle $\mathscr{C}$ we measure the distance $d_R(A, B)$ between two points A and B as the shorter of the lengths of the two arcs of $\mathscr{C}$ joining A to B. (Note $d_R(A, -A) = \pi$.) Prove that d_R is a distance function. Is $\{S^2, \mathscr{L}_R, d_R\}$ a metric geometry?

23. On the Projective Plane (see Problem B26 of Section 2.1) define $d_P([A], [B]) =$ minimum of the two numbers $d_R(A, B)$ and $d_R(A, -B)$. Prove that d_P is a distance function. Is $\{\mathbb{P}, \mathscr{L}_P, d_P\}$ a metric geometry?

Part C. Expository exercises.

24. Compare and contrast the definition of the taxicab metric as given in this section with that of Byrkit [1971].

2.3 Special Coordinate Systems

In this section we shall prove the existence of a special kind of coordinate system. This coordinate system will play an important role in our study of betweenness in Chapter 3. We shall also see that, as a consequence of the Ruler Postulate, every line in a metric geometry must have infinitely many points.

Theorem 2.3.1. *Let f be a coordinate system for the line l in a metric geometry. If $a \in \mathbb{R}$ and ε is ± 1 and if we define $h_{a,\varepsilon}: l \to \mathbb{R}$ by*

$$h_{a,\varepsilon}(P) = \varepsilon(f(P) - a)$$

then $h_{a,\varepsilon}$ is a coordinate system for l.

PROOF. By Lemma 2.2.3 we need only show that $h_{a,\varepsilon}$ is surjective and satisfies the Ruler Equation. If $t \in \mathbb{R}$ is given we know that there is an $R \in l$ with

$f(R) = t/\varepsilon + a$ since f is surjective. But then

$$h_{a,\varepsilon}(R) = \varepsilon(f(R) - a) = \varepsilon\left(\left(\frac{t}{\varepsilon} + a\right) - a\right) = t$$

so that $h_{a,\varepsilon}$ is surjective.

As for the Ruler Equation,

$$\begin{aligned}
|h_{a,\varepsilon}(P) - h_{a,\varepsilon}(Q)| &= |\varepsilon(f(P) - a) - \varepsilon(f(Q) - a)| \\
&= |\varepsilon||f(P) - f(Q)| \\
&= |f(P) - f(Q)| \\
&= d(P, Q)
\end{aligned}$$

since f is a coordinate system for l.   □

Geometrically, when $a = 0$ and $\varepsilon = -1$ the coordinate system of Theorem 2.3.1 interchanges the positive and negative points of l with respect to f. More precisely, if P_0 is that point of l with $f(P_0) = 0$ then $h_{0,-1}$ is the result of reflecting the ruler f about P_0. See Figure 2-8. We may also translate a coordinate system by an element $a \in \mathbb{R}$. This amounts to changing the origin (i.e., the point which corresponds to 0). In Figure 2-9, we assume that $f(P_1) = a$ and $f(P_0) = 0$ so that P_1 corresponds to a and P_0 corresponds to the origin in the coordinate system f. If we apply Theorem 2.3.1 with $\varepsilon = 1$ then P_1 corresponds to the origin and P_0 to $-a$ in the new coordinate system $h_{a,1}$.

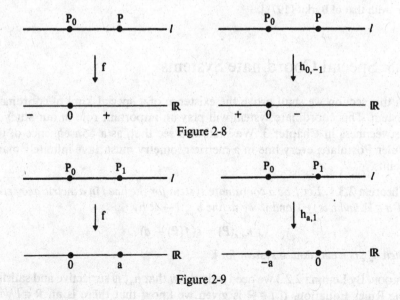

Figure 2-8

Figure 2-9

Theorem 2.3.2 (Ruler Placement Theorem). *Let l be a line in a metric geometry and let A and B be points on the line. There is a coordinate system g on l with* $g(A) = 0$ *and* $g(B) > 0$.

PROOF. Let $f: l \to \mathbb{R}$ be a coordinate system for l and let $a = f(A)$. If $f(B) > a$ let $\varepsilon = +1$. If $f(B) < a$ let $\varepsilon = -1$. By Theorem 2.3.1 $g = h_{a,\varepsilon}$ is a coordinate system for l, and

$$g(A) = h_{a,\varepsilon}(A) = \varepsilon(f(A) - a) = \varepsilon \cdot 0 = 0;$$
$$g(B) = h_{a,\varepsilon}(B) = \varepsilon(f(B) - a) = |f(B) - a| > 0.$$

Thus g is a coordinate system with the desired properties. $\qquad\square$

The special coordinate system of Theorem 2.3.2 is so useful that it merits a special name.

Definition. Let $l = \overleftrightarrow{AB}$. If $g: l \to \mathbb{R}$ is a coordinate system for l with $g(A) = 0$ and $g(B) > 0$, then g is called a **coordinate system with** A **as origin and** B **positive**.

It is reasonable to ask if there are any other operations (besides reflection and translation) that can be done to a coordinate system to get another coordinate system; that is, is every coordinate system of the form $h_{a,\varepsilon}$? The next theorem says the answer is yes. This result will not be used in the rest of the book. It is included for the sake of completeness and is optional.

Theorem 2.3.3. *If l is a line in a metric geometry and if* $f: l \to \mathbb{R}$ *and* $g: l \to \mathbb{R}$ *are both coordinate systems for l, then there is an* $a \in \mathbb{R}$ *and an* $\varepsilon = \pm 1$ *with* $g(P) = \varepsilon(f(P) - a)$ *for all* $P \in l$.

PROOF. Let $P_0 \in l$ be the point with $g(P_0) = 0$. Let $a = f(P_0)$. Since both f and g are rulers for l, we have for each $P \in l$ that

$$|g(P)| = |g(P) - g(P_0)| = d(P, P_0)$$
$$= |f(P) - f(P_0)|$$
$$= |f(P) - a|.$$

Thus for each $P \in l$,

$$g(P) = \pm(f(P) - a). \tag{3-1}$$

We claim we can use the same sign for each value of P.

Suppose to the contrary that there is a point $P_1 \neq P_0$ with $g(P_1) = +(f(P_1) - a)$ and another point $P_2 \neq P_0$ with $g(P_2) = -(f(P_1) - a)$. Then

$$d(P_1, P_2) = |g(P_1) - g(P_2)|$$
$$= |f(P_1) - a + f(P_2) - a|$$
$$= |f(P_1) + f(P_2) - 2a|.$$

But

$$d(P_1, P_2) = |f(P_1) - f(P_2)|.$$

Thus

$$|f(P_1) - f(P_2)| = |f(P_1) + f(P_2) - 2a|$$

and either

$$f(P_1) - f(P_2) = f(P_1) + f(P_2) - 2a$$

or

$$f(P_1) - f(P_2) = -f(P_1) - f(P_2) + 2a.$$

In the first case $f(P_2) = a = f(P_0)$ and in the second case $f(P_1) = a = f(P_0)$. Either way we contradict the fact that f is injective. Thus by Equation (3-1), either

$$g(P) = f(P) - a \qquad \text{for all } P \in l$$

or

$$g(P) = -(f(P) - a) \qquad \text{for all } P \in l.$$

Thus for an appropriate choice of ε (either $+1$ or -1), $g(P) = \varepsilon(f(P) - a)$ for all $P \in l$.   □

A metric geometry always has an infinite number of points (Problem A5). In particular, a finite geometry (Problems A13, B19, and B25 of Section 2.1) cannot be a metric geometry. On the other hand, Problem A6 shows that not every distance on an incidence geometry gives rise to a metric geometry even if it has infinitely many points. The points must "spread out." (Problems B14 and B15 of Section 2.2.) The Ruler Postulate is therefore a very strong restriction to place on an incidence geometry.

PROBLEM SET 2.3

Part A.

1. In the Euclidean Plane find a ruler f with $f(P) = 0$ and $f(Q) > 0$ for the given pair P and Q:
   i. $P = (2, 3), Q = (2, -5)$
   ii. $P = (2, 3), Q = (4, 0)$.

2. In the Poincaré Plane find a ruler f with $f(P) = 0$ and $f(Q) > 0$ for the given pair P and Q:
   i. $P = (2, 3), Q = (2, 1)$
   ii. $P = (2, 3), Q = (-1, 6)$.

3. In the Taxicab Plane find a ruler f with $f(P) = 0$ and $f(Q) > 0$ for the given pair P and Q:

   i. $P = (2, 3), Q = (2, -5)$
   ii. $P = (2, 3), Q = (4, 0)$.

4. Let P and Q be points in a metric geometry. Show that there is a point M such that $M \in \overrightarrow{PQ}$ and $d(P, M) = d(M, Q)$.

5. Prove that a line in a metric geometry has infinitely many points.

6. Let $\{\mathscr{S}, \mathscr{L}, d\}$ be a metric geometry and $Q \in \mathscr{S}$. If l is a line through Q show that for each real number $r > 0$ there is a point $P \in l$ with $d(P, Q) = r$. (This says that the line really extends indefinitely.)

Part B.

7. Let $g : \mathbb{R} \to \mathbb{R}$ by $g(s) = s/(|s| + 1)$. Show that g is injective.

8. Let $\{\mathscr{S}, \mathscr{L}, d\}$ be a metric geometry. For each $l \in \mathscr{L}$ choose a ruler f_l. Define the function $\tilde{d}$ by

$$\tilde{d}(P, Q) = |g(f_l(P)) - g(f_l(Q))|$$

where $l = \overrightarrow{PQ}$ and g is as in Problem B7. Show that $\tilde{d}$ is a distance function.

9. In Problem B8 show that $\{\mathscr{S}, \mathscr{L}, \tilde{d}\}$ is not a metric geometry.

CHAPTER 3
Betweenness and Elementary Figures

3.1 An Alternative Description of the Cartesian Plane

In Chapter 2 we introduced the Cartesian Plane model using ideas from analytic geometry as our motivation. This was useful at that time because it was the most intuitive method and led to simple verification of the incidence axioms. However, treating vertical and non-vertical lines separately does have its drawbacks. By making it necessary to break proofs into two cases, it leads to an artificial distinction between lines that really are not different in any geometric sense. Furthermore, as we develop additional axioms to verify we will need a more tractable notation. For these reasons we introduce an alternative description of the Cartesian Plane, one that is motivated by ideas from linear algebra, especially the notion of a vector.

Definition. If $A = (x_1, y_1)$, $B = (x_2, y_2) \in \mathbb{R}^2$ and $r \in \mathbb{R}$ then

(i) $A + B = (x_1 + x_2, y_1 + y_2) \in \mathbb{R}^2$
(ii) $rA = (rx_1, ry_1) \in \mathbb{R}^2$
(iii) $A - B = A + (-1)B = (x_1 - x_2, y_1 - y_2)$
(iv) $\langle A, B \rangle = x_1 x_2 + y_1 y_2 \in \mathbb{R}$
(v) $\|A\| = \sqrt{\langle A, A \rangle} \in \mathbb{R}$.

For those of you familiar with the ideas, all we are doing is viewing $\mathbb{R}^2$ as a vector space with its standard addition, scalar multiplication, and inner product. (Note that you probably wrote $\langle A, B \rangle$ as $A \cdot B$ before.) The following results are easily verified and are left as Problem A1.

Proposition 3.1.1. *For all $A, B, C \in \mathbb{R}^2$ and $r, s \in \mathbb{R}$*

(i) $A + B = B + A$ (ii) $(A + B) + C = A + (B + C)$
(iii) $r(A + B) = rA + rB$ (iv) $(r + s)A = rA + sA$
(v) $\langle A, B \rangle = \langle B, A \rangle$ (vi) $\langle rA, B \rangle = r \langle A, B \rangle$
(vii) $\langle A + B, C \rangle = \langle A, C \rangle + \langle B, C \rangle$ (viii) $\|rA\| = |r| \|A\|$
(ix) $\|A\| > 0$ if $A \neq (0, 0)$.

Using this notation we make $\mathbb{R}^2$ into an incidence geometry by defining the line through the distinct points A and B to be L_{AB} where

$$L_{AB} = \{X \in \mathbb{R}^2 \,|\, X = A + t(B - A) \text{ for some } t \in \mathbb{R}\}. \tag{1-1}$$

Proposition 3.1.2. *If $\mathscr{L}'$ is the collection of all subsets of $\mathbb{R}^2$ of the form L_{AB}, then $\{\mathbb{R}^2, \mathscr{L}'\}$ is the Cartesian Plane and hence is an incidence geometry.*

PROOF. Let $\mathscr{L}_E$ be the set of Cartesian lines as given in Chapter 2. We will show that $\mathscr{L}_E \subset \mathscr{L}'$ and $\mathscr{L}' \subset \mathscr{L}_E$.

Step 1. Let $l \in \mathscr{L}_E$ be a Cartesian line. If l is the vertical line L_a choose A to be $(a, 0)$ and B to be $(a, 1)$. $A, B \in l$.

$$l = \{(a, t) \,|\, t \in \mathbb{R}\} = \{(a, 0) + t(0, 1) \,|\, t \in \mathbb{R}\} = L_{AB} \in \mathscr{L}'.$$

Thus $l \in \mathscr{L}'$.

If l is the non-vertical line $L_{m,b}$ choose A to be $(0, b)$ and choose B to be $(1, b + m)$. $A, B \in l$.

$$l = \{(x, y) \,|\, y = mx + b\} = \{(x, y) = (t, mt + b) \,|\, t \in \mathbb{R}\}$$
$$= \{(x, y) = (0, b) + t(1, m) \,|\, t \in \mathbb{R}\} = L_{AB} \in \mathscr{L}'$$

Thus $l \in \mathscr{L}'$ and hence $\mathscr{L}_E \subset \mathscr{L}'$.

Step 2. Let $L_{AB} \in \mathscr{L}'$ with $A = (x_1, y_1)$, $B = (x_2, y_2)$, and $A \neq B$. If $x_1 = x_2$, then (since $A \neq B$) $y_2 - y_1 \neq 0$ and

$$L_{AB} = \{(x_1, y_1) + t(0, y_2 - y_1) \,|\, t \in \mathbb{R}\}$$
$$= \{(x_1, y_1 + t(y_2 - y_1)) \,|\, t \in \mathbb{R}\}$$
$$= \{(x, y) \in \mathbb{R}^2 \,|\, x = x_1\} = L_{x_1} \in \mathscr{L}_E.$$

Thus $L_{AB} \in \mathscr{L}_E$.

If $x_1 \neq x_2$ then $x_2 - x_1 \neq 0$, and we let

$$m = \frac{y_2 - y_1}{x_2 - x_1} \quad \text{and} \quad b = y_1 - mx_1.$$

Then

$$L_{AB} = \{(x_1, y_1) + t(x_2 - x_1, y_2 - y_1) | t \in \mathbb{R}\}$$
$$= \{(x_1, mx_1 + b) + t(x_2 - x_1, m(x_2 - x_1)) | t \in \mathbb{R}\}$$
$$= \{(x_1 + t(x_2 - x_1), m(x_1 + t(x_2 - x_1)) + b) | t \in \mathbb{R}\}$$
$$= \{(x, mx + b) | x \in \mathbb{R}\}$$
$$= \{(x, y) \in \mathbb{R} | y = mx + b\} = L_{m,b} \in \mathscr{L}_E.$$

Hence $L_{AB} \in \mathscr{L}_E$ and $\mathscr{L}' \subset \mathscr{L}_E$.

Thus we have shown that $\mathscr{L}_E = \mathscr{L}'$ so that $\{\mathbb{R}^2, \mathscr{L}'\}$ is the Cartesian Plane.   □

In Problem B6 you are asked to prove directly that $\{\mathbb{R}^2, \mathscr{L}'\}$ is an incidence geometry without any reference to the initial model $\{\mathbb{R}^2, \mathscr{L}_E\}$.

In terms of our new notation, the distance function d_E is described slightly differently as you will show in Problem A2. We also have a nice description of some rulers.

Proposition 3.1.3. *If $A, B \in \mathbb{R}^2$ then $d_E(A, B) = \|A - B\|$.*

Proposition 3.1.4. *If L_{AB} is a Cartesian line then $f: L_{AB} \to \mathbb{R}$ defined by*

$$f(A + t(B - A)) = t\|B - A\|$$

is a ruler for $\{\mathbb{R}^2, \mathscr{L}_E, d_E\}$.

PROOF. The function f makes sense only if for each point $P \in L_{AB}$ there is a unique value of t with $P = A + t(B - A)$. This can be seen to be true as follows.

Suppose $P = A + r(B - A)$ and $P = A + s(B - A)$. Then

$$(0, 0) = P - P = (A + r(B - A)) - (A + s(B - A))$$
$$= (r - s)(B - A)$$

so that either $r - s = 0$ or $B - A = (0, 0)$. Since $A \neq B$, $B - A \neq (0, 0)$ and so $r - s = 0$. That is, $r = s$ and there is a unique value of t with $P = A + t(B - A)$. Hence the function f makes sense.

The proof that f actually is a ruler is Problem A3.   □

In a college algebra or linear algebra course you probably learned that the dot product of two vectors is given by the product of the lengths of the vectors and the cosine of the angle in between:

$$\mathbf{a} \cdot \mathbf{b} = \|\mathbf{a}\| \|\mathbf{b}\| \cos \theta.$$

Since $|\cos \theta| \leq 1$, we have $|\mathbf{a} \cdot \mathbf{b}| \leq \|\mathbf{a}\| \|\mathbf{b}\|$. The Cauchy-Schwarz Inequality (Proposition 3.1.5) is a careful statement of this result without any reference to angles or the measurement of angles. It will be used in Chapter 5 when we develop angle measurement. We will apply it in this section to prove a special property of the Euclidean distance function (Proposition 3.1.6).

Proposition 3.1.5 (Cauchy-Schwarz Inequality). *If* $X, Y \in \mathbb{R}^2$ *then*

$$|\langle X, Y \rangle| \leq \|X\| \cdot \|Y\|. \tag{1-2}$$

Furthermore, equality holds in Inequality (1-2) if and only if either $Y = (0, 0)$ *or* $X = tY$ *for some* $t \in \mathbb{R}$.

PROOF. If $Y = (0,0)$ we clearly have $|\langle X, Y \rangle| = 0 = \|X\| \cdot \|Y\|$ and Inequality (1-2) is true. Hence we assume $Y \neq (0,0)$. Consider the function $g: \mathbb{R} \to \mathbb{R}$ by $g(t) = \|X - tY\|^2$. Then

$$g(t) = \langle X - tY, X - tY \rangle = \langle X, X \rangle - 2t\langle X, Y \rangle + t^2\langle Y, Y \rangle.$$

Because $Y \neq (0,0)$, $\langle Y, Y \rangle \neq 0$ and $g(t)$ is a quadratic function. Now $g(t) \geq 0$ for all t so that g cannot have two distinct real zeros. Since a quadratic function $at^2 + 2bt + c$ has distinct real zeros if and only if $b^2 - ac > 0$, it must be that

$$\langle X, Y \rangle^2 - \langle Y, Y \rangle\langle X, X \rangle \leq 0$$

or

$$|\langle X, Y \rangle| \leq \sqrt{\langle X, X \rangle\langle Y, Y \rangle} = \|X\| \cdot \|Y\|.$$

This gives the desired inequality.

When do we get equality in Inequality (1-2)? If $Y \neq 0$ then equality holds only when $g(t) = 0$ has a repeated real root. But $g(t) = 0$ if and only if $\|X - tY\| = 0$, i.e., $X = tY$. Thus equality holds if and only if either $Y = (0,0)$ or $X = tY$ for some $t \in \mathbb{R}$. $\qquad\qquad\square$

So far all our results about distance concerned points on a single line. The more important results of geometry will involve non-collinear points. The first property we will discuss is called the triangle inequality. It is so named because it says that the length of any side of a triangle is less than or equal to the sum of the lengths of the other two sides.

Definition. A distance function d on $\mathcal{S}$ satisfies the **triangle inequality** if

$$d(A, C) \leq d(A, B) + d(B, C) \quad \text{for all } A, B, C \in \mathcal{S}.$$

Proposition 3.1.6. *The Euclidean distance function* d_E *satisfies the triangle inequality.*

PROOF. First we use Proposition 3.1.5 to show that if X, $Y \in \mathbb{R}^2$ then $\|X + Y\| \le \|X\| + \|Y\|$.

$$\begin{aligned}
\|X + Y\|^2 &= \langle X + Y, X + Y \rangle = \langle X, X \rangle + 2\langle X, Y \rangle + \langle Y, Y \rangle \\
&= \|X\|^2 + 2\langle X, Y \rangle + \|Y\|^2 \\
&\le \|X\|^2 + 2|\langle X, Y \rangle| + \|Y\|^2 \\
&\le \|X\|^2 + 2\|X\|\,\|Y\| + \|Y\|^2 \\
&= (\|X\| + \|Y\|)^2
\end{aligned}$$

Hence $$\|X + Y\|^2 \le (\|X\| + \|Y\|)^2$$
or $$\|X + Y\| \le \|X\| + \|Y\|.$$

To complete the proof let $X = A - B$ and $Y = B - C$. $\qquad\square$

We shall see later that the triangle inequality is a consequence of certain other axioms that we will want our geometries to satisfy. In particular, it will hold in the Poincaré Plane. (A direct proof of this fact is for the masochistic.) However, it does not hold in every metric geometry as Problem B9 shows.

PROBLEM SET 3.1

Part A.

1. Prove Proposition 3.1.1.

2. Prove Proposition 3.1.3.

3. Complete the proof of Proposition 3.1.4.

4. Complete the proof of Proposition 3.1.6.

5. Show that the ruler in Proposition 3.1.4 is a coordinate system with A as origin and B positive.

Part B. "Prove" may mean "find a counterexample".

6. Let $\mathcal{L}'$ be the collection of subsets of $\mathbb{R}^2$ of the form given by Equation (1-1). Prove directly that $\{\mathbb{R}^2, \mathcal{L}'\}$ is an incidence geometry without using our previous model of the Cartesian Plane.

7. Prove that the Taxicab distance d_T satisfies the triangle inequality.

8. Prove that the max distance d_S on $\mathbb{R}^2$ satisfies the triangle inequality. (See Problem B18 of Section 2.2.)

9. Define a function d_F for points P and Q in $\mathbb{R}^2$ by

$$d_F(P,Q) = \begin{cases} 0 & \text{if } P = Q \\ d_E(P,Q) & \text{if } L_{PQ} \text{ is not vertical} \\ 3d_E(P,Q) & \text{if } L_{PQ} \text{ is vertical.} \end{cases}$$

a. Prove that d_F is a distance function on $\mathbb{R}^2$ and that $\{\mathbb{R}^2, \mathscr{L}_E, d_F\}$ is a metric geometry.

b. Prove that the triangle inequality is not satisfied for this distance, d_F.

Part C. Expository exercises.

10. What other descriptions of the Cartesian Plane can you find in various mathematic books? Why is it useful to have more than one description of an object such as the Cartesian Plane? (The answer could deal with either technical reasons or the level of the intended audience.)

11. Another example of different descriptions of the same mathematical concept is given by the notion of a "vector". Discuss different definitions of a vector, why they are the same, and what their possible uses are. (Note: a use need not be an application of vectors to another subject—it might be to use vectors to do mathematics.)

3.2 Betweenness

The concept of one point being between two others is an extremely important, yet at the same time, an extremely intuitive idea. It does not appear formally in Euclid, which leads to some logical flaws. (Euclid made certain tacit assumptions about betweenness. These often occurred as he reasoned from a figure—a shaky practice at best!) These flaws were first rectified by Pasch [1882] who axiomatized betweenness. Without a precise definition of between it is possible to produce erroneous "proofs." (What would Euclid have thought of the fallacious "proof" which will appear in Problem Set 6.4B that every triangle is isosceles?) In this section we shall use the distance function to define betweenness. In turn, betweenness will allow us to define elementary figures such as segments, rays, angles, and triangles.

Definition. B is **between** A and C if A, B, and C are distinct collinear points in the metric geometry $\{\mathscr{S}, \mathscr{L}, d\}$ and if

$$d(A,B) + d(B,C) = d(A,C). \tag{2-1}$$

Note that the definition of between requires that the three points all lie on the same line. (See Problem A10.) Because we will be using betweenness and distance constantly throughout the rest of the book we adopt the following simplified notation.

Notation. In a metric geometry $\{\mathscr{S}, \mathscr{L}, d\}$

    (i) A—B—C means B is between A and C

    (ii) AB denotes the distance $d(A, B)$.

Thus in this notation, Equation (2-1) becomes, for distinct collinear points,

$$A—B—C \quad \text{if and only if} \quad AB + BC = AC. \tag{2-2}$$

The axioms of the distance function are written in this notation as

  (i) $PQ \geq 0$;
  (ii) $PQ = 0$ if and only if $P = Q$;
  (iii) $PQ = QP$; and
  (iv) $PQ = |f(P) - f(Q)|$ for a ruler f on $\overleftrightarrow{PQ}$. $\qquad\qquad$ (2-3)

Note that by using PQ for the distance we have dropped all reference to which distance function we are using. Since an incidence geometry may have more than one distance function, whenever we use the notation PQ for $d(P, Q)$ it must be clear which distance is involved. In our basic models we will continue to use the notation d_E, d_H, and d_T.

Example 3.2.1. Let $A = (-\frac{1}{2}, \sqrt{3}/2)$, $B = (0, 1)$, and $C = (\frac{1}{2}, \sqrt{3}/2)$ be points in the Poincaré Plane. Show that A—B—C.

SOLUTION. A, B, and C are on the type II line $_0L_1 = \{(x, y) \in \mathbb{H} \,|\, x^2 + y^2 = 1\}$. From Equation (2-3) of Chapter 2

$$AB = d_H(A, B) = \left| \ln \frac{\frac{-\frac{1}{2}+1}{\sqrt{3}/2}}{(\frac{1}{1})} \right| = \ln \sqrt{3}$$

$$BC = d_H(B, C) = \left| \ln \frac{\sqrt{3}}{3} \right| = \ln\sqrt{3} \quad \text{and} \quad AC = d_H(A, C) = \ln 3.$$

Thus $d_H(A, B) + d_H(B, C) = d_H(A, C)$ and A—B—C. Note that in Figure 3-1 the point B "looks" like it is between A and C. $\qquad\qquad\square$

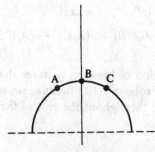

Figure 3-1

Theorem 3.2.2. *If $A-B-C$ then $C-B-A$.*

PROOF. If A, B, and C are distinct and collinear, then so are C, B, and A. Since $A-B-C$, Equation (2-2) shows that $AB + BC = AC$. Since $PQ = QP$ for all P and Q, we have $BA + CB = CA$ or

$$CB + BA = CA$$

which is what we needed to show. $\qquad\qquad\qquad\qquad\qquad\qquad\qquad\square$

If l is a line with a ruler, the next theorem will allow us to interpret betweenness on l in terms of a corresponding notion of betweenness for real numbers. This will be a useful method of proving certain results involving betweenness. Thus we will be using the notion of betweenness on the real line to help us with the betweenness in a metric geometry.

Definition. If x, y, and z are real numbers, then y is **between** x and z (written $x * y * z$) if either

$$x < y < z \quad \text{or} \quad z < y < x.$$

Note that if x, y, and z are distinct real numbers then exactly one is between the other two: one is largest, one is smallest, and the other between them.

Theorem 3.2.3. *Let l be a line and f a coordinate system for l. If A, B, and C are three points of l with coordinates x, y, and z respectively, then $A-B-C$ if and only if $x * y * z$.*

PROOF. Note that if A, B, and C are not distinct then both $A-B-C$ and $x * y * z$ are false. Hence we may assume A, B, and C are distinct. We first prove that $x * y * z$ if $A-B-C$.

We are given that $x = f(A)$, $y = f(B)$, and $z = f(C)$, and that $AB + BC = AC$. The Ruler Equation (2-3) indicates that

$$AB = |f(A) - f(B)| = |x - y|, \qquad BC = |y - z|, \quad \text{and} \quad AC = |x - z|$$

so that

$$|x - y| + |y - z| = |x - z|. \qquad\qquad\qquad (2-4)$$

We shall show that Equation (2-4) implies that either $x < y < z$ or $z < y < x$.

Since A, B, C are distinct then so are x, y, z and exactly one of the following cases must occur:

$$\begin{array}{ll} \text{(i)} \ x < y < z & \text{(ii)} \ z < y < x \\ \text{(iii)} \ y < x < z & \text{(iv)} \ z < x < y \\ \text{(v)} \ x < z < y & \text{(vi)} \ y < z < x \end{array}$$

We will show that case (iii) leads to a contradiction. Similar arguments dispose of cases (iv), (v), and (vi).

Case (iii) implies that

$$|x - y| = x - y, \qquad |y - z| = z - y, \quad \text{and} \quad |x - z| = z - x.$$

If we substitute these equations into Equation (2-4) we obtain

$$x - y + z - y = z - x$$

so that

$$x = y. \tag{2-5}$$

This contradicts the fact that x, y, z are distinct. Hence case (iii) does not hold. By Problem A4 neither do cases (iv), (v), or (vi). Thus $x * y * z$ (cases (i) and (ii)).

We now show that if $x * y * z$ then A—B—C. Assume that $x < y < z$. (The case $z < y < x$ is similar.) In this case $|x - y| = y - x$, $|x - z| = z - x$, and $|y - z| = z - y$ so that

$$|x - y| + |y - z| = |x - z|$$

or

$$|f(A) - f(B)| + |f(B) - f(C)| = |f(A) - f(C)|$$

or

$$AB + BC = AC.$$

Thus since A, B, and C are collinear and distinct, A—B—C. $\qquad\square$

Corollary 3.2.4. *Given three distinct points on a line, one and only one of these points is between the other two.*

PROOF. This is immediate since the corresponding statement is true for three distinct real numbers. $\qquad\square$

Note that this result says that if we have three distinct points on a line, we may name them as A, B, and C with A—B—C. However, if the points are already named A, B, C in some way, then all we can say is that one of A—B—C, B—A—C, or A—C—B is true.

The next result (whose proof is left as Problem A5) gives a useful characterization of betweenness for the Euclidean Plane.

Proposition 3.2.5. *In the Euclidean plane A—B—C if and only if there is a number t with $0 < t < 1$ and $B = A + t(C - A)$.*

Theorem 3.2.6. *If A and B are distinct points in a metric geometry then*

(i) *there is a point C with A—B—C; and*
(ii) *there is a point D with A—D—B.*

PROOF. Let f be a ruler for the line $\overline{AB}$ with $f(A) < f(B)$ and set $x = f(A)$ and $y = f(B)$. To prove (i) let $z = y + 1$ and $C = f^{-1}(z)$. Then A—B—C since $x < y < z$.

To prove (ii), we define $w \in \mathbb{R}$ and $D \in \overrightarrow{AB}$ by $w = (x + y)/2$ and $D = f^{-1}(w)$. Then $A—D—B$ since $x < w < y$. $\qquad\qquad\qquad\square$

In the next section we will define what is meant by a segment. Once we have that terminology we will see that Theorem 3.2.6 (i) says that a segment may be extended (one of Euclid's axioms). Part (ii) says that given two points, there is always at least one point between them. In fact, a careful examination of the proof of part (ii) shows that we could prove that there are infinitely many points between A and B.

Definition. $A—B—C—D$ means that $A—B—C$, $A—B—D$, $A—C—D$, and $B—C—D$.

This definition can be visualized as meaning that A, B, C, and D lie on the same line (see Problem A6) and are in forward or reverse order as in Figure 3-2. To remember the four conditions of the definition merely drop one letter out of $A—B—C—D$. The resulting betweenness relations are the conditions of the definition. Actually only the conditions $A—B—C$ and $B—C—D$ are needed (see Problem A7). Note that this definition does not say that $A—B—C—D$ if A, B, C, and D are collinear and $AB + BC + CD = AD$. The reason for this is that the latter statement is not strong enough to prove that $A—B—C$.

Given four distinct collinear points can we name them A, B, C, D so that $A—B—C—D$? This seems obvious, but a careful proof requires the use of a ruler.

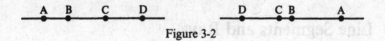

Figure 3-2

PROBLEM SET 3.2

Part A.

1. Let $A = (x_1, y_1)$, $B = (x_2, y_2)$, and $C = (x, y)$ be three collinear points in the Euclidean Plane with $x_1 < x_2$. Prove that $A—C—B$ if and only if $x_1 < x < x_2$.

2. Formulate and prove a condition for $A—C—B$ if A and B are on the same type I line in the Poincaré Plane.

3. If $A = (4, 7)$, $B = (1, 1)$, and $C = (2, 3)$ prove that $A—C—B$ in the Taxicab Plane.

4. Prove that cases (iv), (v), and (vi) of Theorem 3.2.3 also lead to contradictions.

5. Prove Proposition 3.2.5.

6. If A—B—C—D in a metric geometry, prove that $\{A, B, C, D\}$ is a collinear set.

7. Prove that if A—B—C and B—C—D in a metric geometry, then A—B—D and A—C—D also so that A—B—C—D.

8. Let A, B, C be three (not necessarily distinct) collinear points in a metric geometry. Give all possible betweenness relations ($A = B$ or A—B—C or . . .).

9. Let four distinct collinear points be given in a metric geometry. Prove that they can be named A, B, C and D in such a manner that A—B—C—D.

10. In the Taxicab Plane, find three points A, B, C which are not collinear but $d_T(A, C) = d_T(A, B) + d_T(B, C)$. This problem shows why the definition of between requires collinear points.

Part C. Expository exercises.

11. Discuss the problems that can result if one is not careful with the notion of "between". An excellent reference is Chapter 3 of Greenberg [1980]. How would you present this subtlety to high school students? What mistakes did Euclid himself make?

12. What are the Hilbert axioms for betweenness (see Greenberg [1980])? Why do we not have to use all these axioms in our development? Which way do you prefer, and why?

13. Let A, B, and C be three points on a great circle on the Riemann Sphere. Doesn't it look like any one of them is between the other two? How would you explain this apparent contradiction of Corollary 3.2.4? There is a fundamental concept of betweenness involved here.

3.3 Line Segments and Rays

The notion of a line is an integral part of geometry. We are now in a position to talk about parts of a line: line segments and rays. In the next section the concept of a triangle will be defined in terms of line segments, while angles will be defined in terms of rays. In this section we will develop some of the basic properties of segments and rays and will introduce the idea of congruence of segments. This concept will be needed for the study of congruence of triangles and is fundamental in geometry.

Definition. If A and B are distinct points in a metric geometry $\{\mathscr{S}, \mathscr{L}, d\}$ then the **line segment** from A to B is the set

$$\overline{AB} = \{C \in \mathscr{S} \mid A\text{—}C\text{—}B \text{ or } C = A \text{ or } C = B\}.$$

Example 3.3.1. Let $A = (x_1, y_1)$ and $B = (x_2, y_2)$ lie on the type II line $_cL_r$ in the Poincaré Plane. If $x_1 < x_2$ show that

$$\overline{AB} = \{C = (x, y) \in {}_cL_r | x_1 \leq x \leq x_2\}.$$

SOLUTION. Since $x = x_1$ corresponds to $C = A$ and $x = x_2$ corresponds to $C = B$ we must show that A—C—B if and only if $x_1 < x < x_2$, where $C = (x, y) \in {}_cL_r$. Recall that for $l = {}_cL_r$, the standard ruler $f : l \rightarrow \mathbb{R}$ is given by

$$f(x, y) = \ln\left(\frac{x - c + r}{y}\right)$$

as in Equation (2-12) of Chapter 2. Now Theorem 3.2.3 says that if A—C—B then

$$f(A) * f(C) * f(B). \tag{3-1}$$

The inverse of f is the function $g : \mathbb{R} \rightarrow {}_cL_r$ given by

$$g(t) = (c + r\tanh(t), r\operatorname{sech}(t)).$$

(See Equation (2-10) of Section 2.2.) If we let $f(A) = t_1, f(B) = t_2$, and $f(C) = t_3$ then Equation (3-1) becomes

$$t_1 * t_3 * t_2. \tag{3-2}$$

Since $\tanh(t)$ is a strictly increasing function, Equation (3-2) implies that

$$(c + r\tanh(t_1)) * (c + r\tanh(t_3)) * (c + r\tanh(t_2))$$

or

$$x_1 * x * x_2. \tag{3-3}$$

Since $x_1 < x_2$ by hypothesis, we have $x_1 < x < x_2$. Hence if $C \in \overline{AB}$ with $C = (x, y)$, then $x_1 \leq x \leq x_2$. The other direction of the proof ($x_1 \leq x \leq x_2$ implies $C \in \overline{AB}$) is left to Problem A1. See Figure 3-3 for a sketch of $\overline{AB}$. $\qquad\square$

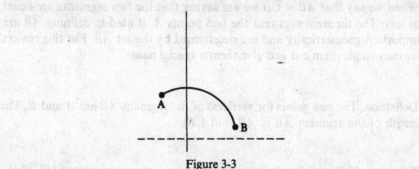

Figure 3-3

Definition. Let $\mathcal{A}$ be a subset of a metric geometry. A point $B \in \mathcal{A}$ is a **passing point** of $\mathcal{A}$ if there exists points $X, Y \in \mathcal{A}$ with X—B—Y. Otherwise B is an **extreme point** of $\mathcal{A}$.

The concept of extreme points and the next result allow us to define the end points of a segment.

Theorem 3.3.2. *If A and B are two points in a metric geometry then the only extreme points of the segment $\overline{AB}$ are A and B themselves. In particular, if $\overline{AB} = \overline{CD}$ then $\{A, B\} = \{C, D\}$.*

PROOF. We use proof by contradiction to show that A is not a passing point of $\overline{AB}$. Suppose that A is between two points X, Y of $\overline{AB}$ so that $X—A—Y$. The proof hinges on the fact there is then no place for B. There are six possibilities: $B—X—A—Y$, $B = X$, $X—B—A—Y$, $X—A—B—Y$, $B = Y$, or $X—A—Y—B$. The first three cases imply that $B—A—Y$ (so that $Y \notin \overline{AB}$) and the last three cases imply that $X—A—B$ (so that $X \notin \overline{AB}$). Either way, we have a contradiction of $X, Y \in \overline{AB}$. Thus A is not between two points of $\overline{AB}$. Similarly, B is not between two points of $\overline{AB}$. Thus A and B are extreme points.

We next show that any other point of $\overline{AB}$ is a passing point of $\overline{AB}$. If $Z \in \overline{AB}$ and $Z \neq A$, $Z \neq B$, then $A—Z—B$. Hence Z is between two points of $\overline{AB}$ and is a passing point. Thus A and B are the only extreme points of $\overline{AB}$.

Finally, suppose that $\overline{AB} = \overline{CD}$. Then

$$\{A, B\} = \{Z \in \overline{AB} | Z \text{ is an extreme point of } \overline{AB}\}$$
$$= \{Z \in \overline{CD} | Z \text{ is an extreme point of } \overline{CD}\}$$
$$= \{C, D\}. \qquad \square$$

The importance of this result is the following. $\overline{AB}$ is defined as a set. When we say that $\overline{AB} = \overline{CD}$ we are saying that the two segments are equal as sets. The theorem says that the two points A, B used in defining $\overline{AB}$ are important geometrically and are determined by the set $\overline{AB}$. For this reason we may single them out and give them a special name.

Definition. The **end points** (or **vertices**) of the segment $\overline{AB}$ are A and B. The **length** of the segment $\overline{AB}$ is $AB = d(A, B)$.

Definition. If A and B are distinct points in a metric geometry $\{\mathcal{S}, \mathcal{L}, d\}$ then the **ray** from A toward B is the set

$$\overrightarrow{AB} = \overline{AB} \cup \{C \in \mathcal{S} | A—B—C\}. \qquad (3\text{-}4)$$

Note the ray $\overrightarrow{AB}$ is a subset of the line $\overleftrightarrow{AB}$. Rays in the Euclidean Plane and the Poincaré Plane are illustrated in Figures 3-4 and 3-5.

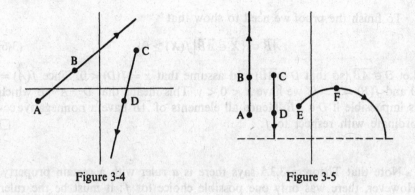

Figure 3-4   Figure 3-5

The next two results are left as exercises.

Proposition 3.3.3. *In the Euclidean Plane, line segments and rays are given by*

$$\overline{AB} = \{C \in \mathbb{R}^2 \,|\, C = A + t(B - A) \text{ for some } t \text{ with } 0 \le t \le 1\}$$
$$\overrightarrow{AB} = \{C \in \mathbb{R}^2 \,|\, C = A + t(B - A) \text{ for some } t \ge 0\}.$$

Theorem 3.3.4. *In a metric geometry*

(i) *if* $C \in \overrightarrow{AB}$ *and* $C \ne A$, *then* $\overrightarrow{AC} = \overrightarrow{AB}$;
(ii) *if* $\overrightarrow{AB} = \overrightarrow{CD}$ *then* $A = C$.

Theorem 3.3.4 (i) tells us that a given ray can be named in many ways. Part (ii) says that one point of $\overrightarrow{AB}$ is special. (A can be shown to be the only extreme point of $\overrightarrow{AB}$.) For this reason we can give it a special name.

Definition. The **vertex** (or **initial point**) of the ray $\overrightarrow{AB}$ is the point A.

Theorem 3.3.5. *If A and B are distinct points in a metric geometry then there is a ruler $f : \overrightarrow{AB} \to \mathbb{R}$ such that*

$$\overrightarrow{AB} = \{X \in \overleftrightarrow{AB} \,|\, f(X) \ge 0\}.$$

PROOF. Let f be the special coordinate system with origin A and B positive. We claim that this ruler f is the one we desire. We first show that

$$\{X \in \overleftrightarrow{AB} \,|\, f(X) \ge 0\} \subset \overrightarrow{AB}. \tag{3-5}$$

Suppose $X \in \overleftrightarrow{AB}$ with $f(X) \ge 0$. Let $x = f(X)$ and let $f(B) = y$, which is positive by assumption. If $x = 0$ then $X = A$ and $X \in \overrightarrow{AB}$. If $x = y$ then $X = B$ and $X \in \overrightarrow{AB}$. There are only two possibilities left. Either $0 < x < y$, in which case $A—X—B$ so that $X \in \overline{AB}$ and $X \in \overrightarrow{AB}$, or $0 < y < x$, in which case $A—B—X$ and so $X \in \overrightarrow{AB}$ by Equation (3-4). In all cases, $X \in \overrightarrow{AB}$ and Condition (3-5) is proved.

To finish the proof we need to show that

$$\overrightarrow{AB} \subset \{X \in \overleftrightarrow{AB} \mid f(X) \geq 0\}. \tag{3-6}$$

Let $D \in \overrightarrow{AB}$ (so that $D \in \overleftrightarrow{AB}$) and assume that $x = f(D) < 0$. Since $f(A) = 0$ and $f(B) = y > 0$, we have $x < 0 < y$. This means that $D—A—B$ which is impossible if $D \in \overrightarrow{AB}$. Hence all elements of $\overrightarrow{AB}$ have a nonnegative coordinate with respect to f. $\qquad\qquad\qquad\square$

Note that Theorem 3.3.5 says there is *a* ruler with a certain property. However, there was only one possible choice for f: it must be the ruler which is zero at the (unique) vertex and positive elsewhere on the ray.

One of the most familiar (and most basic) topics in geometry is the study of congruences, especially the congruence of triangles. In order to reach the point where we can develop this rigorously, we must consider the congruence of line segments and the congruence of angles. We consider the former here and the latter in Chapter 5.

Definition. Two line segments $\overline{AB}$ and $\overline{CD}$ in a metric geometry are **congruent** (written $\overline{AB} \simeq \overline{CD}$) if their lengths are equal; that is

$$\overline{AB} \simeq \overline{CD} \quad \text{if } AB = CD.$$

The next result will be used continually when dealing with congruence in triangles. It allows us to mark off (or construct) on a ray a unique segment which is congruent to a given segment.

Theorem 3.3.6 (Segment Construction). *If $\overrightarrow{AB}$ is a ray and $\overline{PQ}$ is a line segment in a metric geometry, then there is a unique point $C \in \overrightarrow{AB}$ with $\overline{PQ} \simeq \overline{AC}$.*

PROOF. Let f be a special coordinate system for the line $\overleftrightarrow{AB}$ with A as origin and B positive. Then $f(A) = 0$ and $\overrightarrow{AB} = \{X \in \overleftrightarrow{AB} \mid f(X) \geq 0\}$. Let $r = PQ$ and set $C = f^{-1}(r)$. Since $r = PQ > 0$, we have $C \in \overrightarrow{AB}$. Furthermore,

$$AC = |f(A) - f(C)| = |0 - r| = r = PQ$$

so that $\overline{AC} \simeq \overline{PQ}$. Thus we have at least one point C on $\overrightarrow{AB}$ with $\overline{AC} \simeq \overline{PQ}$.

Now suppose $C' \in \overrightarrow{AB}$ with $\overline{AC'} \simeq \overline{PQ}$. Then since $C' \in \overrightarrow{AB}$, $f(C') > 0$ and

$$f(C') = f(C') - f(A) = |f(C') - f(A)|$$
$$= AC' = PQ = f(C).$$

Since f is injective, $C' = C$ and so there is exactly one point $C \in \overrightarrow{AB}$ with $\overline{AC} \simeq \overline{PQ}$.   □

Example 3.3.7. In the Poincaré Plane let $A = (0, 2)$, $B = (0, 1)$, $P = (0, 4)$, $Q = (7, 3)$. Find $C \in \overrightarrow{AB}$ so that $\overline{AC} \simeq \overline{PQ}$.

SOLUTION. First we must determine PQ. Both P and Q lie on $_3L_5$ so that

$$PQ = d_H(P, Q) = \left| \ln \left(\frac{\left(\dfrac{-3 + 5}{4} \right)}{\dfrac{4 + 5}{3}} \right) \right| = \left| \ln \frac{1}{6} \right| = \ln 6.$$

Since $C = (0, y)$ is on the type I line $\overleftrightarrow{AB}$, $d_H(A, C) = |\ln y/2|$. In order that $\overline{AC} \simeq \overline{PQ}$ we need $\ln y/2 = \pm \ln 6$. Hence

$$\frac{y}{2} = 6 \quad \text{or} \quad \frac{y}{2} = \frac{1}{6}.$$

Thus

$$y = 12 \quad \text{or} \quad y = \frac{1}{3}.$$

Since we want $C \in \overrightarrow{AB}$ we must take $C = (0, \frac{1}{3})$. See Figure 3-6.   □

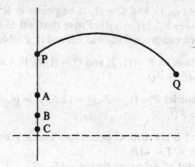

Figure 3-6

Note that in this example a segment from a type I line is congruent to a segment from a type II line. Of course, two such segments could never be equal but they can be congruent. Do not confuse "congruent" and "equal"! Congruence of segments means "equal in length" whereas equality of segments means "equal as sets."

Theorem 3.3.8 (Segment Addition). *In a metric geometry, if A—B—C, P—Q—R, $\overline{AB} \simeq \overline{PQ}$, and $\overline{BC} \simeq \overline{QR}$, then $\overline{AC} \simeq \overline{PR}$.*

Theorem 3.3.8 says that we create congruent segments by "adding" congruent segments. The following theorem, whose proof is also left as an exercise, says we may also "subtract" congruent segments.

Theorem 3.3.9 (Segment Subtraction). *In a metric geometry, if* A—B—C, P—Q—R, $\overline{AB} \simeq \overline{PQ}$, *and* $\overline{AC} \simeq \overline{PR}$, *then* $\overline{BC} \simeq \overline{QR}$.

PROBLEM SET 3.3

Part A.

1. Complete the solution of Example 3.3.1.

2. Prove Proposition 3.3.3.

3. Prove Theorem 3.3.4.

*4. Prove that "congruence" is an equivalence relation on the set of all line segments in a metric geometry.

5. Prove Theorem 3.3.8.

6. Prove Theorem 3.3.9.

7. In the Taxicab Plane show that if $A = (-\frac{5}{2}, 2)$, $B = (\frac{1}{2}, 2)$, $C = (2, 2)$, $P = (0, 0)$, $Q = (2, 1)$ and $R = (3, \frac{3}{2})$then A—B—C and P—Q—R. Show that $\overline{AB} \simeq \overline{PQ}$, $\overline{BC} \simeq \overline{QR}$ and $\overline{AC} \simeq \overline{PR}$. Sketch an appropriate picture.

8. Let $A = (0, 0)$, $B = (\frac{1}{10}, 1)$, and $C = (1, 1)$ be points in $\mathbb{R}^2$ with the max distance $d_S(P, Q) = \max\{|x_1 - x_2|, |y_1 - y_2|\}$. Prove that $\overline{AB} \simeq \overline{AC}$. Sketch the two segments. Do they look congruent? (d_S was defined in Problem B18 of Section 2.2.)

9. In the Poincaré Plane let $P = (1, 2)$ and $Q = (1, 4)$. If $A = (0, 2)$ and $B = (1, \sqrt{3})$, find $C \in \overrightarrow{AB}$ with $\overline{AC} \simeq \overline{PQ}$.

10. In the Taxicab Plane let $P = (1, -2)$, $Q = (2, 5)$, $A = (4, -1)$ and $B = (3, 2)$. Find $C \in \overrightarrow{AB}$ with $\overline{AC} \simeq \overline{PQ}$.

*11. Suppose that A and B are distinct points in a metric geometry. $M \in \overrightarrow{AB}$ is called a **midpoint** of $\overline{AB}$ if $AM = MB$.
    a. If M is a midpoint of $\overline{AB}$ prove that A—M—B.
    b. If $A = (0, 4)$ and $B = (0, 1)$ are points in the Poincaré Plane find a midpoint, M, of $\overline{AB}$. Sketch A, B and M on a graph. Does M look like a midpoint?

*12. If A and B are distinct points of a metric geometry, prove that
    a. the segment $\overline{AB}$ has a midpoint M. (See Problem A11.)
    b. the midpoint M of $\overline{AB}$ is unique.

13. Prove that $\overline{AB} = \overline{BA}$ for any distinct points A and B in a metric geometry.

14. If $D \in \overrightarrow{AB}$—$\overline{AB}$ in a metric geometry, prove that $\overrightarrow{AB} = \overline{AB} \cup \overrightarrow{AD}$.

15. If $A \neq B$ in a metric geometry, prove that $\overleftrightarrow{AB} = \overrightarrow{AB} \cup \overrightarrow{BA}$ and $\overline{AB} = \overrightarrow{AB} \cap \overrightarrow{BA}$.

Part B. "Prove" may mean "find a counterexample".

16. Prove that in a metric geometry, $\overrightarrow{AB}$ is the set of all points $C \in \overleftrightarrow{AB}$ such that A is not between C and B.

17. Prove that in a metric geometry any segment can be divided into n congruent parts for any $n > 0$. More formally: Let A and B be distinct points in a metric geometry.
   a. Prove there are points $X_0, X_1, \ldots, X_n$ on $\overrightarrow{AB}$ such that $X_0 = A$, $X_n = B$, $X_i{-}X_{i+1}{-}X_{i+2}$ for $i = 0, 1, \ldots, n - 2$; $X_i X_{i+1} = AB/n$, and $\overline{AB} = \bigcup_{i=0}^{n-1} \overline{X_i X_{i+1}}$.
   b. Prove that the points X_i given by the above are unique.

18. If $\overline{AB} = \overline{CD}$ in a metric geometry, prove that $A = C$ and $B = D$.

19. If $D \notin \overleftrightarrow{AB}$ in a metric geometry, prove that $\overrightarrow{AD} \cap \overrightarrow{AB} = \{A\}$.

20. If $D \in \overleftrightarrow{AB}$ in a metric geometry, prove that either $\overrightarrow{AD} = \overrightarrow{AB}$ or $\overrightarrow{AD} \cup \overrightarrow{AB} = \overleftrightarrow{AB}$.

21. In a metric geometry suppose that $A{-}B{-}C$, $\overline{AB} \simeq \overline{PQ}$, $\overline{AC} \simeq \overline{PR}$, $\overline{BC} \simeq \overline{QR}$. Prove that $P{-}Q{-}R$.

Part C. Expository exercises.

22. Rays, segments, and points can be quite beautiful. Go to an art book such as Feldman [1981] or McCall [1970] and identify pictures with significant geometric content. The work of artists such as Seurat, Mondrian, and Kandinski show manifold geometric ideas. See Millman-Speranza [1990] for a presentation of these ideas at the elementary or middle school level.

23. Discuss the statement "Congruent triangles are the same" on both a mathematical and a philosophical level.

3.4 Angles and Triangles

In this section we will define angles and triangles in an arbitrary metric geometry. Just as in the case of segments and rays, they will be defined as sets using the concept of betweenness. We will also show that the idea of a vertex of an angle or a triangle is well defined. It is important to note that an angle is a set, not a number like 45°. We will view numbers as properties of angles when we define angle measure in Chapter 5.

For us an angle will consist of two rays which are not collinear but have the same initial point.

Definition. If A, B and C are noncollinear points in a metric geometry then the angle $\angle ABC$ is the set

$$\angle ABC = \overrightarrow{BA} \cup \overrightarrow{BC}.$$

Note that a line is not permitted to be an angle nor is a ray since $\{A, B, C\}$ in the definition must be noncollinear. This is for convenience: If "straight angles" or "zero angles" were allowed, we would have to make assumptions to deal with those special cases when stating theorems. Some

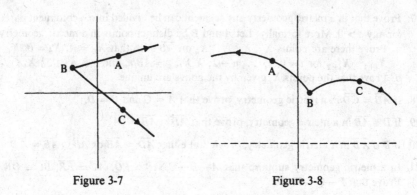

Figure 3-7   Figure 3-8

angles in the Euclidean and Poincaré Planes are sketched in Figures 3-7 and 3-8.

It is customary to talk about the vertex of $\angle ABC$ as the point B. However, *a priori*, it is not clear that the point B is well defined. After all, it might be possible for $\angle ABC = \angle DEF$ without $B = E$. (Of course, we will prove that B does in fact equal E.) If this seems unnecessarily pedantic, ask someone if $\angle ABC = \angle DEF$ implies that $A = D$ and $C = F$. Theorem 3.4.2 below contains the "well defined" result referred to above and is similar in spirit to Theorem 3.3.2. Its proof needs a preliminary lemma.

Lemma 3.4.1. *In a metric geometry, B is the only extreme point of $\angle ABC$.*

PROOF. We first show that if $Z \in \angle ABC$ and $Z \neq B$ then Z is a passing point of $\angle ABC$. If $Z \in \angle ABC$ and $Z \neq B$ then either $Z \in \overrightarrow{BA}$ or $Z \in \overrightarrow{BC}$. Since the two cases are similar we may assume $Z \in \overrightarrow{BA}$. Since $Z \neq B$, Theorem 3.3.4 implies that $\overrightarrow{BA} = \overrightarrow{BZ}$. There exists a $D \in \overrightarrow{BZ}$ such that $B—Z—D$. (Why?) Thus $D \in \overrightarrow{BA}$ and Z is between two points of $\angle ABC$, namely B and D.

Next we show that B is not a passing point of $\angle ABC$. We do this with a proof by contradiction. Suppose that $X—B—Y$ with $X, Y \in \angle ABC$. X belongs to either $\overrightarrow{BA}$ or $\overrightarrow{BC}$. Both cases are similar so that we may assume that $X \in \overrightarrow{BA}$. Since $X \neq B$, $\overrightarrow{BA} = \overrightarrow{BX}$ by Theorem 3.3.4. Since $Y—B—X$, $Y \notin \overrightarrow{BX} = \overrightarrow{BA}$. Since $Y \in \angle ABC$, this means that $Y \in \overrightarrow{BC}$ and $\overrightarrow{BC} = \overrightarrow{BY}$. But then $A \in \overrightarrow{BA} = \overrightarrow{BX} \subset \overleftrightarrow{XY}$, $B \in \overline{XY} \subset \overleftrightarrow{XY}$, and $C \in \overrightarrow{BC} = \overrightarrow{BY} \subset \overleftrightarrow{XY}$. Thus A, B, and C are collinear, which is impossible since we are given $\angle ABC$. Thus B is not between two points of $\angle ABC$ and is the only extreme point of $\angle ABC$.   ☐

Theorem 3.4.2. *In a metric geometry, if $\angle ABC = \angle DEF$ then $B = E$.*

PROOF.

$$\{B\} = \{Z \in \angle ABC \mid Z \text{ is an extreme point of } \angle ABC\}$$
$$= \{Z \in \angle DEF \mid Z \text{ is an extreme point of } \angle DEF\}$$
$$= \{E\}. \qquad\qquad\qquad\qquad\qquad\qquad\qquad\qquad\quad ☐$$

After Theorem 3.4.2 we may make the following definition without any ambiguity.

Definition. The **vertex** of the angle $\angle ABC$ in a metric geometry is the point B.

Definition. If $\{A, B, C\}$ are noncollinear points in a metric geometry then the **triangle** ABC is the set

$$\triangle ABC = \overline{AB} \cup \overline{BC} \cup \overline{CA}.$$

Triangles in the Euclidean Plane and the Poincaré Plane are given in Figures 3-9 and 3-10. The Poincaré triangles certainly do not look standard!

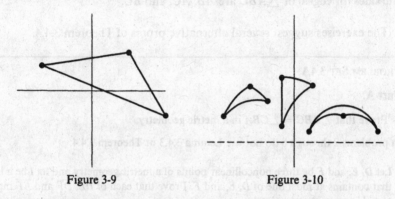

Figure 3-9   Figure 3-10

We now know that the vertex of a ray, the pair of endpoints of a segment, and the vertex of an angle are all uniquely determined by the ray, segment or angle. We shall show that the points A, B and C of $\triangle ABC$ are also uniquely determined in Theorem 3.4.4. This also needs a preliminary result.

Lemma 3.4.3. *In a metric geometry, if A, B, and C are not collinear then A is an extreme point of $\triangle ABC$.*

PROOF. Our proof is by contradiction. Suppose that $D\text{---}A\text{---}E$ with D, $E \in \triangle ABC$. We show that this implies that both D and E are in $\overline{BC}$, which leads to a contradiction.

If $D \in \overline{AB}$ then either $D = B$ so that $E\text{---}A\text{---}B$ or $D \neq B$ so that $E\text{---}A\text{---}D\text{---}B$ and $E\text{---}A\text{---}B$ ($D \neq A$ because $D\text{---}A\text{---}E$). Either way $E \notin \overline{AB}$. If E belongs to $\overline{AC}$ or $\overline{BC}$ then either $C\text{---}E\text{---}A\text{---}B$ or $C = E$ so that $C\text{---}A\text{---}B$. But A, B, C are not collinear. Hence we cannot have $D \in \overline{AB}$ because E must belong to one of $\overline{AB}$, $\overline{AC}$, or $\overline{BC}$.

In a similar fashion $D \notin \overline{AC}$. Since $D \in \triangle ABC$ it must be that $D \in \overline{BC}$. A similar proof shows that $E \in \overline{BC}$ also. Thus $D, E \in \overline{BC}$.

But $D\text{---}A\text{---}E$ implies $A \in \overleftrightarrow{BC}$ also, which is contrary to the hypothesis that A, B, C are noncollinear. Hence it cannot be that A is between two points of $\triangle ABC$.   □

Theorem 3.4.4. *In a metric geometry, if* $\triangle ABC = \triangle DEF$ *then* $\{A, B, C\} = \{D, E, F\}$.

PROOF. If $X \in \triangle ABC$ and $X \notin \{A, B, C\}$ then X is in one the segments $\overline{AB}$, $\overline{BC}$, or $\overline{AC}$ but is not an end point. Then X is a passing point of that segment and hence a passing point of $\triangle ABC$. By Lemma 3.4.3 we have

$$\{A, B, C\} = \{X \in \triangle ABC | X \text{ is an extreme point of } \triangle ABC\}$$
$$= \{X \in \triangle DEF | X \text{ is an extreme point of } \triangle DEF\}$$
$$= \{D, E, F\}. \qquad \qquad \square$$

Definition. In a metric geometry the **vertices** of $\triangle ABC$ are the points A, B, C. The **sides** (or **edges**) of $\triangle ABC$ are $\overline{AB}$, $\overline{AC}$, and $\overline{BC}$.

The exercises suggest several alternative proofs of Theorem 3.4.4.

PROBLEM SET 3.4A

Part A.

1. Prove that $\angle ABC = \angle CBA$ in a metric geometry.

In problems 2 through 8 do not use Lemma 3.4.3 or Theorem 3.4.4.

2. Let D, E, and F be three noncollinear points of a metric geometry and let l be a line that contains at most one of D, E, and F. Prove that each of $\overleftrightarrow{DE}$, $\overleftrightarrow{DF}$ and $\overleftrightarrow{EF}$ intersects l in at most one point.

3. Prove that if $\triangle ABC = \triangle DEF$ in a metric geometry then $\overleftrightarrow{AB}$ contains exactly two of the points D, E and F.

4. Use Problem A3 to give an alternative proof of Theorem 3.4.4.

5. In a metric geometry, prove that if A, B and C are not collinear then $\overline{AB} = \overleftrightarrow{AB} \cap \triangle ABC$.

6. Use Problem A5 to prove Theorem 3.4.4.

7. Prove that, in a metric geometry, if $\triangle ABC = \triangle DEF$ then $\overleftrightarrow{AB}$ contains two of the three points D, E, and F.

8. Use Problem A7 to prove Theorem 3.4.4.

Part C. Expository exercises.

9. Prior to Lemma 3.4.1 there is a discussion of the idea of what it means for a concept to be "well defined". What examples do you know about from your previous mathematics courses where a concept needed to be well defined? What concepts in this course need to be well defined? Explain what the notion of "well defined" is in your own words.

CHAPTER 4
Plane Separation

4.1 The Plane Separation Axiom

The Plane Separation Axiom is a careful statement of the very intuitive idea that every line has "two sides." Such an idea seems so natural that we might expect it to be a consequence of our present axiom system. However, as we shall see in Section 4.3, there are models of a metric geometry that do not satisfy this new axiom. Thus the Plane Separation Axiom does not follow from the axioms of a metric geometry, and it is therefore necessary to add it to our list of axioms if we wish to use it. In this section we will introduce the concept of convexity, use it to state the Plane Separation Axiom, and develop some of the very basic results coming from the new axiom. In the second section we will show that our two basic models, the Euclidean Plane and the Poincaré Plane, do satisfy this new axiom. In the third section we will introduce an alternative formulation of plane separation in terms of triangles. This substitute for the Plane Separation Axiom is called Pasch's Postulate. We shall see that it is equivalent to the Plane Separation Axiom: any metric geometry that satisfies one of these axioms satisfies the other.

Definition. Let $\{\mathscr{S}, \mathscr{L}, d\}$ be a metric geometry and let $\mathscr{S}_1 \subset \mathscr{S}$. $\mathscr{S}_1$ is said to be **convex** if for every two points $P, Q \in \mathscr{S}_1$, the segment $\overline{PQ}$ is a subset of $\mathscr{S}_1$.

In Figure 4-1 each of the individual subsets of $\mathbb{R}^2$ is convex while each of those in Figure 4-2 is not convex. In Figure 4-2 the segment $\overline{PQ}$ is contained in $\mathscr{S}_1$ but the segment $\overline{P'Q'}$ is not. This means that $\mathscr{S}_1$ is not convex: convexity requires that the segment between *any* two points of $\mathscr{S}_1$ be in $\mathscr{S}_1$,

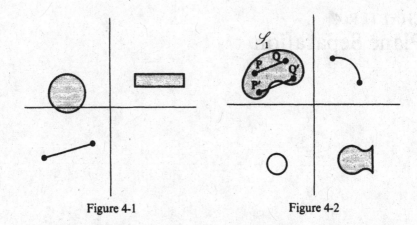

Figure 4-1   Figure 4-2

not just some. To show that a set is convex we must show that for every pair of points in the set, the segment joining them is contained in the set. To show a set is not convex, we need only find one pair of points such that the line segment joining them is not entirely contained in the set.

We should also note that the concept of convexity depends on the metric geometry. This is because convexity involves line segments, which in turn involve betweenness, which is defined in terms of distance. Thus a change in the distance function affects which sets are convex. For example, consider the set of ordered pairs (x, y) with $(x - 1)^2 + y^2 = 9$, $0 < x < 4$ and $0 < y$ (see Figure 4-3). This set is not convex as a subset of the Euclidean Plane. However it is convex as a subset of the Poincaré Plane. (This is because a line in a metric geometry is always convex. See Problem A2.) In this example we changed both the set of lines and the distance function. See Problem B20 for an example where a set $\mathscr{S}_1$ is convex in $\{\mathscr{S}, \mathscr{L}, d\}$ but not in $\{\mathscr{S}, \mathscr{L}, d_N\}$.

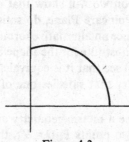

Figure 4-3

Definition. A metric geometry $\{\mathscr{S}, \mathscr{L}, d\}$ satisfies the **plane separation axiom** (PSA) if for every $l \in \mathscr{L}$ there are two subsets H_1 and H_2 of $\mathscr{S}$ (called

half planes determined by l) such that

(i) $\mathscr{S} - l = H_1 \cup H_2$;
(ii) H_1 and H_2 are disjoint and each is convex;
(iii) If $A \in H_1$ and $B \in H_2$ then $\overline{AB} \cap l \neq \varnothing$.

The definition demands that the line l have two sides (H_1 and H_2) both of which are convex. Further, condition (iii) says that you cannot get from one side to the other without cutting across l. Of course $\overline{AB} \cap l$ can only have one point in it, otherwise $\overline{AB} = l$. In Figure 4-4 we see the situation in the Euclidean Plane whereas Figure 4-5 gives the picture for a type II line in the Poincaré Plane. Note there is no way to distinguish H_1 from H_2. They are distinct, but no geometric property makes one different from the other.

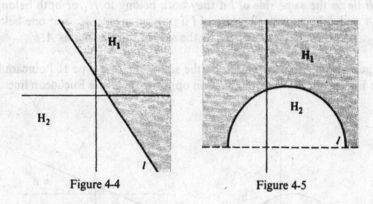

Figure 4-4   Figure 4-5

To be pedagogically correct, we should now prove that both the Euclidean Plane and Poincaré Plane satisfy PSA. However, to do this requires a bit of work, which is left to the next section. Thus we will illustrate our theorems and definitions in this section by using the Euclidean and Poincaré Planes and assume that they satisfy PSA. In the next section, we will prove that they really do. In Section 4.3 we will give an example, the Missing Strip Plane, which does not satisfy PSA.

First, we want to show that a pair of half planes determined by l is unique. Then it will make sense to talk about *the* pair of half planes determined by l.

Theorem 4.1.1. *Let l be a line in a metric geometry. If both H_1, H_2 and H_1', H_2' satisfy the conditions of* PSA *for the line l then either $H_1 = H_1'$ (and $H_2 = H_2'$) or $H_1 = H_2'$ (and $H_2 = H_1'$).*

PROOF. Let $A \in H_1$. Since $A \notin l$, either $A \in H_1'$ or $A \in H_2'$. Suppose that $A \in H_1'$. We will show that in this case $H_1 = H_1'$. The case where $A \in H_2'$ yields $H_1 = H_2'$ in a manner similar to what follows.

To show that $H_1 \subset H'_1$ let $B \in H_1$. We must prove that $B \in H'_1$. This is obvious if $B = A$. Suppose that $B \neq A$. If $B \notin H'_1$ then $B \in H'_2$ since $B \notin l$. This means that $\overline{AB} \cap l \neq \varnothing$ since $A \in H'_1$ and $B \in H'_2$. On the other hand, A and B belong to the convex set H_1 so that $\overline{AB} \subset H_1$ and $\overline{AB} \cap l = \varnothing$. This contradiction shows that $B \in H'_1$ and thus $H_1 \subset H'_1$.

The proof that $H'_1 \subset H_1$ is similar. Thus $H_1 = H'_1$. Finally $H_2 = \mathscr{S} - l - H_1 = \mathscr{S} - l - H'_1 = H'_2$. $\square$

We may now formalize the idea of two points being on the same side of a line.

Definition. Let $\{\mathscr{S}, \mathscr{L}, d\}$ be a metric geometry which satisfies PSA, let $l \in \mathscr{L}$, and let H_1 and H_2 be the half planes determined by l. Two points A and B **lie on the same side of** l if they both belong to H_1 or both belong to H_2. A and B **lie on opposite sides of** l if one belongs to H_1 and one belongs to H_2. If $A \in H_1$, we say that H_1 is the **side of** l **that contains** A.

Figure 4-6 shows two points on the same side of a type II Poincaré line while Figure 4-7 shows two points on opposite sides of a Euclidean line.

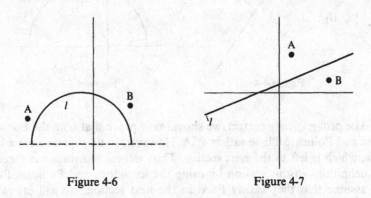

Figure 4-6   Figure 4-7

An alternate but useful interpretation of "same side"/"opposite side" is contained in the next result whose proof is left to Problem A6.

Theorem 4.1.2. *Let $\{\mathscr{S}, \mathscr{L}, d\}$ be a metric geometry which satisfies PSA. Let A and B be two points of $\mathscr{S}$ not on a given line l. Then*

(i) *A and B are on opposite sides of l if and only if $\overline{AB} \cap l \neq \varnothing$.*
(ii) *A and B are on the same side of l if and only if either $A = B$ or $\overline{AB} \cap l = \varnothing$.*

The next two results will be used frequently. Their proofs are left as Problems A9 and A10.

Theorem 4.1.3. *Let l be a line in a metric geometry which satisfies PSA. If P and Q are on opposite sides of l and if Q and R are on opposite sides of l then P and R are on the same side of l.*

Theorem 4.1.4. *Let l be a line in a metric geometry which satisfies PSA. If P and Q are on opposite sides of l and if Q and R are on the same side of l then P and R are on opposite sides of l.*

In Theorem 4.1.1 we showed that a line l had a unique pair of half planes associated with it. Now we will show that a given half plane comes from just one line.

Theorem 4.1.5. *Let l be a line in a metric geometry with PSA. Assume that H_1 is a half plane determined by the line l. If H_1 is also a half plane determined by the line l', then l = l'.*

PROOF. Suppose that $l \neq l'$. Then $l \cap l'$ has at most one point. Since every line has at least two points, there must be a point $P \in l—l'$ and a point $Q \in l'—l$. Note that $\overline{PQ} \neq l$ and $\overline{PQ} \neq l'$ since $Q \notin l$ and $P \notin l'$. Thus $\overline{PQ} \cap l = \{P\}$ and $\overline{PQ} \cap l' = \{Q\}$. Choose points A and B on $\overline{PQ}$ with $A—P—Q$ and $P—Q—B$, so that $A—P—Q—B$. See Figure 4-8. Since $A—P—B, P \in \overline{AB} \cap l$. On the other hand, $\overline{AB} \cap l \subset \overline{PQ} \cap l = \{P\}$. Hence $\overline{AB} \cap l = \{P\}$. By Theorem 4.1.2, A and B are on opposite sides of l. Hence either $A \in H_1$ or $B \in H_1$. We assume $A \in H_1$ and leave the case that $B \in H_1$ to Problem A7.

Now $\overline{AP} \cap l' \subset \overline{PQ} \cap l' = \{Q\}$. Since $A—P—Q, Q \notin \overline{AP}$ and $\overline{AP} \cap l' = \varnothing$. Thus by Theorem 4.1.2, A and P are on the same side of l', and that side is the half plane H_1 since $A \in H_1$. But this means $P \in H_1$ which is impossible since $P \in l$. This contradiction implies $l = l'$. $\square$

Definition. If H_1 is a half plane determined by the line l, then the **edge** of H_1 is l.

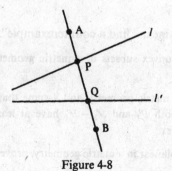

Figure 4-8

The point of Theorem 4.1.5 is that a half plane has *exactly* one edge. Thus we have shown that a line uniquely determines its half planes and a half plane uniquely determines its edge.

You should note that nowhere have we said that H_1 and H_2 are nonempty. However, this is true as you will prove in Problem A4.

Martin [1975] uses the suggestive term "scissors geometry" to refer to a metric geometry which satisfies PSA. The idea is that a line "cuts" the plane into two parts.

PROBLEM SET 4.1

Part A.

1. If $\mathscr{S}_1$ and $\mathscr{S}_2$ are convex subsets of a metric geometry, prove that $\mathscr{S}_1 \cap \mathscr{S}_2$ is convex.

2. If l is a line in a metric geometry, prove that l is convex.

3. If H_1 is a half plane determined by l prove that $H_1 \cup l$ is convex.

4. If H_1 and H_2 are the half planes determined by the line l, prove that neither H_1 nor H_2 is empty.

5. If H_1 is a half plane determined by the line l, prove that H_1 has at least three non-collinear points.

6. Prove Theorem 4.1.2.

7. Complete the proof of Theorem 4.1.5 in the case $B \in H_1$.

*8. Let l be a line in a metric geometry $\{\mathscr{S}, \mathscr{L}, d\}$ which satisfies PSA. We write $P \sim Q$ if P and Q are on the same side of l. Prove that $\sim$ is an equivalence relation on $\mathscr{S} - l$. How many equivalence classes are there and what are they?

9. Prove Theorem 4.1.3.

10. Prove Theorem 4.1.4.

Part B. "Prove" may mean "find a counterexample".

11. If $\mathscr{S}_1$ and $\mathscr{S}_2$ are convex subsets of a metric geometry, prove that $\mathscr{S}_1 \cup \mathscr{S}_2$ is convex.

12. Let $\{\mathscr{S}, \mathscr{L}, d\}$ be a metric geometry and assume that $\mathscr{S}_1 \subset \mathscr{S}_2 \subset \mathscr{S}$, that $\mathscr{S}_2$ is convex, and that both $\mathscr{S}_1$ and $\mathscr{S}_2 - \mathscr{S}_1$ have at least two points. Prove that $\mathscr{S}_2 - \mathscr{S}_1$ is *not* convex.

13. If A, B, C are noncollinear in a metric geometry, prove that $\triangle ABC$ is convex.

14. Let $\mathscr{S}_1$ be a subset of a metric geometry which satisfies PSA. $\mathscr{S}_1$ is a **passing set** if every point of $\mathscr{S}_1$ is a passing point of $\mathscr{S}_1$. Prove that a line is a passing set.

15. If $\mathscr{S}_1$ and $\mathscr{S}_2$ are passing sets in a metric geometry which satisfies PSA, prove that $\mathscr{S}_1 \cup \mathscr{S}_2$ is a passing set.

16. If $\mathscr{S}_1$ and $\mathscr{S}_2$ are passing sets in a metric geometry which satisfies PSA, prove that $\mathscr{S}_1 \cap \mathscr{S}_2$ is a passing set.

17. Prove that if $\mathscr{S}_1$ is a convex subset of a metric geometry which satisfies PSA and has more than one point, then $\mathscr{S}_1$ is a passing set.

18. Let $\mathscr{P}(\mathscr{S}_1)$ denote the set of all passing points of $\mathscr{S}_1$ where $\mathscr{S}_1$ is a subset of a metric geometry which satisfies PSA. Prove that $\mathscr{P}(\mathscr{S}_1)$ is convex.

19. We define a new distance on the Cartesian Plane $\{\mathbb{R}^2, \mathscr{L}_E\}$ as follows. Let $f: L_0 \to \mathbb{R}$ by

$$f(0, y) = \begin{cases} y & \text{if } y \text{ is not an integer} \\ -y & \text{if } y \text{ is an integer.} \end{cases}$$

a. Prove that f is a bijection.

For every other line of $\mathbb{R}^2$ choose a Euclidean ruler. By Theorem 2.2.8 this collection of bijections determines a distance function d_N which makes $\{\mathbb{R}^2, \mathscr{L}_E, d_N\}$ into a metric geometry. Now

b. Prove that $\{(0, y) | \frac{1}{2} \leq y \leq \frac{3}{2}\}$ is convex in the Euclidean Plane but not in $\{\mathbb{R}^2, \mathscr{L}_E, d_N\}$.

c. In $\{\mathbb{R}^2, \mathscr{L}_E, d_N\}$, what is the segment from $(0, \frac{1}{2})$ to $(0, \frac{3}{2})$? Show that this set is convex in $\{\mathbb{R}^2, \mathscr{L}_E, d_N\}$ but not in the Euclidean Plane.

d. Show that $\{\mathbb{R}^2, \mathscr{L}_E, d_N\}$ does not satisfy PSA. (Hint: Consider the line $l = L_{0,1}$ and the three points $(0, \frac{1}{2}), (0, \frac{3}{2}), (1, \frac{1}{2})$. Use Problems A8, A9, A10.)

Part C. Expository exercises.

20. Perform the following experiment and then write up what happened. If you have access to middle or high school students, show it to them and record their reactions. What conclusions can you draw from the way they responded? The purpose of what follows is to show the sort of non-intuitive difficulties that can occur if concepts are not defined carefully or if geometries do not satisfy some of the axioms we would like.

Take a long narrow strip of paper and draw a line down the middle of it on both sides. On one side mark both ends with the letter X. Tape or glue the two ends together, twisting the strip so that the two Xs are touching and not visible. Note that you have a line drawn on the new figure (which is called a Möbius Strip). With a pair of scissors carefully cut the strip along the line and see what happens. Are you surprised? What happens when you cut the Euclidean or Poincaré Plane along a line? Is there a difference in the qualitative behavior of these constructions?

4.2 PSA for the Euclidean and Poincaré Planes

In this section we will show that both the Euclidean and Poincaré Planes satisfy PSA, as was promised in the last section. As a homework problem (Problem A5), you will show that the Taxicab Plane also satisfies PSA.

For the Euclidean Plane, it is useful to have the following notation.

Notation. If $X = (x, y) \in \mathbb{R}^2$ then $X^\perp$ (read "X perp") is the element

$$X^\perp = (-y, x) \in \mathbb{R}^2. \tag{2-1}$$

Intuitively, $X^\perp$ is obtained by rotating the vector X counterclockwise about the origin 90°. Note that this is just an intuitive idea—we have *not* defined what is meant by "90°." The reason for the name is that $X^\perp$ is *perpendicular* to X in the following sense.

Lemma 4.2.1. (a) *If* $X \in \mathbb{R}^2$ *then* $\langle X, X^\perp \rangle = 0$.
(b) *If* $X \in \mathbb{R}^2$ *and* $X \neq (0,0)$ *then* $\langle Z, X^\perp \rangle = 0$ *implies that* $Z = tX$ *for some* $t \in \mathbb{R}$.

PROOF. We leave part (a) to Problem A1. For part (b) we proceed as follows. Let $X = (x, y)$ and $Z = (z, w)$ so that $X^\perp = (-y, x)$. Then $\langle Z, X^\perp \rangle = 0$ means

$$-zy + wx = 0. \tag{2-2}$$

Since $X \neq (0,0)$ one of x and y is not zero. If $x \neq 0$ then we may solve Equation (2-2) for $w = zy/x$ so that $Z = tX$ with $t = z/x$. If $y \neq 0$ then $z = xw/y$ so that $Z = tX$ with $t = w/y$. Either way $Z = tX$ for some $t \in \mathbb{R}$. $\square$

Using $X^\perp$ we can give an alternative description of a line in $\mathbb{R}^2$. Our motivation here is the idea from linear algebra that a line can be described by giving one point on the line and a vector normal to it. See Figure 4-9.

Proposition 4.2.2. *If* P *and* Q *are distinct points in* $\mathbb{R}^2$ *then*

$$\overline{PQ} = \{A \in \mathbb{R}^2 \mid \langle A - P, (Q - P)^\perp \rangle = 0\}. \tag{2-3}$$

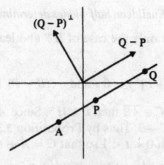

Figure 4-9

PROOF. First we show that $\overleftrightarrow{PQ} \subset \{A \in \mathbb{R}^2 | \langle A - P, (Q - P)^\perp \rangle = 0\}$. Let $A \in \overleftrightarrow{PQ}$ so that $A = P + t(Q - P)$ for some $t \in \mathbb{R}$. Then

$$\langle A - P, (Q - P)^\perp \rangle = t\langle Q - P, (Q - P)^\perp \rangle = 0.$$

Thus $\overleftrightarrow{PQ} \subset \{A \in \mathbb{R}^2 | \langle A - P, (Q - P)^\perp \rangle = 0\}$.

To prove the reverse containment assume that $A \in \mathbb{R}^2$ with $\langle A - P, (Q - P)^\perp \rangle = 0$. Now $Q - P \neq (0,0)$ since $Q \neq P$. Thus by Lemma 4.2.1 there is a real number t with $A - P = t(Q - P)$. Hence

$$A = P + t(Q - P) \in \overleftrightarrow{PQ}.$$

Thus $\{A \in \mathbb{R}^2 | \langle A - P, (Q - P)^\perp \rangle = 0\} \subset \overleftrightarrow{PQ}$. We now have containment in both directions so the sets are equal. $\square$

Definition. Let $l = \overleftrightarrow{PQ}$ be a Euclidean line. The **Euclidean half planes** determined by l are

$$H^+ = \{A \in \mathbb{R}^2 | \langle (A - P), (Q - P)^\perp \rangle > 0\}$$
$$H^- = \{A \in \mathbb{R}^2 | \langle (A - P), (Q - P)^\perp \rangle < 0\} \tag{2-4}$$

(See Figure 4-10.)

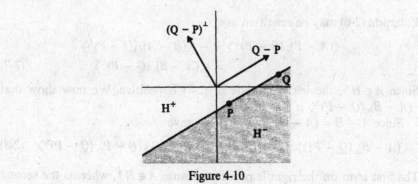

Figure 4-10

Proposition 4.2.3. *The Euclidean half planes determined by* $l = \overline{PQ}$ *are convex.*

PROOF. We will handle only the case of H^+ and leave H^- as Problem A2. Let $A, B \in H^+$ so that

$$\langle (A - P), (Q - P)^\perp \rangle > 0 \quad \text{and} \quad \langle (B - P), (Q - P)^\perp \rangle > 0 \qquad (2\text{-}5)$$

We must show that if $C \in \overline{AB}$ then $C \in H^+$. Since A, $B \in H^+$ we need only consider the case A—C—B. Thus by Proposition 3.3.3 we may assume that there is a number t with $0 < t < 1$ so that $C = A + t(B - A)$. Therefore

$$C = (1 - t)A + tB \quad \text{with } 0 < t < 1$$

Then

$$\begin{aligned}
\langle (C - P), (Q - P)^\perp \rangle &= \langle ((1 - t)A + tB - P), (Q - P)^\perp \rangle \\
&= \langle ((1 - t)(A - P) + t(B - P)), (Q - P)^\perp \rangle \\
&= (1 - t)\langle (A - P), (Q - P)^\perp \rangle + t\langle (B - P), (Q - P)^\perp \rangle
\end{aligned}$$

Since $0 < t < 1$, Inequalities (2-5) show that each term on the right of the last equality is positive. Thus $\langle (C - P), (Q - P)^\perp \rangle > 0$ and $C \in H^+$. $\qquad\square$

Proposition 4.2.4. *The Euclidean Plane satisfies PSA.*

PROOF. Let $l = \overline{PQ}$ be a line. If $A \in \mathbb{R}^2$ then $\langle (A - P), (Q - P)^\perp \rangle$ is either positive (so that $A \in H^+$), zero (so that $A \in l$ by Proposition 4.2.2), or negative (so that $A \in H^-$). Thus $\mathbb{R}^2 - l = H^+ \cup H^-$. Since H^+ and H^- are clearly disjoint and Proposition 4.2.3 says they are convex, we need only show that condition (iii) of PSA holds.

Let $A \in H^+$ and $B \in H^-$. To show that $\overline{AB} \cap l \neq \varnothing$ we must find t with $0 < t < 1$ and $A + t(B - A) \in l$. We could at this point write down the explicit formula for t. (It is given by Equation (2-9).) However, it is more illuminating to see how the formula for t is derived.

According to Proposition 4.2.2, $A + t(B - A) \in l$ if and only if

$$\langle (A + t(B - A) - P), (Q - P)^\perp \rangle = 0 \qquad (2\text{-}6)$$

Equation (2-6) may be rewritten as

$$\begin{aligned}
\langle (A - P), (Q - P)^\perp \rangle &= -t\langle (B - A), (Q - P)^\perp \rangle \\
&= t\langle (A - B), (Q - P)^\perp \rangle
\end{aligned} \qquad (2\text{-}7)$$

Since $A \in H^+$, the left hand side of (2-7) is positive. We now show that $\langle (A - B), (Q - P)^\perp \rangle$ is also positive.

Since $A - B = (A - P) - (B - P)$ we have

$$\langle (A - B), (Q - P)^\perp \rangle = \langle (A - P), (Q - P)^\perp \rangle - \langle (B - P), (Q - P)^\perp \rangle \qquad (2\text{-}8)$$

The first term on the right is positive because $A \in H^+$, whereas the second term is negative because $B \in H^-$. Thus the difference is positive. Hence we may divide Equation (2-7) by $\langle (A - B), (Q - P)^{\perp} \rangle$ to obtain

$$t = \frac{\langle (A - P), (Q - P)^{\perp} \rangle}{\langle (A - B), (Q - P)^{\perp} \rangle} > 0 \qquad (2-9)$$

To finish the proof we must show that the value of t in Equation (2-9) is less than one. Note that Equation (2-8) implies that the numerator of t is less than the denominator. Hence $t < 1$. With the value of t given by Equation (2-9) we have a point $X = A + t(B - A) \in \overline{AB} \cap l$.   □

An alternate proof that condition (iii) is satisfied is given in Problem A3. Now we turn our attention to the Poincaré Plane. In this proof we shall use calculus. A reader who has not had calculus should skip the proof and go on to Section 4.3. The results we need from calculus are

(i) if $f'(t) > 0$ for all t then $f(t)$ is an increasing function;
(ii) the Intermediate Value Theorem, which says that if $f(t)$ is a continuous function and $f(a) < r < f(b)$ then there is a point c between a and b with $f(c) = r$.

Definition. If $l = {}_aL$ is a type I line in the Poincaré Plane then the **Poincaré half planes** determined by l are

$$H_+ = \{(x, y) \in \mathbb{H} \,|\, x > a\}$$
$$H_- = \{(x, y) \in \mathbb{H} \,|\, x < a\}. \qquad (2-10)$$

If $l = {}_cL_r$ is a type II line then the **Poincaré half planes** determine by l are

$$H_+ = \{(x, y) \in \mathbb{H} \,|\, (x - c)^2 + y^2 > r^2\}$$
$$H_- = \{(x, y) \in \mathbb{H} \,|\, (x - c)^2 + y^2 < r^2\}. \qquad (2-11)$$

The half planes for a type II line were sketched in Figure 4-5.

Proposition 4.2.5. *The Poincaré Plane satisfies* PSA.

PROOF. Let l be a line in $\mathcal{H}$. Let H_+ and H_- be the Poincaré half planes determined by l. Clearly from the definition of Poincaré half planes, $\mathbb{H} - l = H_+ \cup H_-$ and $H_+ \cap H_- = \varnothing$. We must show that each half plane is convex and that condition (iii) of the definition of PSA holds.

Case 1. l is a type I line. This is left to Problem A4.

Case 2. Let $l = {}_cL_r$ be a type II line and suppose that A and B are distinct points of $\mathbb{H} - l$. We shall parametrize the line segment $\overline{AB}$. The form of this parametrization depends on whether $\overline{AB}$ is type I or type II. This parametrization is used to show H_+ and H_- are convex. We write $A = (x_1, y_1)$ and $B = (x_2, y_2)$. If $\overline{AB}$ is a type I line then we will assume that $y_1 < y_2$. If $\overline{AB}$ is a type II line then we will assume that $x_1 < x_2$. Our plan is to show that, no matter where A and B are, a certain function $g(t)$ is either always increasing, always decreasing, or constant. This function will be zero at points of l.

If $\overrightarrow{AB} = {}_aL$ is a type I line then $\overrightarrow{AB}$ may be parametrized by $(x, y) \in \overrightarrow{AB}$ if and only if $(x, y) = (a, e^t)$ for some $t \in \mathbb{R}$. Define $f_I(t) = (a, e^t)$. f_I is the inverse of the standard ruler for ${}_aL$. We let

$$g_I(t) = (x - c)^2 + y^2 - r^2 = (a - c)^2 + e^{2t} - r^2.$$

Since $g_I'(t) = 2e^{2t} > 0$, g_I is always increasing.

If $\overrightarrow{AB} = {}_aL_s$ is a type II line we parametrize $\overrightarrow{AB}$ by $(x, y) \in \overrightarrow{AB}$ if and only if

$$(x, y) = (d + s \tanh(t), s \operatorname{sech}(t)) = f_{II}(t).$$

f_{II} is the inverse of the standard ruler for ${}_aL_s$. Again we let

$$\begin{aligned}
g_{II}(t) &= (x - c)^2 + y^2 - r^2 \\
&= (d - c + s \tanh(t))^2 + (s \operatorname{sech}(t))^2 - r^2
\end{aligned}$$

and find that

$$\begin{aligned}
g_{II}'(t) &= 2(d - c + s \tanh(t))s \operatorname{sech}^2(t) \\
&\quad + 2(s \operatorname{sech}(t))(-s \operatorname{sech}(t)\tanh(t)) \\
&= 2(d - c)s \operatorname{sech}^2(t),
\end{aligned}$$

so that g_{II} is either increasing $(d > c)$ or decreasing $(d < c)$ or constant $(d = c)$.

We let $f = f_I$ and $g = g_I$ if $\overrightarrow{AB}$ is a type I line and we let $f = f_{II}$ and $g = g_{II}$ if $\overrightarrow{AB}$ is a type II line. Thus we have a function f from the real numbers to $\overrightarrow{AB}$ and real numbers $t_1 < t_2$ with $f(t_1) = A$ and $f(t_2) = B$. We also have a continuous real valued function g such that $g(t) > 0$ if $f(t) \in H_+$, $g(t) < 0$ if $f(t) \in H_-$, and $g(t) = 0$ if $f(t) \in l$.

We now prove that H_+ is convex. Suppose $A, B \in H_+$ and let f, g be as above so that $A = f(t_1)$ and $B = f(t_2)$ with $t_1 < t_2$. If $A{-}C{-}B$ then $C = f(t_3)$ with $t_1 < t_3 < t_2$. Since g is strictly increasing, strictly decreasing, or constant, this means $g(t_3)$ is between $g(t_1)$ and $g(t_2)$ (or equal to both if g is constant). Since $A, B \in H_+$, $g(t_1)$ and $g(t_2)$ are both positive. Thus $g(t_3)$ is positive and $C \in H_+$. Thus H_+ is convex. Similarly H_- is convex.

Finally suppose that one of A and B is in H_+ and the other is in H_-. We show that $\overline{AB} \cap l \neq \varnothing$. Using f and g as before, $A = f(t_1)$, $B = f(t_2)$ and one of $g(t_1)$ and $g(t_2)$ is positive while the other is negative. Since g is continuous, the Intermediate Value Theorem implies that there is a number t_3 between t_1 and t_2 with $g(t_3) = 0$. But then $C = f(t_3)$ is a point on $\overline{AB}$ since $t_1 < t_3 < t_2$ and is a point on l since $g(t_3) = 0$. Thus $C \in \overline{AB} \cap l \neq \varnothing$. $\qquad\square$

PROBLEM SET 4.2

Part A.

1. Prove Lemma 4.2.1(a).

2. Prove that the Euclidean half plane H^- is convex.

3. Let l be a line in the Euclidean Plane and suppose that $A \in H^+$ and $B \in H^-$. Show that $\overline{AB} \cap l \neq \emptyset$ in the following way. Let

$$g(t) = \langle A + t(B - A) - P, (Q - P)^\perp \rangle \quad \text{if } t \in \mathbb{R}.$$

Evaluate $g(0)$ and $g(1)$, show that g is continuous, and then prove that $\overline{AB} \cap l \neq \emptyset$.

4. If $l = {}_aL$ is a type I line in the Poincaré Plane then prove that
   a. H_+ and H_- as defined in Equation (2-10) are convex.
   b. If $A \in H_+$ and $B \in H_-$ then $\overline{AB} \cap l \neq \emptyset$.

   (Hint: One way to do this is to mimic what was done when l was a type II line, but use a different g. Another way involves Example 3.3.1.)

5. For the Taxicab Plane $\{\mathbb{R}^2, \mathscr{L}_E, d_T\}$ prove that
   a. If $A = (x_1, y_1)$, $B = (x_2, y_2)$ and $C = (x_3, y_3)$ are collinear but do not lie on a vertical line then $A—B—C$ if and only if $x_1 * x_2 * x_3$.
   b. The Taxicab Plane satisfies PSA.

4.3 Pasch Geometries

We are now ready to prove our first important theorem. Roughly, it will say that in a metric geometry which satisfies PSA, a line which intersects one side of a triangle must intersect one of the other two sides. Despite the simplicity of its statement and proof, this result is remarkably powerful. Its importance was initially noticed by Pasch [1882] when he gave the first modern axiomatization of geometry.

One remark is in order about Pasch's Theorem. It is easy not to realize that there really is something to prove here—after all, the situation is so obvious geometrically! In fact it is this kind of result (which involves betweenness) that Euclid "forgot" to prove. We will use Pasch's Theorem to end this section with an example of a metric geometry which doesn't satisfy PSA.

Definition. A metric geometry satisfies **Pasch's Postulate** (PP) if for any line l, any triangle $\triangle ABC$, and any point $D \in l$ such that $A—D—B$, then either $l \cap \overline{AC} \neq \emptyset$ or $l \cap \overline{BC} \neq \emptyset$. (See Figure 4-11.)

Theorem 4.3.1 (Pasch's Theorem). *If a metric geometry satisfies PSA then it also satisfies PP.*

PROOF. Let $\triangle ABC$ and a line l be given. Assume that there is a point $D \in l$ with $A—D—B$. We will show that either $l \cap \overline{AC} \neq \emptyset$ or $l \cap \overline{BC} \neq \emptyset$. See Figure 4-11.

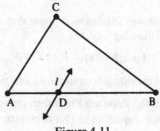

Figure 4-11

Suppose that $\overline{AC} \cap l = \emptyset$. We will show that $\overline{BC} \cap l \neq \emptyset$. Now $l \neq \overleftrightarrow{AB}$ since $A \in \overline{AC} \cap \overleftrightarrow{AB}$. Thus A and B are not on l and must be on opposite sides of l since $\overline{AB} \cap l = \{D\} \neq \emptyset$. A and C are on the same side of l ($\overline{AC} \cap l = \emptyset$). By Theorem 4.1.4, B and C are on opposite sides of l. Hence $\overline{BC} \cap l \neq \emptyset$. Thus either $\overline{AC} \cap l \neq \emptyset$ or $\overline{BC} \cap l \neq \emptyset$.   □

Note that another way of stating Pasch's Theorem is that if a line intersects a triangle, then it intersects two sides of the triangle, possibly at the common vertex. Even when a line intersects the interior of a side, the second point of intersection could be a vertex. See Figure 4-12.

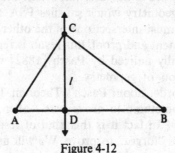

Figure 4-12

We now turn to a kind of result which we have not discussed before. By assuming the Plane Separation Axiom we got Pasch's Postulate for free. We now show that the reverse is true: If a metic geometry satisfies Pasch's Postulate then it also satisfies PSA. Logically, PSA and PP are equivalent and either may be assumed as an axiom with the other then becoming a theorem. Before we explore the logical equivalence involved we make the following definition.

Definition. A **Pasch Geometry** is a metric geometry which satisfies PSA.

Theorem 4.3.2. *Let* $\{\mathscr{S}, \mathscr{L}, d\}$ *be a metric geometry which satisfies* PP. *If* A, B, C *are noncollinear and if the line* l *does not contain any of the points* A, B, C, *then* l *cannot intersect all three sides of* $\triangle ABC$.

PROOF. In search of a contradiction, we assume that l *does* intersect all three sides in points other than vertices, i.e., $\overline{AB} \cap l = \{D\}$, $\overline{AC} \cap l = \{E\}$, $\overline{BC} \cap l = \{F\}$ with $A—D—B$, $A—E—C$, and $B—F—C$. Since D, E, and F all lie on l, one is between the other two. We assume $D—E—F$. (The other cases are similar.) The situation is illustrated in Figure 4-13 (which looks funny because it *is* an impossible situation!).

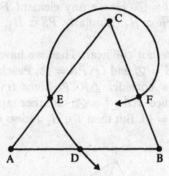

Figure 4-13

Now B, D and F are not collinear (or else A, B, C are collinear). Thus we have a triangle, $\triangle BDF$. Since $\overrightarrow{AC} \cap \overline{DF} = \{E\}$, we know that $\overleftrightarrow{AC}$ intersects either $\overline{BD}$ or $\overline{BF}$ by Pasch's Postulate applied to $\triangle BDF$.

First note that

$$\overleftrightarrow{AC} \cap \overline{BD} \subset \overleftrightarrow{AC} \cap \overline{BA} = \{A\}.$$

Since $A \notin \overline{BD}$ (because $A—D—B$) we have $\overleftrightarrow{AC} \cap \overline{BD} = \varnothing$.

On the other hand,

$$\overleftrightarrow{AC} \cap \overline{BF} \subset \overleftrightarrow{AC} \cap \overline{BC} = \{C\}.$$

Since $C \notin \overline{BF}$ we have $\overleftrightarrow{AC} \cap \overline{BF} = \varnothing$. This contradicts Pasch's Postulate (applied to $\triangle BDF$) which says that $\overleftrightarrow{AC} \cap \overline{BF} \neq \varnothing$ or $\overleftrightarrow{AC} \cap \overline{BD} \neq \varnothing$. Hence, our original assumption that l intersects all three sides at points other than the vertices is erroneous and the theorem is proved. □

Theorem 4.3.3. *If a metric geometry satisfies PP then it also satisfies PSA.*

PROOF. Let l be a line and let P be a point not on l. (P exists because a metric geometry has at least three noncollinear points!) We shall define two sets H_1 and H_2 and show they satisfy the axioms for half planes. Define sets H_1 and H_2 as follows:

$$H_1 = \{Q \in \mathscr{S} \,|\, Q = P \text{ or } \overline{QP} \cap l = \varnothing\}$$
$$H_2 = \{Q \in \mathscr{S} \,|\, Q \notin l \text{ and } \overline{QP} \cap l \neq \varnothing\}. \tag{3-1}$$

Clearly $H_1 \cap H_2 = \varnothing$ and $\mathscr{S} - l = H_1 \cup H_2$. We need to show that H_1 and H_2 are convex and that condition (iii) of PSA holds.

Step 1. First we show H_1 is convex. Let $R, S \in H_1$ and suppose $R—T—S$. We must show $T \in H_1$. We shall need two cases.

Case 1a. R, S, P are collinear. In this situation either $R = P$, $S = P$, $R—S—P$, $S—R—P$, or $R—P—S$. In all of the possibilities

$$\overline{RS} \subset \overline{PR} \cup \overline{PS}. \tag{3-2}$$

Since $R \in H_1$, $\overline{PR} \cap l = \emptyset$. Hence any element $F \in \overline{PR}$ has the property $\overline{PF} \cap l = \emptyset$. Thus $\overline{PR} \subset H_1$. Similarly $\overline{PS} \subset H_1$, and thus by Equation (3-2), $\overline{RS} \subset H_1$.

Case 1b. $R, S. P$ are not collinear. Then we have a triangle $\triangle RSP$ (Figure 4-14). Since $l \cap \overline{PS} = \emptyset$ and $l \cap \overline{PR} = \emptyset$, Pasch's Postulate tells us that $l \cap \overline{RS} = \emptyset$ also. Now consider $\triangle RTP$. Since $l \cap \overline{RS} = \emptyset$ and $\overline{RT} \subset \overline{RS}$, $\overline{RT} \cap l = \emptyset$ also. Since $\overline{RP} \cap l = \emptyset$, another application of Pasch (to $\triangle RTP$) yields $\overline{PT} \cap l = \emptyset$. But then $T \in H_1$ and so H_1 is convex as required.

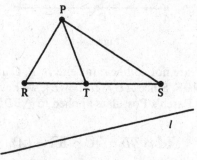

Figure 4-14

Step 2. Next we show that H_2 is convex. Let $R, S \in H_2$ so that $\overline{RP} \cap l \ne \emptyset$ and $\overline{SP} \cap l \ne \emptyset$.

Case 2a. R, S, P are collinear (and distinct). Then $\overline{RP} \cap l = \overline{SP} \cap l = \{Q\}$ for some Q and either $P—Q—R—S$ or $P—Q—S—R$. Hence if $S—T—R$ then $P—Q—T$ and $\overline{TP} \cap l \ne \emptyset$. Hence $\overline{RS} \subset H_2$.

Case 2b. R, S, P are not collinear. Then if $R—T—S$, Theorem 4.3.2 says that $T \notin l$ (or else l intersects all three sides of $\triangle PRS$). Another application of Theorem 4.3.2 shows that $\overline{RS} \cap l = \emptyset$, so that $\overline{RT} \cap l = \emptyset$. But $\overline{PR} \cap l \ne \emptyset$ so PP (applied to $\triangle RPT$) implies $\overline{PT} \cap l \ne \emptyset$. Thus $T \in H_2$ and $\overline{RS} \subset H_2$. Hence H_2 is convex.

Step 3. Finally, suppose $R \in H_1$ and $S \in H_2$. We must show $\overline{RS} \cap l \ne \emptyset$. If $R = P$ then $\overline{RS} \cap l = \overline{PS} \cap l \ne \emptyset$ and we are done. Hence we assume $R \ne P$.

Case 3a. R, S, P are not collinear. Then since $\overline{RP} \cap l = \emptyset$ and $\overline{SP} \cap l \ne \emptyset$, PP implies $\overline{RS} \cap l \ne \emptyset$.

Case 3b. R, S, P are collinear. Then since $\overline{SP} \cap l \ne \emptyset$ we may let $\overline{SP} \cap l = \{Q\}$ with $P—Q—S$. Since $R \in \overleftrightarrow{SP}$ and $R \ne P$, $R \ne S$, $R \ne Q$, either

$$P—Q—R, \quad R—P—Q, \quad \text{or} \quad P—R—Q.$$

The first situation cannot occur since $R \in H_1$ implies $\overline{PR} \cap l = \varnothing$. If the second situation $(R—P—Q)$ occurs then $R—P—Q—S$ and $\overline{RS} \cap l = \{Q\}$. In last situation, $P—R—Q$, we have $P—R—Q—S$ and again $\overline{RS} \cap l = \{Q\}$. Hence $\overline{RS} \cap l \neq \varnothing$.

We have thus shown that H_1 and H_2 are convex in steps 1 and 2 and that any segment from a member of H_1 to a member of H_2 must intersect l in step 3. Hence the geometry satisfies PSA. □

We now give an example of a metric geometry where PSA fails to hold. What we will really show is that the geometry does not satisfy PP, which is equivalent to PSA. This geometry can be found in Martin [1975].

Definition. The **Missing Strip Plane** is the abstract geometry $\{\mathscr{S}, \mathscr{L}\}$ given by

$$\mathscr{S} = \{(x, y) \in \mathbb{R}^2 \,|\, x < 0 \text{ or } 1 \leq x\}$$
$$\mathscr{L} = \{l \cap \mathscr{S} \,|\, l \text{ is a Cartesian line and } l \cap \mathscr{S} \neq \varnothing\}.$$

You are asked to prove in Problem A4 that the Missing Strip Plane is an incidence geometry. To make the Missing Strip Plane into a metric geometry we need to define a ruler for each line. If $l = L_{m,b}$ is a nonvertical line, recall that $f_l : l \to \mathbb{R}$ which is given by

$$f_l(x, y) = x\sqrt{1 + m^2}$$

is the standard Euclidean ruler for l. We cannot use f_l as a ruler for the line $l \cap \mathscr{S}$ of the Missing Strip Plane because f_l is not a bijection. ($f_l(l \cap \mathscr{S})$ omits the half open interval $[0, \sqrt{1 + m^2})$.) We remedy this by defining a new function g_l which is f_l before the strip (i.e., $x < 0$) and which moves every point after the strip to the left by $f_l(1, m + b) = \sqrt{1 + m^2}$. More precisely, define $g_l : l \cap \mathscr{S} \to \mathbb{R}$ by

$$g_l(x, y) = \begin{cases} f_l(x, y) & \text{if } x < 0 \\ f_l(x, y) - \sqrt{1 + m^2} & \text{if } x \geq 1. \end{cases}$$

The next result is Problem A5.

Proposition 4.3.4. *If $\{\mathscr{S}, \mathscr{L}\}$ is the Missing Strip Plane and $l = L_{m,b}$ then $g_l : l \cap \mathscr{S} \to \mathbb{R}$ is a bijection.*

The coordinates of several points on the lines $L_{0,3} \cap \mathscr{S}$ and $L_{1,-2} \cap \mathscr{S}$ are shown in Figure 4-15.

For each vertical line l in $\mathscr{S}$ let g_l be any Euclidean ruler. By Theorem 2.2.8 this collection of rulers g_l determines a distance d' on $\mathscr{S}$ that makes $\{\mathscr{S}, \mathscr{L}, d'\}$ a metric geometry.

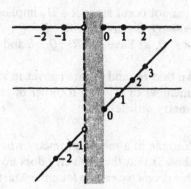

Figure 4-15

Proposition 4.3.5. *The Missing Strip Plane is not a Pasch geometry.*

PROOF. Consider $\triangle ABC$ where $A = (2,0)$, $B = (2,3)$ and $C = (-2,0)$. The line $l \cap \mathscr{S}$, where $l = L_{0,2}$, intersects $\overline{AB}$ at $D = (2,2)$. However, $(l \cap \mathscr{S}) \cap \overline{AC} \neq \emptyset$ and $(l \cap \mathscr{S}) \cap \overline{BC} = \emptyset$, contradicting PP. See Figure 4-16. $\square$

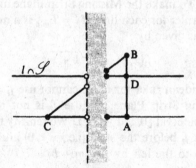

Figure 4-16

Note that if PSA were to follow from the axioms of a metric geometry then every model of a metric geometry would satisfy PSA. Proposition 4.3.5 gives a model of a metric geometry which doesn't satisfy PSA. Thus, Proposition 4.3.5 shows us that PSA really is an addition to our list of axioms and cannot be deduced from the previous ones. In Problems B6 and B8 there are two more examples where PSA is not satisfied.

PROBLEM SET 4.3

Part A.

1. (Peano's Axiom) Given a triangle $\triangle ABC$ in a metric geometry which satisfies PSA and points D, E with $B—C—D$ and $A—E—C$, prove there is a point $F \in \overleftrightarrow{DE}$ with $A—F—B$, and $D—E—F$.

2. Given $\triangle ABC$ in a metric geometry which satisfies PSA and points D, F with B—C—D, A—F—B, prove there exists $E \in \overrightarrow{DF}$ with A—E—C and D—E—F.

3. Given $\triangle ABC$ and a point P in a metric geometry which satisfies PSA prove there is a line through P that contains exactly two points of $\triangle ABC$.

4. Prove that the Missing Strip Plane is an incidence geometry.

5. Prove Proposition 4.3.4.

Part B. "Prove" may mean "find a counterexample".

6. Let $\{\mathbb{R}^2, \mathscr{L}_E, d_N\}$ be the metric geometry of Problem B20 of Section 4.1. Prove that PP is not satisfied.

7. Given $\triangle ABC$ in a metric geometry and points D, E with A—D—B and C—E—B, prove $\overrightarrow{AE} \cap \overrightarrow{CD} \neq \varnothing$.

8. Let $\mathbb{R}^3 = \{(x, y, z) | x, y, z \in \mathbb{R}\}$. If $A, B \in \mathbb{R}^3$ define $L_{AB} = \{A + t(B - A) | t \in \mathbb{R}\}$. Let $\mathscr{L} = \{L_{AB} | A, B \in \mathbb{R}^3, A \neq B\}$. If $A, B \in \mathbb{R}^3$ let $d(A, B) = \|A - B\|$. Prove that $\{\mathbb{R}^3, \mathscr{L}, d\}$ is a metric geometry but that it does not satisfy PSA.

Part C. Expository exercises.

9. We have just shown (Theorems 4.3.1 and 4.3.3) that two axioms are equivalent. Write a short essay on the equivalence of axiom systems using as a reference an appropriate book on mathematical logic.

10. Find some middle or high school students, ask them to draw a triangle and to pick a point on the triangle. Then ask them to draw a line through the point. They will probably construct it so that it crosses one of the other sides. Ask them how they know it would cross the side and write up their reactions. (The answer that Euclidean geometry satisfies Pasch's Postulate will be too subtle for them.) After the experiment be sure to tell them not to feel bad about not knowing the answer—neither did Euclid!

4.4 Interiors and the Crossbar Theorem

In this section we will be interested in interiors—the interior of a ray, of a segment, of an angle, and of a triangle. These concepts will aid us in proving the main theorem of this section which says that a ray starting at a vertex of a triangle and which passes through a point in the interior of the angle at that vertex must intersect the opposite side; that is, it must "cross the bar."

Theorem 4.4.1. *In a Pasch geometry if $\mathscr{A}$ is a nonempty convex set that does not intersect the line l, then all points of $\mathscr{A}$ lie on the same side of l.*

PROOF. Let $A \in \mathscr{A}$ and let B be any other point of $\mathscr{A}$. Since $\mathscr{A}$ is convex, $\overline{AB} \subset \mathscr{A}$. Since $\mathscr{A} \cap l = \emptyset$, $\overline{AB} \cap l = \emptyset$. Thus A and B are on the same side of l. Thus every point of $\mathscr{A}$ is on the same side of l as A is. $\square$

Definition. The **interior of the ray** $\overrightarrow{AB}$ in a metric geometry is the set

$$\mathrm{int}(\overrightarrow{AB}) = \overrightarrow{AB} - \{A\}.$$

The **interior of the segment** $\overline{AB}$ in a metric geometry is the set

$$\mathrm{int}(\overline{AB}) = \overline{AB} - \{A, B\}.$$

In Problem A1 you will show that the interior of a ray or a segment is convex. Theorem 4.4.1 can then be applied in a number of interesting cases. The proof of the next result is left to Problem A2.

Theorem 4.4.2. *Let $\mathscr{A}$ be a line, ray, segment, the interior of a ray, or the interior of a segment in a Pasch geometry. If l is a line with $\mathscr{A} \cap l = \emptyset$ then all of $\mathscr{A}$ lies on one side of l. If there is a point B with $A{-}B{-}C$ and $\overrightarrow{AC} \cap l = \{B\}$ then $\mathrm{int}(\overrightarrow{BA})$ and $\mathrm{int}(\overline{BA})$ both lie on the same side of l while $\mathrm{int}(\overline{BA})$ and $\mathrm{int}(\overline{BC})$ lie on opposite sides of l.*

Theorem 4.4.3 (Z Theorem). *In a Pasch geometry, if P and Q are on opposite sides of the line $\overleftrightarrow{AB}$ then $\overrightarrow{BP} \cap \overrightarrow{AQ} = \emptyset$. In particular, $\overline{BP} \cap \overline{AQ} = \emptyset$.*

PROOF. See Figure 4-17. By Theorem 4.4.2, $\mathrm{int}(\overrightarrow{BP})$ lies on one side of $\overleftrightarrow{AB}$ and $\mathrm{int}(\overrightarrow{AQ})$ lies on the other (since P and Q are on opposite sides). Thus $\mathrm{int}(\overrightarrow{BP}) \cap \mathrm{int}(\overrightarrow{AQ}) = \emptyset$. Since $B \notin \overrightarrow{AQ}$ (because A, B, Q are not collinear) we have $\overrightarrow{BP} \cap \mathrm{int}(\overrightarrow{AQ}) = \emptyset$. Since $A \notin \overrightarrow{BP}$, we have $\overrightarrow{BP} \cap \overrightarrow{AQ} = \emptyset$. The rest follows from $\overline{BP} \subset \overrightarrow{BP}$ and $\overline{AQ} \subset \overrightarrow{AQ}$. $\square$

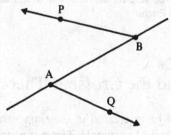

Figure 4-17

Theorem 4.4.3 is surprisingly useful. The key to using it is to recognize a "Z configuration" in the picture you have sketched. With a little imagination Figure 4-17 looks like a Z. The name of the theorem comes from this observation. The Z Theorem will be used repeatedly in the proof of the Crossbar Theorem.

Definition. In a Pasch geometry the **interior** of $\angle ABC$, written int($\angle ABC$), is the intersection of the side of $\overrightarrow{AB}$ that contains C with the side of $\overrightarrow{BC}$ that contains A. (See Figure 4-18 for a picture in $\mathscr{E}$.)

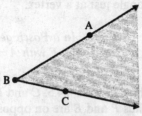

Figure 4-18

Note that the definition of int($\angle ABC$) uses the points A and C. However, $\angle ABC$ can be named in more than one way, say as $\angle A'BC'$. How do we know we get the same interior when we use A, C as when we use A', C'? This is a question of "well-defined." Does the definition depend on our choice of name? The answer is no, as the next theorem shows. Philosophically this is similar to Theorems 3.3.2 and 3.4.2.

Theorem 4.4.4. *In a Pasch geometry, if* $\angle ABC = \angle A'B'C'$ *then* int($\angle ABC$) = int($\angle A'B'C'$).

PROOF. By Theorem 3.4.2, $B = B'$ and $\overrightarrow{BA}$ is either $\overrightarrow{B'A'}$ or $\overrightarrow{B'C'}$. Assume that $\overrightarrow{BA} = \overrightarrow{B'A'}$ (so that $\overrightarrow{BC} = \overrightarrow{B'C'}$). Then A, $A' \in \text{int}(\overrightarrow{BA})$ and by Theorem 4.4.2 both A and A' lie on the same side of $l = \overrightarrow{BC} = \overrightarrow{B'C'}$. Thus the side of $\overrightarrow{BC} = l$ containing A is the same as the side of $\overrightarrow{B'C'} = l$ containing A'. Similarly the side of $\overrightarrow{BA}$ containing C is the same as the side of $\overrightarrow{B'A'}$ containing C'. Hence the intersection of the sides giving int($\angle ABC$) is the same as the intersection of the sides giving int($\angle A'B'C'$).

If $\overrightarrow{BA} = \overrightarrow{B'C'}$ then we may repeat the above argument with A' and C' interchanged. □

The following two theorems, whose proofs are left as exercises, as well as Problems A9 and A10, give tests for when $P \in \text{int}(\angle ABC)$.

Theorem 4.4.5. *In a Pasch geometry,* $P \in \text{int}(\angle ABC)$ *if and only if* A *and* P *are on the same side of* $\overrightarrow{BC}$ *and* C *and* P *are on the same side of* $\overrightarrow{BA}$.

Theorem 4.4.6. *Given* $\triangle ABC$ *in a Pasch geometry, if* $A - P - C$ *then* $P \in \text{int}(\angle ABC)$ *and therefore* int($\overline{AC}$) $\subset$ int($\angle ABC$).

We can now prove the theorem we have been working towards, the Crossbar Theorem. Both Pasch's Theorem and the Crossbar Theorem deal with what happens after a line enters a triangle. Pasch's Theorem can be thought of as saying that when a line enters through a non-vertex, it must pass out one of the other two sides (possibly at a vertex). The Crossbar Theorem tells us what happens if the line enters through a vertex—it must pass out the opposite side. Of course, "enters" must be made precise because a line may intersect a triangle just at a vertex.

Theorem 4.4.7 (Crossbar Theorem). *In a Pasch geometry if $P \in \text{int}(\angle ABC)$ then $\overrightarrow{BP}$ intersects $\overline{AC}$ at a unique point F with $A—F—C$.*

PROOF. Let E be a point such that $E—B—C$ (see Figure 4-19). P and C are on the same side of $\overleftrightarrow{AB}$ by Theorem 4.4.5. C and E are on opposite sides of $\overleftrightarrow{AB}$ by Theorem 4.4.2. Thus P and E are on opposite sides of $\overleftrightarrow{AB}$. By the Z Theorem, $\overrightarrow{BP} \cap \overline{AE} = \varnothing$. Let Q be a point such that $P—B—Q$. Then Q and A are on opposite sides of $\overleftrightarrow{BC} = \overleftrightarrow{EC}$. By the Z Theorem again $\overrightarrow{BQ} \cap \overline{AE} = \varnothing$. Hence $\overleftrightarrow{BP} \cap \overline{AE} = \varnothing$. Applying Pasch's Postulate to $\triangle ECA$ we see that $\overleftrightarrow{BP} \cap \overline{AC} \neq \varnothing$. Since A, B, C are not collinear, $\overleftrightarrow{BP} \cap \overline{AC} = \{F\}$ for some F.

$F \neq A$ (or else $\overleftrightarrow{BP} \cap \overline{AE} \neq \varnothing$) and $F \neq C$ (or else B, P, C are collinear). Thus $F \in \text{int}(\overline{AC})$. Finally P, A, and F are all on the same side of $\overleftrightarrow{BC}$ so that $F \in \overleftrightarrow{BP}$ implies $F \in \overrightarrow{BP}$. Hence $\overrightarrow{BP}$ intersects $\overline{AC}$ at a unique point F with $A—F—C$. $\qquad\square$

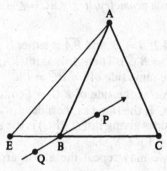

Figure 4-19

The following two results will be used frequently to verify that a point is in the interior of an angle. Their proofs are left to Problems A9 and A10.

Theorem 4.4.8. *In a Pasch geometry, if $\overline{CP} \cap \overrightarrow{AB} = \varnothing$ then $P \in \text{int}(\angle ABC)$ if and only if A and C are on opposite sides of $\overleftrightarrow{BP}$.*

Theorem 4.4.9. *In a Pasch geometry, if $A—B—D$ then $P \in \text{int}(\angle ABC)$ if and only if $C \in \text{int}(\angle DBP)$.*

Definition. In a Pasch geometry, the **interior of** $\triangle ABC$, written $\text{int}(\triangle ABC)$, is the intersection of the side of $\overleftrightarrow{AB}$ which contains C, the side of $\overleftrightarrow{BC}$ which contains A, and the side of $\overleftrightarrow{CA}$ which contains B.

The next result, as well as two other characterizations of int($\triangle ABC$), is left to the problem set.

Theorem 4.4.10. In a Pasch geometry int($\triangle ABC$) *is convex.*

PROBLEM SET 4.4

Part A.

1. Prove that in a metric geometry, int($\overrightarrow{AB}$) and int($\overline{AB}$) are convex sets.

2. Prove Theorem 4.4.2.

3. Prove Theorem 4.4.5.

4. Prove Theorem 4.4.6.

5. In a Pasch geometry, if $P \in \text{int}(\angle ABC)$ prove int($\overrightarrow{BP}$) $\subset$ int($\angle ABC$).

*6. In a Pasch geometry, given $\triangle ABC$ and points D, E, F such that B—C—D, A—E–-C, and B—E—F, prove that $F \in \text{int}(\angle ACD)$.

7. Prove the strong form of Pasch's Theorem: In a metric geometry which satisfies PSA, if A—D—B and C and E are on the same side of $\overleftrightarrow{AB}$, then $\overrightarrow{DE} \cap \overline{AC} \neq \varnothing$ or $\overrightarrow{DE} \cap \overline{BC} \neq \varnothing$. How is this different from Pasch's Theorem?

8. In a Pasch geometry, if $P \in \text{int}(\angle ABC)$ and if $D \in \overrightarrow{AP} \cap \overline{BC}$, then prove that A—P—D.

9. Prove Theorem 4.4.8.

10. Prove Theorem 4.4.9.

11. In a Pasch geometry, if $\overrightarrow{CP} \cap \overrightarrow{AB} = \varnothing$, prove that either $\overrightarrow{BC} = \overrightarrow{BP}$, or $P \in \text{int}(\angle ABC)$, or $C \in \text{int}(\angle ABP)$.

12. Given $\angle ABC$ and a point P in a Pasch geometry, prove that if $\overrightarrow{BP} \cap \text{int}(\overline{AC}) \neq \varnothing$ then $P \in \text{int}(\angle ABC)$. (This is the converse of the Crossbar Theorem.)

13. In a Pasch geometry, if $\angle ABC = \angle DBE$ and $\overrightarrow{BF} \cap \text{int}(\overline{AC}) \neq \varnothing$ then prove $\overrightarrow{BF} \cap \text{int}(\overline{DE}) \neq \varnothing$.

14. In a Pasch geometry, if int($\angle ABC$) = int($\angle DEF$), prove $\angle ABC = \angle DEF$.

15. Prove that in a Pasch geometry, int($\angle ABC$) is convex.

16. Prove that in a Pasch geometry, int($\triangle ABC$) is convex.

17. Prove that if l is a line in a Pasch geometry and $l \cap \text{int}(\triangle ABC) \neq \varnothing$ then $l \cap \triangle ABC$ has exactly two points.

18. In a Pasch geometry, prove int($\triangle ABC$) = int($\angle ABC$) $\cap$ int($\angle BCA$) $\cap$ int($\angle CAB$).

19. In a Pasch geometry, prove int($\triangle ABC$) = $\{P \mid \text{there exists a } D \text{ with } B$—$D$—$C \text{ and } A$—$P$—$D\}$.

20. In a Pasch geometry, prove that $\triangle ABC \cup \text{int}(\triangle ABC)$ is convex.

21. Show that the conclusion of the Crossbar Theorem is false in the Missing Strip Plane. Explain.

Part B. "Prove" may mean "find a counterexample".

22. Prove that $\text{int}(\overrightarrow{AB}) \subset \text{int}(\overline{AB})$ in a metric geometry.

23. Prove that in a Pasch geometry if $l \cap \text{int}(\angle ABC) \neq \varnothing$ then $l \cap \angle ABC \neq \varnothing$.

24. In a Pasch geometry, given $\triangle ABC$ and two points P, Q with A—P—B and B—Q—C, prove that if $R \in \overrightarrow{PQ} \cap \text{int}(\triangle ABC)$ then P—R—Q.

25. In a metric geometry define the **crossbar interior** of $\angle ABC$ to be $\text{cint}(\angle ABC) = \{P | D$—$P$—$E$ for some $D \in \text{int}(\overline{BA})$ and some $E \in \text{int}(\overline{BC})\}$. Prove that $\text{cint}(\angle ABC) = \text{int}(\angle ABC)$ in a Pasch geometry.

26. In a Pasch geometry if $P \in \text{int}(\angle ABC)$ prove that there is a line through P which intersects both $\overrightarrow{BA}$ and $\overrightarrow{BC}$ but which does not pass through B.

4.5 Convex Quadrilaterals

We end this chapter with a short section giving an application of the Crossbar Theorem. The main result will be that the diagonals of a convex quadrilateral intersect in a Pasch geometry. The preliminary results are left to the exercises. This material will not be needed until Chapter 7 and may be postponed until then.

Definition. Let $\{A, B, C, D\}$ be a set of four points in a metric geometry no three of which are collinear. If no two of $\text{int}(\overline{AB})$, $\text{int}(\overline{BC})$, $\text{int}(\overline{CD})$ and $\text{int}(\overline{DA})$ intersect, then

$$\square ABCD = \overline{AB} \cup \overline{BC} \cup \overline{CD} \cup \overline{DA}$$

is a **quadrilateral**.

In Figure 4-20, parts (a) and (b) represent quadrilaterals while part (c) does not. Note that although we use a square as the symbol for a quadrilateral, you should not think that $\square ABCD$ is a square. In fact we don't even know what a square is yet!

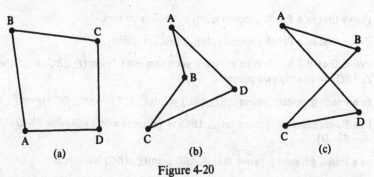

(a)   (b)   (c)

Figure 4-20

Theorem 4.5.1. *Given a quadrilateral* $\square ABCD$ *in a metric geometry then*

$$\square ABCD = \square BCDA = \square CDAB = \square DABC = \square ADCB$$
$$= \square DCBA = \square CBAD = \square BADC.$$

If both $\square ABCD$ *and* $\square ABDC$ *exist, they are not equal.*

Definition. In the quadrilateral $\square ABCD$, the **sides** are $\overline{AB}$, $\overline{BC}$, $\overline{CD}$, and $\overline{DA}$; the **vertices** are A, B, C, and D; the **angles** are $\angle ABC$, $\angle BCD$, $\angle CDA$, and $\angle DAB$; and the **diagonals** are $\overline{AC}$ and $\overline{BD}$. The endpoints of a diagonal are called **opposite vertices**. If two sides contain a common vertex, the sides are **adjacent**; otherwise they are **opposite**. If two sides contain a common side, the angles are **adjacent**; otherwise they are **opposite**.

Just as for earlier geometric forms (segments, angles, triangles) we must show that the angles, sides, vertices, etc. of a quadrilateral are well defined.

Theorem 4.5.2. *In a metric geometry, if* $\square ABCD = \square PQRS$ *then* $\{A, B, C, D\} = \{P, Q, R, S\}$. *Furthermore, if* $A = P$ *then* $C = R$ *and either* $B = Q$ *or* $B = S$ *so that the sides, angles, and diagonals of* $\square ABCD$ *are the same as those of* $\square PQRS$.

Definition. A quadrilateral $\square ABCD$ in a Pasch geometry is a **convex quadrilateral** if each side lies entirely in a half plane determined by its opposite side.

In Figure 4-20 (a) is convex while (b) is not.

Theorem 4.5.3. *In a Pasch geometry, a quadrilateral is a convex quadrilateral if and only if the vertex of each angle is contained in the interior of the opposite angle.*

Theorem 4.5.4. *In a Pasch geometry, the diagonals of a convex quadrilateral intersect.*

PROOF. Let $\square ABCD$ be a convex quadrilateral. We must show that $\overline{AC} \cap \overline{BD} \neq \emptyset$. By Theorem 4.5.3, $D \in \text{int}(\angle ABC)$. By the Crossbar Theorem, $\overrightarrow{BD}$ intersects $\overline{AC}$ at a unique point E with $A-E-C$. (See Figure 4-21.) We must show that $E \in \overline{BD}$ (not just $E \in \overrightarrow{BD}$). $C \in \text{int}(\angle DAB)$ so by the Crossbar Theorem, $\overrightarrow{AC}$ intersects $\overline{DB}$ at a unique point F with $B-F-D$. Then $\{E\} = \overrightarrow{AC} \cap \overrightarrow{BD} = \overrightarrow{AC} \cap \overrightarrow{BD} = \overrightarrow{AC} \cap \overrightarrow{BD} = \{F\}$ since $\overrightarrow{AC} \neq \overrightarrow{BD}$. Thus $E = F$ and $\overline{AC} \cap \overline{BD} = \{E\}$. □

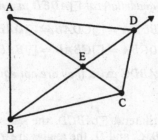

Figure 4-21

The last result will be extremely useful in Chapter 7. It says that a "trapezoid" is a convex quadrilateral.

Theorem 4.5.5. *Let* □*ABCD be a quadrilateral in a Pasch geometry. If* $\overleftrightarrow{BC} \| \overrightarrow{AD}$ *then* □*ABCD is a convex quadrilateral.*

PROOF. Since $\overleftrightarrow{BC} \| \overrightarrow{AD}$, $\overleftrightarrow{BC}$ lies on one side of $\overleftrightarrow{AD}$ and $\overleftrightarrow{AD}$ lies on one side of $\overleftrightarrow{BC}$. See Figure 4-22. We now show that $\overline{AB}$ lies on one side of $\overleftrightarrow{CD}$. Suppose to the contrary that $\overline{AB}$ does not all lie on one side of $\overleftrightarrow{CD}$. Then int$(\overline{AB}) \cap \overleftrightarrow{CD} \neq \varnothing$. Let H_1 be the side of $\overleftrightarrow{BC}$ that contains A and let H_1^* be the side of $\overleftrightarrow{AD}$ that contains B. Int$(\overline{AB}) \subset H_1 \cap H_1^*$ by Theorem 4.4.2. Therefore $\varnothing \neq$ int$(\overline{AB}) \cap \overleftrightarrow{CD} \subset H_1 \cap H_1^* \cap \overleftrightarrow{CD} =$ int$(\overline{CD})$. Hence int$(\overline{AB}) \cap$ int$(\overline{CD}) \neq \varnothing$, which contradicts the definition of a quadrilateral. Thus $\overline{AB}$ lies all on one side of $\overleftrightarrow{CD}$. Similarly $\overline{CD}$ lies all on one side of $\overleftrightarrow{AB}$. Therefore □*ABCD* is a convex quadrilateral.   □

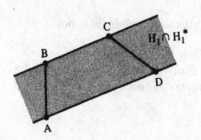

Figure 4-22

PROBLEM SET 4.5

Part A.

1. Prove Theorem 4.5.1.

2. Prove Theorem 4.5.2.

3. Prove Theorem 4.5.3.

4. Sketch two quadrilaterals in the Poincaré Plane, one of which is a convex quadrilateral and the other of which is not.

5. Prove that the quadrilateral $\square ABCD$ in a Pasch geometry is a convex quadrilateral if and only if each side does not intersect the line determined by its opposite side.

6. Give a "proper" definition of the interior of a convex quadrilateral. Then prove that the interior of a convex quadrilateral is a convex set.

7. Prove that in a Pasch geometry if the diagonals of a quadrilateral intersect then the quadrilateral is a convex quadrilateral.

Part B. "Prove" may mean "find a counterexample".

8. Prove that a convex quadrilateral in a Pasch geometry is a convex set.

9. Prove that for any quadrilateral $\square ABCD$ in a Pasch geometry either $\overrightarrow{AB} \cap \overleftrightarrow{CD} = \emptyset$ or $\overleftrightarrow{AB} \cap \overrightarrow{CD} = \emptyset$.

10. Prove that in a Pasch geometry at least one vertex of a quadrilateral is in the interior of the opposite angle.

11. Prove that in a Pasch geometry the lines containing the diagonals of a quadrilateral intersect. How does this differ from Theorem 4.5.4?

12. If the quadrilateral $\square ABCD$ in a Pasch geometry is not a convex quadrilateral, prove that either $\overrightarrow{BC} \cap \overrightarrow{AD} \neq \emptyset$ or $\overrightarrow{BC} \cap \overrightarrow{AD} \neq \emptyset$.

13. Let $\square ABCD$ be a quadrilateral in a Pasch geometry with $\overleftrightarrow{AB} \cap \overleftrightarrow{CD} = \{E\}$, $\overleftrightarrow{AC} \cap \overleftrightarrow{BD} = \{F\}$, and $\overleftrightarrow{AD} \cap \overleftrightarrow{BC} = \{G\}$. Prove that E, F, G are *not* collinear.

14. Suppose that $\square ABCD$ and $\square ABDC$ are both quadrilaterals in a Pasch geometry. Prove that neither one is a convex quadrilateral.

CHAPTER 5
Angle Measure

5.1 The Measure of an Angle

As we have progressed through the first four chapters, we have built up an axiom system piece by piece. Starting with the underlying set structure in an abstract geometry, we added the incidence axioms, the ruler postulate, and the plane separation axiom. Now we add another axiom to make our geometry look more like the geometry that is studied in high school. In this section we shall define what is meant by an angle measure and indicate how angle measures are defined in our two basic models. In the second section we shall develop a new model with some very interesting properties. In the third section, some of the basic results associated with angle measure are discussed. The last section is devoted to the technical details of verifying the existence of angle measure on $\mathbb{R}^2$ and $\mathbb{H}$.

After we have an angle measure our geometry will look very much like "high school geometry." However, we will still be missing one important assumption. That assumption is the Side-Angle-Side (SAS) congruence axiom. Without it, some results can occur which are very unusual for someone accustomed only to Euclidean geometry. In particular, we will see in examples that without SAS it is possible for the sum of the measures of the angles of a triangle to be greater than 180 degrees, or for the length of one side of a triangle to be longer than the sum of the other two. The SAS axiom will be formally introduced in Chapter 6.

Definition. Let r_0 be a fixed positive real number. In a Pasch geometry, an **angle measure** (or **protractor**) based on r_0 is a function m from the set $\mathscr{A}$ of all angles to the set of real numbers such that

(i) If $\angle ABC \in \mathscr{A}$ then $0 < m(\angle ABC) < r_0$;

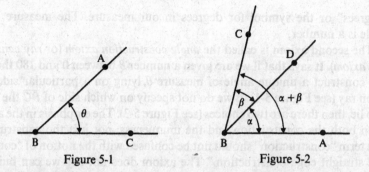

Figure 5-1   Figure 5-2

(ii) If $\overrightarrow{BC}$ lies in the edge of the half plane H_1 and if θ is a real number with $0 < \theta < r_0$, then there is a unique ray $\overrightarrow{BA}$ with $A \in H_1$ and $m(\angle ABC) = \theta$ (see Figure 5-1);

(iii) If $D \in \text{int}(\angle ABC)$ then (see Figure 5-2)

$$m(\angle ABD) + m(\angle DBC) = m(\angle ABC). \tag{1-1}$$

We should note that the definition of an angle measure does not even make sense unless we have PSA (or equivalently, Pasch) since we must use the idea of the interior of an angle. If $r_0 = 180$, m is called *degree measure*. If $r_0 = \pi$, then m is called *radian measure*. If $r_0 = 200$, then m is called *grade measure*. Traditionally, degree measure is used in geometry. Radian measure is used in calculus because it makes the differentiation formulas most natural. The origin of degree measure is not really known. Perhaps it was chosen because a year has (almost) 360 days and in degree measure, a complete revolution is one of 360 degrees. Another possible explanation is that 360 has many factors so that there are many "natural" subdivisions of a circle into equal parts. Although the precise value of r_0 is not crucial (see Problem A4) we adopt the following convention.

> **Convention.** Except in Section 5.4, we shall always use degree measure ($r_0 = 180$) without further assumption.

Definition. A **protractor geometry** $\{\mathscr{S}, \mathscr{L}, d, m\}$ is a Pasch geometry $\{\mathscr{S}, \mathscr{L}, d\}$ together with an angle measure m.

The various axioms contained in the definition of an angle measure are worth discussing. One of the consequences of the first axiom is that an angle *cannot* have measure 0 or measure 180. What we might intuitively think of as an angle of measure 0 is not an angle—it is a ray. Likewise, what we might think of as an angle of measure 180 is also not an angle—it is a line. $\angle ABC$ makes sense only if A, B, C are not collinear. Note we never use the word "degrees" or the symbol for degrees in our measure. The measure of an angle is a number.

The second axiom is called the *angle construction axiom* (or *ray construction axiom*). It says that if we are given a number θ between 0 and 180 then we can construct a unique angle of measure θ lying on a particular side of a given ray (see Figure 5-1). If we do not specify on which side of $\overrightarrow{BC}$ the angle is to lie, then there are two choices (see Figure 5-3). The emphasis in the axiom is on both the construction and the uniqueness, not just the construction. The term "construction" should not be confused with the notion of "compass and straight edge construction." The axiom does not say we can build an angle in that fashion. Rather, it is really postulating the existence of an angle of any particular measure between 0 and 180 with no indication as to how it is found.

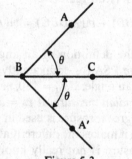

Figure 5-3

The third axiom is called the *angle addition axiom* and reflects the familiar statement that the whole is the sum of its parts. Note in Figure 5-2 we illustrated this by marking the various angles with lower case Greek letters which denote their measure. This will be our standard procedure.

The Euclidean Plane has an angle measure which we have been accustomed to using since high school. Yet to carefully show that there really is an angle measure in $\mathbb{R}^2$ can be quite involved. The basic idea is to use the familiar formula for the scalar product of two vectors

$$\langle \mathbf{a}, \mathbf{b} \rangle = \|\mathbf{a}\| \|\mathbf{b}\| \cos \theta \tag{1-2}$$

when θ is the measure of the angle between the vectors. If we solve Equation (1-2) for θ we have

$$\theta = \cos^{-1} \frac{\langle \mathbf{a}, \mathbf{b} \rangle}{\|\mathbf{a}\| \|\mathbf{b}\|}.$$

This is fine, *provided* we know what the function $\cos^{-1}(x)$ is. Unfortunately, $\cos(x)$ (and $\cos^{-1}(x)$) is usually defined in terms of angle measures and so cannot be used to define angle measure. The main work involved in defining an angle measure for $\mathbb{R}^2$ is to develop the function $\cos(x)$ without reference to angles at all. This will be done in Section 5.4. However, those readers who are willing to accept the existence and standard properties of $\cos^{-1}(x)$ may omit Section 5.4 and assume Proposition 5.1.2.

Definition. In the Euclidean Plane, the **Euclidean angle measure** of $\angle ABC$ is

$$m_E(\angle ABC) = \cos^{-1}\left(\frac{\langle A - B, C - B \rangle}{\|A - B\| \cdot \|C - B\|}\right). \tag{1-3}$$

Example 5.1.1. In $\mathscr{E}$ what is $m_E(\angle ABC)$ if $A = (0,3)$, $B = (0,1)$ and $C = (\sqrt{3}, 2)$?

SOLUTION. $A - B = (0,2)$ and $C - B = (\sqrt{3}, 1)$ so that

$$\frac{\langle A - B, C - B \rangle}{\|A - B\| \cdot \|C - B\|} = \frac{2}{2 \cdot 2} = \frac{1}{2}$$

$$m_E(\angle ABC) = \cos^{-1}(\tfrac{1}{2}) = 60. \qquad \square$$

The following proposition will be proved carefully in Section 5.4.

Proposition 5.1.2. m_E is an angle measure on $\{\mathbb{R}^2, \mathscr{L}_E, d_E\}$.

Since the Poincaré Plane as a set is a subset of the Euclidean Plane, and since its lines are defined in terms of Euclidean lines and circles, it should not be surprising that we define Poincaré angle measure in terms of Euclidean angle measure. The basic idea is to replace the Poincaré rays that make up the angle by Euclidean rays that are tangent to the given Poincaré rays. The measure of the angle formed by the Euclidean rays will be used as the measure of the Poincaré angle.

The Euclidean tangent rays are determined by finding tangent vectors of the given Poincaré rays when those rays are viewed as curves. In the case of a type I ray $\overrightarrow{BA}$, there is a natural choice for the tangent vector to $\overrightarrow{BA}$ at B: If $A = (x_A, y_A)$ and $B = (x_B, y_B)$ belong to a type I line, then $x_A = x_B$ and $A - B = (0, y_A - y_B)$ is the tangent vector. See Figure 5-4.

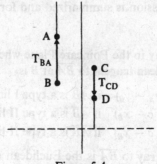

Figure 5-4

We now investigate what the tangent to a type II ray should be in order to motivate the definition of Poincaré angle measure. Suppose that $\overrightarrow{BA}$ is part of the type II line $_cL_r$ so that if $(x, y) \in \overrightarrow{BA}$ then

$$(x - c)^2 + y^2 = r^2. \tag{1-4}$$

The slope of the tangent to the curve in $\mathbb{R}^2$ whose equation is given by Equation (1-4) at the point $B = (x_B, y_B)$ should be given by the derivative $(dy/dx)(B)$. This is found by implicit differentiation of Equation (1-4):

$$2(x - c) + 2y \frac{dy}{dx} = 0$$

or

$$\frac{dy}{dx} = \frac{c - x}{y}.$$

Thus the slope of the tangent at B should be $(c - x_B)/y_B$. The vectors $\pm(y_B, c - x_B)$ have this as their slope and so are prime candidates for the tangent vector. The $\pm$ sign reflects whether A is to the right $(+)$ or left $(-)$ of B. See Figure 5-5.

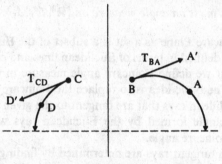

Figure 5-5

If T_{BA} denotes the tangent vector to $\overrightarrow{BA}$ as found above and if $A' = B + T_{BA}$, then the Euclidean ray $\overrightarrow{BA'}$ is parallel to the vector T_{BA} and is the desired Euclidean tangent ray.

The preceding discussion is summarized and formalized in the following definition.

Definition. If $\overrightarrow{BA}$ is a ray in the Poincaré Plane where $B = (x_B, y_B)$ and $A = (x_A, y_A)$, then the **Euclidean tangent** to $\overrightarrow{BA}$ at B is

$$T_{BA} = \begin{cases} (0, y_A - y_B) & \text{if } \overleftrightarrow{AB} \text{ is a type I line} \\ (y_B, c - x_B) & \text{if } \overleftrightarrow{AB} \text{ is a type II line } _cL_r,\ x_B < x_A \\ -(y_B, c - x_B) & \text{if } \overleftrightarrow{AB} \text{ is a type II line } _cL_r,\ x_B > x_A. \end{cases}$$

The **Euclidean tangent ray** to $\overrightarrow{BA}$ is the Euclidean ray $\overrightarrow{BA'}$ where $A' = B + T_{BA}$.

Definition. The **measure** of the Poincaré angle $\angle ABC$ in $\mathbb{H}$ is

$$m_H(\angle ABC) = m_E(\angle A'BC') = \cos^{-1}\left(\frac{\langle T_{BA}, T_{BC}\rangle}{\|T_{BA}\|\|T_{BC}\|}\right) \qquad (1\text{-}5)$$

where $A' = B + T_{BA}$ and $C' = B + T_{BC}$, and $m_E(\angle A'BC')$ is the Euclidean measure of the Euclidean angle $\angle A'BC'$. (See Figure 5-6.)

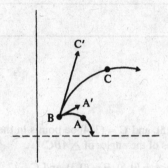

Figure 5-6

Note that in Equation (1-5), we do not really need A' and C' to compute $m_H(\angle ABC)$, only T_{BA} and T_{BC}.

Example 5.1.3. In the Poincaré Plane find the measure of $\angle ABC$ where $A = (0, 1)$, $B = (0, 5)$, and $C = (3, 4)$.

SOLUTION. It is easy to see that $\overrightarrow{BC} = {}_0L_5$, that $\overrightarrow{CA} = {}_4L_{\sqrt{17}}$, and that $\overrightarrow{BA} = {}_0L$. Hence

$$T_{BA} = (0, -4) \quad \text{and} \quad T_{BC} = (5, 0).$$

Thus $m_H(\angle ABC) = \theta$ where

$$\cos\theta = \frac{\langle T_{BA}, T_{BC}\rangle}{\|T_{BA}\|\|T_{BC}\|} = \frac{0}{20}.$$

Hence

$$\cos(\theta) = 0 \quad \text{and} \quad m_H(\angle ABC) = 90.$$

In Problem A1 you will find the measure of the other angles of $\triangle ABC$ and will show that the sum of the measures of all three angles is approximately 155. Thus the angle sum of a triangle in the Poincaré Plane need not be 180! $\square$

The proof of the following Proposition will be left to Section 5.4.

Proposition 5.1.4. m_H *is an angle measure on* $\{\mathbb{H}, \mathscr{L}_H, d_H\}$.

> **Convention.** From now on the terms Euclidean Plane, Poincaré Plane, and Taxicab Plane will refer to the protractor geometries
> $$\mathcal{E} = \{\mathbb{R}^2, \mathcal{L}_E, d_E, m_E\}$$
> $$\mathcal{H} = \{\mathbb{H}, \mathcal{L}_H, d_H, m_H\}$$
> $$\mathcal{T} = \{\mathbb{R}^2, \mathcal{L}_E, d_T, m_E\}.$$

PROBLEM SET 5.1

Part A.

1. Let $A = (0, 1)$, $B = (0, 5)$, and $C = (3, 4)$ be points in the Poincaré Plane $\mathcal{H}$. Find the sum of the measures of the angles of $\triangle ABC$.

2. Repeat Problem 1 with $A = (0, 5)$, $B = (0, 3)$, and $C = (2, \sqrt{21})$.

3. Repeat Problem 1 with $A = (5, 1)$, $B = (8, 4)$, and $C = (1, 3)$.

4. Let m be an angle measure for $\{\mathcal{S}, \mathcal{L}, d\}$ based on α. Let $t > 0$ and define m_t by
   $$m_t(\angle ABC) = t \cdot m(\angle ABC).$$
   Prove that m_t is an angle measure for $\{\mathcal{S}, \mathcal{L}, d\}$. What value is m_t based on?

5. Assume that m_E is an angle measure for Euclidean metric geometry $\{\mathbb{R}^2, \mathcal{L}_E, d_E\}$. Prove that m_E is an angle measure for the Taxicab Plane $\{\mathbb{R}^2, \mathcal{L}_E, d_T\}$.

6. Show that Euclidean angle measure is well defined; i.e., if $\angle ABC = \angle A'BC'$ prove that $m_E(\angle ABC) = m_E(\angle A'BC')$ by using Equation (1-3).

Part B. "Prove" may mean "find a counterexample".

7. Let θ be a real number with $0 < \theta < 180$ and let $\overrightarrow{BC}$ lie in the edge of a half plane H_1 in a protractor geometry. Prove that there is a unique point $A \in H_1$ with $m(\angle ABC) = \theta$.

8. Assume that m_E is an angle measure for the Euclidean Plane. Prove that $\{\mathbb{R}^2, \mathcal{L}_E, d_S, m_E\}$ is a protractor geometry, where d_S is the max distance defined in Problem B18 of Section 2.2.

Part C. Expository exercises.

9. Write an essay contrasting degree, radian, and grade measure. You should include information on their practical use, history, and development.

5.2 The Moulton Plane

In this section whose details are optional we shall develop another model of a protractor geometry—the Moulton Plane. This model was introduced by the American mathematician Forest Moulton [1902] and is an important example in the study of projective geometry. It will have some strange characteristics. In fact, we will eventually see (Problem A10) that the sum of the measures of the angles of a triangle in the Moulton Plane may be more than 180.

The underlying set of the Moulton Plane will be $\mathbb{R}^2$, but the set of lines will not be $\mathscr{L}_E$. Some lines will be Euclidean and the rest will be in the form $M_{m,b}$ as follows.

$$M_{m,b} = \left\{(x, y) \in \mathbb{R}^2 \,\middle|\, \begin{array}{l} y = mx + b \text{ if } x \le 0 \\ y = \tfrac{1}{2}mx + b \text{ if } x > 0 \end{array} \right\}.$$

The Moulton line $M_{1,2}$ is sketched in Figure 5-7. One way to view a Moulton line $M_{m,b}$ is as the path of a ray of light that is bent or refracted as it crosses the y-axis. (Not every line of the form $M_{m,b}$ will be used, only those with $m > 0$.)

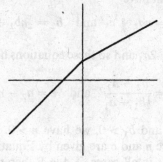

Figure 5-7

Definition. The **Moulton Plane** is the collection $\mathcal{M} = \{\mathbb{R}^2, \mathscr{L}_M\}$ where

$$\mathscr{L}_M = \{L_a \in \mathscr{L}_E\} \cup \{L_{m,b} \in \mathscr{L}_E | m \le 0\} \cup \{M_{m,b} | m > 0\}.$$

Proposition 5.2.1. *The Moulton Plane is an incidence geometry.*

PROOF. It is clear that $\{(1, 0), (0, 0), (0, 1)\}$ is a noncollinear set and that each line has at least two points. We need only prove that there is a unique line between any two distinct points A and B. We shall show that there is at least one line and leave the proof of uniqueness to Problem A2. The proof proceeds by examining cases which depend on the relative positions of A and B.

If $A \neq B$ then A and B lie on a unique *Euclidean* line $l \in \mathscr{L}_E$. If $l = L_a$ or $l = L_{m,b}$ with $m \leq 0$ then $l \in \mathscr{L}_M$. Hence we need only consider the case where $l = L_{m,b}$ with $m > 0$.

Let $A = (a_1, a_2)$ and $B = (b_1, b_2)$. We may assume that $a_1 < b_1$. Note that since $m > 0$, $a_2 < b_2$. See Figure 5-8 for three possible situations.

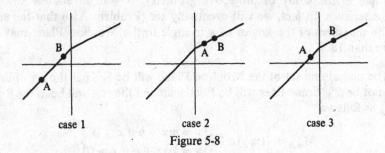

case 1   case 2   case 3

Figure 5-8

Case 1. $a_1 < b_1 < 0$. In this case $A, B \in M_{m,b}$.

Case 2. $0 < a_1 < b_1$. In this case $A, B \in M_{2m,b}$.

Case 3. $a_1 < 0 < b_1$. In this case we must work harder to find what Moulton line A and B lie on. We want to show that $A, B \in M_{n,c}$ for some n, c.

If $A, B \in M_{n,c}$ we must have

$$a_2 = na_1 + c \quad \text{and} \quad b_2 = \tfrac{1}{2}nb_1 + c.$$

Since $a_1 < 0 < b_1$, $b_1 \neq 2a_1$ and so these equations have a unique solution

$$n = \frac{b_2 - a_2}{\tfrac{1}{2}b_1 - a_1} \quad \text{and} \quad c = a_2 - na_1. \tag{2-1}$$

Since $a_2 < b_2$, $a_1 < 0$ and $b_1 > 0$, we have $n > 0$. Thus in the given case $A, B \in M_{n,c} \in \mathscr{L}_M$ where n and c are given by Equations (2-1).

We have shown that in all cases, if $A \neq B$ then there is at least one line $l \in \mathscr{L}_M$ with $A, B \in l$. Hence $\mathscr{M}$ is an abstract geometry. Once you show that the line through A and B is unique, we will know that $\mathscr{M}$ is an incidence geometry.   ☐

The next step is to make $\mathscr{M}$ into a metric geometry. We will define the distance between two points to be the Euclidean distance unless the two points lie on opposite sides of the y-axis on a "bent line". In this case we will view the Moulton segment joining the points as the union of two Euclidean segments and add their lengths. Note the condition $x_1 x_2 < 0$ in the next definition means that (x_1, y_1) and (x_2, y_2) lie on opposite sides of the y-axis.

Definition. The **Moulton distance** between the points $P = (x_1, y_1)$ and $Q = (x_2, y_2)$ in $\mathbb{R}^2$ is given by

$$d_M(P,Q) = \begin{cases} d_E(P,(0,b)) + d_E((0,b),Q) & \text{if } P, Q \in M_{m,b} \text{ with } x_1 x_2 < 0 \\ d_E(P,Q) & \text{otherwise.} \end{cases}$$

Example 5.2.2. Find the lengths of the sides of $\triangle ABC$ in the Moulton Plane, where $A = (-1,0)$, $B = (2,-1)$, and $C = (2,2)$.

SOLUTION.

$$d_M(A,B) = d_E(A,B) = \sqrt{10}$$
$$d_M(B,C) = d_E(B,C) = 3$$

A, C lie on $M_{1,1}$ by Equation (2-1). Hence if $D = (0,1)$ then

$$d_M(A,C) = d_E(A,D) + d_E(D,C) = \sqrt{2} + \sqrt{5}.$$

See Figure 5-9. $\square$

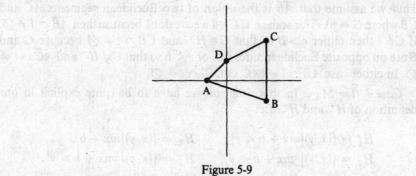

Figure 5-9

Proposition 5.2.3. $\{\mathbb{R}^2, \mathscr{L}_M, d_M\}$ is a metric geometry.

PROOF. In Problem A3 you will show that d_M actually is a distance function. Thus we need only show that each line has a ruler. We may use a Euclidean ruler for each line of the form L_a or $L_{m,b}$. Hence we need only consider the lines $M_{m,b}$.

Define $f: M_{m,b} \to \mathbb{R}$ by

$$f(x,y) = \begin{cases} x\sqrt{1 + m^2} & \text{if } x \le 0 \\ x\sqrt{1 + \dfrac{m^2}{4}} & \text{if } x > 0. \end{cases} \qquad (2\text{-}2)$$

In Problem A4 you will show that f is a ruler. $\square$

Proposition 5.2.4. $\{\mathbb{R}^2, \mathscr{L}_M, d_M\}$ satisfies the plane separation axiom.

PROOF. We must verify the three parts of PSA for every line l. The proof breaks into three cases depending on the line $l \in \mathscr{L}_M$.

Case 1. $l = L_a \in \mathscr{L}_E$. In this case the half planes determined by l are the Euclidean half planes H^+ and H^- determined by l. Clearly $\mathbb{R}^2 - l = H^+ \cup H^-$. Also, we see that $H^+ \cap H^- = \varnothing$ and that both of H^+ and H^- are Moulton convex.

All that remains is to let $A \in H^+$ and $B \in H^-$ and show that $\overline{AB} \cap L_a \neq \varnothing$. If $\overline{AB}$ is a segment of a Euclidean line then this follows from the fact that Euclidean geometry satisfies PSA. Suppose that $\overline{AB} = M_{m,b}$. If $a \leq 0$ then $X = (a, ma + b) \in \overline{AB} \cap L_a$. If $a > 0$ then $X = (a, \frac{1}{2}ma + b) \in \overline{AB} \cap L_a$. In either case, $\overline{AB} \cap L_a \neq \varnothing$ so that H^+ and H^- are half planes for L_a.

Case 2. $l = L_{m,b}$ with $m \leq 0$. Again we let the half plane determined by l be the Euclidean half planes H^+ and H^- determined by l. Clearly $\mathbb{R}^2 - l = H^+ \cup H^-$ and $H^+ \cap H^- = \varnothing$. We leave the proof that H^+ and H^- are Moulton convex to Problem A5.

Suppose that $A \in H^+$ and $B \in H^-$ where $H^+ = \{(x, y) \mid mx + b < y\}$. If the Moulton segment $\overline{AB}$ is actually a Euclidean segment then $\overline{AB} \cap l \neq \varnothing$. Thus we assume that $\overline{AB}$ is the union of two Euclidean segments $\overline{AC}$ and $\overline{CB}$ where $C = (0, r)$ for some r. If $C \in l$ we are done because then $\overline{AB} \cap l \neq \varnothing$. If $C \notin l$ then either $r > b$ so that $C \in H^+$ and $\overline{CB} \cap l \neq \varnothing$ because C and B are on opposite Euclidean sides of l, or $r < b$ so that $C \in H^-$ and $\overline{AC} \cap l \neq \varnothing$. In either case $\overline{AB} \cap l = (\overline{AC} \cup \overline{CB}) \cap l \neq \varnothing$.

Case 3. $l = M_{m,b}$. In this situation we have to be quite explicit in our definition of H^+ and H^-. Let

$$H_1^+ = \{(x, y) \mid mx + b < y\}, \qquad H_1^- = \{(x, y) \mid mx + b > y\}$$
$$H_2^+ = \{(x, y) \mid \tfrac{1}{2}mx + b < y\}, \qquad H_2^- = \{(x, y) \mid \tfrac{1}{2}mx + b > y\}.$$

We define $H^+ = H_1^+ \cup H_2^+$ and $H^- = H_1^- \cap H_2^-$ (H^+ is shaded in Figure 5-10). By Problem A6, $\mathbb{R}^2 - l = H^+ \cup H^-$ and $H^+ \cap H^- = \varnothing$.

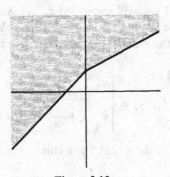

Figure 5-10

We must show that H^+ and H^- are convex. Suppose that $A, B \in H^+$. If A and B lie on the same (Euclidean = Moulton) side of the y-axis or on the y-axis then A and B either both belong to H_1^+ or both belong to H_2^+. In either case the Moulton segment $\overline{AB}$ is actually Euclidean so that $\overline{AB}$ is either in H_1^+ or in H_2^+. Hence $\overline{AB} \subset H^+$.

If A and B lie on opposite sides of the y-axis, L_0, we may assume that A is in the left half plane of L_0 and hence in H_1^+ while B is in the right half plane of L_0 and hence in H_2^+. The Moulton segment $\overline{AB}$ intersects L_0 at some point C by Case 1. If $C \in H^+$ then the Euclidean segment $\overline{AC} \subset H_1^+$ and $\overline{CB} \subset H_2^+$ so that the Moulton segment $\overline{AB} = \overline{AC} \cup \overline{CB} \subset H^+$.

We will now show that $C \notin H^-$. $M_{m,b}$ is the union of two Euclidean rays $\overrightarrow{PQ}$ and $\overrightarrow{PR}$ with Q to the right of L_0 and R to the left. If $C \in H^-$, the Euclidean Crossbar Theorem applied to $\triangle APC$ and the ray $\overrightarrow{PR}$ shows that $\overrightarrow{PR} \cap$ int $\overline{AC} \neq \emptyset$. Likewise $\overrightarrow{PQ} \cap$ int $\overline{CB} \neq \emptyset$. Hence the Moulton line $\overleftrightarrow{RQ} = L_{m,b}$ intersects the Moulton segment $\overline{AB} = \overline{AC} \cup \overline{CB}$ in two points, which is impossible. See Figure 5-11. Hence $C \notin H^-$. We leave the proof that $C \notin M_{m,b}$ to Problem A7.

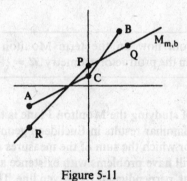

Figure 5-11

The proof that H^- is convex is left to Problem A8. Finally, the proof that if $A \in H^+$ and $B \in H^-$ then $\overline{AB}$ intersects $M_{m,b}$ is left to Problem A9. $\square$

Next we want to define an angle measure for the Moulton Plane using the Euclidean angle measure. If B is not on the line L_0, then given $\angle ABC$ we may choose $A' \in \text{int}(\overrightarrow{BA})$ and $C' \in \text{int}(\overrightarrow{BC})$ so that A', B, and C' all lie on the same side of L_0. Then we set $m_M(\angle ABC) = m_E(\angle A'BC')$. See Figure 5-12. If $B \in L_0$ we proceed as follows. For each $b \in \mathbb{R}$ and each $P = (x, y)$ let

$$P_b = \begin{cases} (x, 2y - b) & \text{if } x > 0 \text{ and } y > b \\ (x, y) & \text{otherwise.} \end{cases}$$

Then if $B = (0, b) \in L_0$ set

$$m_M(\angle ABC) = m_E(\angle A_bBC_b).$$

See Figure 5-13. Note that if $B = (0, b) \in L_0$ what we are doing is "unbending" $\overline{AB}$ before we compute the angle measure.

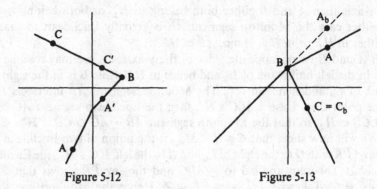

Figure 5-12   Figure 5-13

The detailed proof that m_M actually is an angle measure is lengthy but not hard. It is left to Problem B11.

Proposition 5.2.5. $\mathcal{M} = \{\mathbb{R}^2, \mathcal{L}_M, d_M, m_M\}$ *is a protractor geometry.*

> **Convention.** From now on the term Moulton Plane and the symbol $\mathcal{M}$ mean the protractor geometry $\mathcal{M} = \{\mathbb{R}^2, \mathcal{L}_M, d_M, m_M\}$.

The main value of studying the Moulton Plane is that it supplies us with counterexamples of familiar results in Euclidean geometry. In Problem A10 we have a triangle for which the sum of the measures of the angles is greater than 180. We also will have problems with existence and uniqueness of lines through a given point, perpendicular to a given line. This illustrates our need for an additional axiom in order to obtain familiar results. This triangle congruence axiom will be introduced in Chapter 6.

PROBLEM SET 5.2

Part A.

1. Find the Moulton lines through the following pairs of points:
   a. $(2, 3)$ and $(3, -1)$
   b. $(1, 4)$ and $(2, 6)$
   c. $(-1, 3)$ and $(-3, -2)$
   d. $(-1, 4)$ and $(2, 7)$
   e. $(-4, -4)$ and $(4, 4)$

2. Complete the proof of Proposition 5.2.1 by showing that for every pair of points $A \neq B$ there is *exactly* one line $l \in \mathcal{L}_M$ through A and B.

3. Prove that d_M is a distance function for $\{\mathbb{R}^2, \mathcal{L}_M\}$.

4. Complete the proof of Proposition 5.2.3 by showing that f as defined by Equation (2-2) is a ruler.

5. Complete the proof of Proposition 5.2.4 in Case 2 by proving that H^+ and H^- are Moulton convex.

6. In Case 3 of the proof of Proposition 5.2.4 show that $\mathbb{R}^2 - l = H^+ \cup H^-$ and that $H^+ \cap H^- = \varnothing$.

7. In Case 3 of the proof of Proposition 5.2.4 show that $C \notin M_{m,b}$.

8. In Case 3 of the proof of Proposition 5.2.4 show that H^- is convex.

9. Complete the proof of Proposition 5.2.4 in Case 3 by showing that if $A \in H^+$ and $B \in H^-$ then the Moulton segment $\overline{AB}$ intersects $M_{m,b}$.

*10. In the Moulton Plane let $A = (-5, 0)$, $B = (0, 5)$, $C = (10, 10)$, $D = (-5, 10)$, $E = (5, 0)$, and $F = (2, 6)$ as in Figure 5-14.
   a. Show that $A—B—C$ and $D—B—E$.
   b. Find the sum of the measures of the angles of $\triangle BFE$.
   c. Repeat part (b) for $\triangle BFD$.

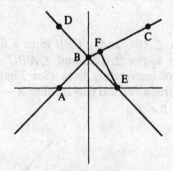

Figure 5-14

Part B. "Prove" may mean "find a counterexample".

11. Prove that m_M is an angle measure for $\mathcal{M}$.

12. Prove that d_M does not satisfy the triangle inequality. (Hint: Consider $\triangle ABC$ where $A = (-1, 0)$, $B = (2, 2)$ and $C = (0, \frac{2}{3})$.)

13. In the Moulton Plane show that it is possible to have three points A, B, C which are NOT collinear but $AC = AB + BC$. (This illustrates again why we insist on collinearity in the definition of between.) If you have access to a computer algebra system (program) such as DERIVE, MAPLE, or MATHEMATICA, let $A = (-1, 0)$, $B = (0, b)$, and $C = (2, 2)$. Find the exact value of b so that $AC = AB + BC$ and A, B, C are not collinear. (The calculation may be done by hand if you so desire and have the time.)

14. Show that the Moulton Plane satisfies Pasch's Postulate directly. This gives an alternative proof that $\mathcal{M}$ is a Pasch geometry.

Part C. Expository exercises.

15. Write a short essay which gives a description of the Moulton Plane and describe
its important properties. What is relevant here is what *you* think is important
about the model.

5.3 Perpendicularity and Angle Congruence

In terms of angle measure there is some standard terminology for angles. This,
in turn, can be related to certain configurations of angles. This section will
deal especially with one such configuration—that of a right angle.

Definition. An **acute angle** is an angle whose measure is less than 90. A **right
angle** is an angle whose measure is 90. An **obtuse angle** is an angle whose
measure is greater than 90. Two angles are **supplementary** if the sum of their
measures is 180. Two angles are **complementary** if the sum of their measures
is 90.

Definition. Two angles $\angle ABC$ and $\angle CBD$ form a **linear pair** if $A—B—D$
(see Figure 5-15). Two angles $\angle ABC$ and $\angle A'BC'$ form a **vertical pair** if
their union is a pair of intersecting lines. (See Figure 5-16. Alternatively,
$\angle ABC$ and $\angle A'BC'$ form a vertical pair if either $A—B—A'$ and $C—B—C'$,
or $A--B—C'$ and $C—B—A'$.)

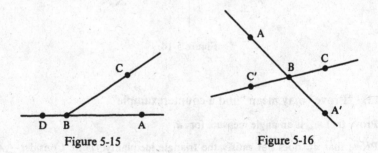

Figure 5-15   Figure 5-16

Theorem 5.3.1. *If C and D are points of a protractor geometry and are on the
same side of* $\overleftrightarrow{AB}$ *and* $m(\angle ABC) < m(\angle ABD)$, *then* $C \in \text{int}(\angle ABD)$.

PROOF. Either A and C are on the same side of $\overrightarrow{BD}$, or $C \in \overrightarrow{BD}$, or A and C
are on opposite sides of $\overrightarrow{BD}$. We eliminate the latter two cases as follows. If
$C \in \overrightarrow{BD}$, then since C and D are on the same side of $\overleftrightarrow{AB}$, then $C \in \text{int}(\overrightarrow{BD})$.
Hence $\angle ABC = \angle ABD$ and $m(\angle ABC) = m(\angle ABD)$ which is a contradiction.

If A and C are on opposite sides of $\overrightarrow{BD}$, then by Problem A9 of Section 4.4,
$D \in \text{int}(\angle ABC)$. This means that

$$m(\angle ABD) + m(\angle DBC) = m(\angle ABC) < m(\angle ABD).$$

Hence $m(\angle DBC) < 0$, which is impossible. Thus the only possibility is that

A and C are on the same side of $\overleftrightarrow{BD}$ and so $C \in \text{int}(\angle ABD)$. □

The next theorem is sometimes taken as an axiom (for example, in Moise [1990]). However, as we shall now see, it is a consequence of the other axioms.

Theorem 5.3.2 (Linear Pair Theorem). *If $\angle ABC$ and $\angle CBD$ form a linear pair in a protractor geometry then they are supplementary.*

PROOF. Let $m(\angle ABC) = \alpha$ and $m(\angle CBD) = \beta$. We must show that $\alpha + \beta = 180$. We do this by showing that both $\alpha + \beta < 180$ and $\alpha + \beta > 180$ lead to contradictions.

Suppose $\alpha + \beta < 180$. By the Angle Construction Axiom, there is a unique ray $\overrightarrow{BE}$ with E on the same side of $\overleftrightarrow{AB}$ as C and with $m(\angle ABE) = \alpha + \beta$. See Figure 5-17. By Theorem 5.3.1, $C \in \text{int}(\angle ABE)$ so that $m(\angle ABC) + m(\angle CBE) = m(\angle ABE)$. Thus

$$\alpha + m(\angle CBE) = \alpha + \beta \quad \text{or} \quad m(\angle CBE) = \beta.$$

On the other hand, $E \in \text{int}(\angle CBD)$ (Why?) so that $m(\angle CBE) + m(\angle EBD) = m(\angle CBD)$. Thus

$$\beta + m(\angle EBD) = \beta \quad \text{or} \quad m(\angle EBD) = 0$$

which is impossible. Thus $\alpha + \beta < 180$ cannot occur.

Now suppose $\alpha + \beta > 180$. Since both α and β are less than 180, $\alpha + \beta < 360$ and $0 < \alpha + \beta - 180 < 180$. Then there exists a unique ray $\overrightarrow{BF}$ with F on the same side of $\overleftrightarrow{AB}$ as C and $m(\angle ABF) = \alpha + \beta - 180$. See Figure 5-18. Since $\beta < 180$, $\alpha + \beta - 180 < \alpha$ and so $F \in \text{int}(\angle ABC)$. Hence $m(\angle ABF) + m(\angle FBC) = m(\angle ABC)$. Hence

$$\alpha + \beta - 180 + m(\angle FBC) = \alpha \quad \text{or} \quad m(\angle FBC) = 180 - \beta.$$

On the other hand, $C \in \text{int}(\angle FBD)$ (Why?) so that $m(\angle FBC) + m(\angle CBD) = m(\angle FBD)$. Thus

$$180 - \beta + \beta = m(\angle FBD) \quad \text{or} \quad m(\angle FBD) = 180$$

which is also impossible. Hence $\alpha + \beta > 180$ is false.

Thus the only possibility is that $\alpha + \beta = 180$. □

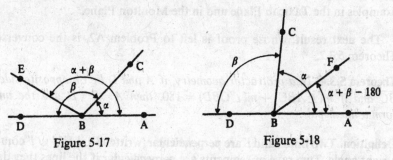

Figure 5-17   Figure 5-18

Now we can prove the converse of axiom (iii) for an angle measure.

Theorem 5.3.3. *In a protractor geometry, if* $m(\angle ABC) + m(\angle CBD) = m(\angle ABD)$*, then* $C \in \text{int}(\angle ABD)$*.*

PROOF. We shall show that C and D are on the same side of $\overleftrightarrow{AB}$ by contradiction. Suppose that C and D are on opposite sides of $\overleftrightarrow{AB}$. Now neither A nor D lies on $\overleftrightarrow{BC}$. If A and D lie on the same side of $\overleftrightarrow{BC}$ then $A \in \text{int}(\angle CBD)$. (Why? See Figure 5-19.) But then

$$m(\angle CBA) + m(\angle ABD) = m(\angle CBD) < m(\angle ABD)$$

which is impossible. Hence A and D are on opposite sides of $\overleftrightarrow{BC}$ (see Figure 5-20). Choose E with E—B—A and note that E and D are on the same side of $\overleftrightarrow{BC}$. Then $E \in \text{int}(\angle CBD)$ (Why?) and so

$$m(\angle CBE) + m(\angle EBD) = m(\angle CBD).$$

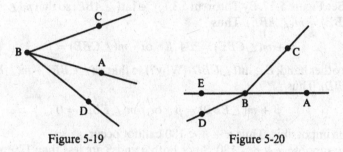

Figure 5-19   Figure 5-20

Since $\angle ABC$ and $\angle CBE$ form a linear pair, $m(\angle CBE) = 180 - m(\angle ABC)$ and therefore

$$180 - m(\angle ABC) + m(\angle EBD) = m(\angle CBD)$$

or

$$180 + m(\angle EBD) = m(\angle ABC) + m(\angle CBD) = m(\angle ABD).$$

But this means $m(\angle ABD) > 180$, which is impossible. Thus C and D cannot be on opposite sides of $\overleftrightarrow{AB}$, so they must be on the same side. The result then follows from Theorem 5.3.1. ☐

Note that the result about distance that corresponds to Theorem 5.3.3 is false. If $AB + BC = AC$ it need not be true that $B \in \text{int}(\overline{AB})$. We have seen examples in the Taxicab Plane and in the Moulton Plane.

The next result, whose proof is left to Problem A2, is the converse of Theorem 5.3.2.

Theorem 5.3.4. *In a protractor geometry, if* A *and* D *lie on opposite sides of* $\overleftrightarrow{BC}$ *and if* $m(\angle ABC) + m(\angle CBD) = 180$*, then* A—B—D *and the angles form a linear pair.*

Definition. Two lines l and l' are **perpendicular** (written $l \perp l'$) if $l \cup l'$ contains a right angle. Two rays or segments are **perpendicular** if the lines they determine are perpendicular.

The existence and uniqueness of a perpendicular to a line through a point on the line is guaranteed in the next theorem, whose proof is left to Problem A3.

Theorem 5.3.5. *Given a line l and a point B ∈ l in a protractor geometry, there exists a unique line l' that contains B such that l ⊥ l'.*

Example 5.3.6. In the Poincaré Plane, find the line through $B = (3, 4)$ that is perpendicular to the line

$$_0L_5 = \{(x, y) \in \mathbb{H} \,|\, x^2 + y^2 = 25\}.$$

SOLUTION. We use our knowledge of analytic geometry as motivation. Clearly the solution is a type II line. Its tangent at B must be perpendicular to the tangent to $_0L_5$ at B. Thought of as a semicircle in $\mathbb{R}^2$, the desired line must have its radius through B tangent to $_0L_5$. Hence the slope must be $-\frac{3}{4}$. The Euclidean line of slope $-\frac{3}{4}$ through $(3, 4)$ has equation

$$(y - 4) = -\tfrac{3}{4}(x - 3).$$

This line crosses the x-axis at $x = \frac{25}{3}$. Hence the desired line should be $_cL_r$ where $c = \frac{25}{3}$ and $r = \sqrt{(\frac{16}{3})^2 + 4^2} = \frac{20}{3}$. See Figure 5-21. We leave the verification that $_cL_r$ actually is the desired line as Problem A8.   □

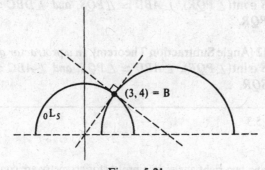

Figure 5-21

We should note at this time that if $B \notin l$ we do not know if there is a unique line through B perpendicular to l. In fact, there may not be any such line unless we add another axiom relating protractors and rulers. Also note that the familiar Pythagorean Theorem may not be true in this setting. You will show in Problem A9 that the triangle $\triangle ABC$ of Example 5.1.3 has a right angle at B but $(AC)^2 \neq (AB)^2 + (BC)^2$.

The remaining results of this section are left to the problems.

Corollary 5.3.7. *In a protractor geometry, every line segment $\overline{AB}$ has a unique* **perpendicular bisector;** *that is, a line $l \perp \overline{AB}$ with $l \cap \overline{AB} = \{M\}$ where M is the midpoint of $\overline{AB}$.*

Theorem 5.3.8. *In a protractor geometry, every angle* $\angle ABC$ *has a unique* **angle bisector** *that is, a ray* $\overrightarrow{BD}$ *with* $D \in \text{int}(\angle ABC)$ *and* $m(\angle ABD) = m(\angle DBC)$.

Recall that two line segments in a metric geometry are said to be congruent if they have the same length. We mimic this idea to define the congruence of angles.

Definition. In a protractor geometry $\{\mathscr{S}, \mathscr{L}, d, m\}$, $\angle ABC$ is **congruent** to $\angle DEF$ (written as $\angle ABC \simeq \angle DEF$) if $m(\angle ABC) = m(\angle DEF)$.

Many of the results of this section can be stated in terms of the notion of congruence and are very easy to prove. Be careful, however, not to confuse "congruent" with "equal."

Theorem 5.3.9 (Vertical Angle Theorem). *In a protractor geometry, if* $\angle ABC$ *and* $\angle A'BC'$ *form a vertical pair then* $\angle ABC \simeq \angle A'BC'$.

Theorem 5.3.10 (Angle Construction Theorem). *In a protractor geometry, given* $\angle ABC$ *and a ray* $\overrightarrow{ED}$ *which lies in the edge of a half plane* H_1, *then there exists a unique ray* $\overrightarrow{EF}$ *with* $F \in H_1$ *and* $\angle ABC \simeq \angle DEF$.

Theorem 5.3.11 (Angle Addition Theorem). *In a protractor geometry, if* $D \in \text{int}(\angle ABC)$, $S \in \text{int}(\angle PQR)$, $\angle ABD \simeq \angle PQS$, *and* $\angle DBC \simeq \angle SQR$, *then* $\angle ABC \simeq \angle PQR$.

Theorem 5.3.12 (Angle Subtraction Theorem). *In a protractor geometry, if* $D \in \text{int}(\angle ABC)$, $S \in \text{int}(\angle PQR)$, $\angle ABD \simeq \angle PQS$, *and* $\angle ABC \simeq \angle PQR$, *then* $\angle DBC \simeq \angle SQR$.

PROBLEM SET 5.3

Part A.

1. Prove that any two right angles in a protractor geometry are congruent.

2. Prove Theorem 5.3.4.

3. Prove Theorem 5.3.5.

4. Prove Corollary 5.3.7.

5. Prove Theorem 5.3.8.

6. Prove Theorem 5.3.9.

7. Let $\angle ABC$ and $\angle A'BC'$ form a vertical pair in a protractor geometry. Prove that if $\angle ABC$ is a right angle so are $\angle A'BC$, $\angle A'BC'$ and $\angle ABC'$.

8. Verify that the line found in Example 5.3.6 really is perpendicular to $_0L_5$ at $(3,4)$.

9. Show that if $\triangle ABC$ is as given in Example 5.1.3, then $(AC)^2 \neq (AB)^2 + (BC)^2$. Thus the Pythagorean Theorem need not be true in a protractor geometry.

10. Prove Theorem 5.3.10.

11. In $\mathscr{H}$ find the angle bisector of $\angle ABC$ if $A = (0,5)$, $B = (0,3)$ and $C = (2,\sqrt{21})$.

12. Repeat Problem 11 with $A = (1,3)$, $B = (1,\sqrt{3})$ and $C = (\sqrt{3},1)$.

13. Prove Theorem 5.3.11.

14. Prove Theorem 5.3.12.

15. Prove that in a protractor geometry $\angle ABC$ is a right angle if and only if there exists a point D with $D—B—C$ and $\angle ABC \simeq \angle ABD$.

16. In the Taxicab Plane let $A = (0,2)$, $B = (0,0)$, $C = (2,0)$, $Q = (-2,1)$, $R = (-1,0)$ and $S = (0,1)$. Show that $\overline{AB} \simeq \overline{QR}$, $\angle ABC \simeq \angle QRS$, and $\overline{BC} \simeq \overline{RS}$. Is $\overline{AC} \simeq \overline{QS}$?

Part B. "Prove" may mean "find a counterexample".

17. A **trisector** of $\angle ABC$ is a ray $\overrightarrow{BD}$ with $D \in \mathrm{int}(\angle ABC)$ such that either $m(\angle ABD) = \frac{1}{3}m(\angle ABC)$ or $m(\angle CBD) = \frac{1}{3}(\angle ABC)$. Prove that for every angle $\angle ABC$ in a protractor geometry there are exactly two trisectors.

18. Let $\triangle ABC$ be a triangle in a protractor geometry with $\overline{AB} \simeq \overline{CB}$. Prove $\angle BAC \simeq \angle BCA$.

19. Suppose that in the Poincaré Plane the line l is perpendicular to the line $_aL$. Prove that l is a type II line and that its "c" parameter is equal to a.

20. Prove that if two angles in a protractor geometry are supplementary then they form a linear pair.

21. Prove that any two right angles in a protractor geometry are equal.

22. In a protractor geometry assume that $D \in \mathrm{int}(\angle ABC)$ and that $\angle ABD \simeq \angle PQS$, $\angle DBC \simeq \angle SQR$, and $\angle ABC \simeq \angle PQR$. Prove that $S \in \mathrm{Int}(\angle PQR)$.

23. In the Moulton Plane $\mathscr{M}$ find a line l and two points P, Q such that there are two lines through P perpendicular to l and no lines through Q perpendicular to l. (Hint: Figure 5-14.)

Part C. Expository exercises.

24. To do this exercise you will need the computer program POINCARE which is described in the preface. Prepare a demonstration of geometry in the Poincaré plane for high school students using the program. As the written part of the work, first describe what you would hope to accomplish, how you would go about it, and then how you would determine if you were successful.

5.4 Euclidean and Poincaré Angle Measure

In this optional section we shall carefully verify that the Euclidean and Poincaré angle measures defined in Section 5.1 actually satisfy the axioms of an angle measure. The key step will be the construction of an inverse cosine function. This will involve techniques quite different from those of the rest of the book. As a result, you may choose to omit this section knowing that the only results that we will use in the sequel are that m_E and m_H are angle measures and that the cosine function is injective. On the other hand, it is interesting to see a variety of mathematical techniques tied together to develop one concept as is done in this section. The material on the construction of Euclidean angle measure is taken from Parker [1980].

Precisely what are we assuming in this section? We are assuming the standard facts about differentiation and integration but nothing about trigonometric functions. This will force us to consider the notion of an improper integral in order to define the inverse cosine function. Since general results about differential equations may not be familiar to the reader, we shall need to develop some very specific theorems regarding the solutions of $y'' = -y$. (In calculus we learned that both $\sin(x)$ and $\cos(x)$ are solutions of this differential equation. That is why we are interested in this equation.)

Definition. Let $f(t)$ be a function which is continuous for $c \leq t < d$ and which may not be defined at $t = d$. Then the **improper integral** $\int_c^d f(t)\,dt$ **converges** if $\lim_{b \to d^-} \int_c^b f(t)\,dt$ exists. In this case, we say $\lim_{b \to d^-} \int_c^b f(t)\,dt = \int_c^d f(t)\,dt$.

Lemma 5.4.1. *The improper integral* $\int_0^1 dt/\sqrt{1-t^2}$ *converges.*

PROOF. Since we are trying to develop the trigonometric functions and we are assuming nothing about them, we cannot use the "fact" that $\sin^{-1}(t)$ is an antiderivative of $1/\sqrt{1-t^2}$. Instead, we note that

$$\int_0^b \frac{dt}{\sqrt{1-t^2}} = \int_0^{1/2} \frac{dt}{\sqrt{1-t^2}} + \int_{1/2}^b \frac{dt}{\sqrt{1-t^2}}$$

for every choice of b with $\frac{1}{2} < b < 1$, so that to show that $\int_0^1 dt/\sqrt{1-t^2}$ converges we need only show that $\int_{1/2}^1 dt/\sqrt{1-t^2}$ converges. We proceed by integrating by parts:

$$\int_{1/2}^b \frac{dt}{\sqrt{1-t^2}} = \int_{1/2}^b \frac{1}{t} \frac{t\,dt}{\sqrt{1-t^2}}$$

$$= \frac{-\sqrt{1-t^2}}{t} \bigg|_{1/2}^b - \int_{1/2}^b \frac{\sqrt{1-t^2}}{t^2}\,dt$$

$$= \sqrt{3} - \frac{\sqrt{1-b^2}}{b} - \int_{1/2}^b \frac{\sqrt{1-t^2}}{t^2}\,dt$$

Now $\sqrt{1-t^2}/t^2$ is continuous on the interval $[\frac{1}{2}, 1]$ (including at $t = 1$) so its integral over that interval exists. Thus

$$\lim_{b \to 1^-} \int_{1/2}^{b} \frac{dt}{\sqrt{1-t^2}} = \sqrt{3} - 0 - \int_{1/2}^{1} \frac{\sqrt{1-t^2}}{t^2} dt$$

exists. Hence the improper integrals $\int_{1/2}^{1} dt/\sqrt{1-t^2}$ and $\int_{0}^{1} dt/\sqrt{1-t^2}$ both converge.   □

A similar argument shows that the improper integral $\int_{-1}^{0} dt/\sqrt{1-t^2}$ converges so that $\int_{0}^{-1} dt/\sqrt{1-t^2}$ also exists.

We define a number p to be twice the value of the integral in Lemma 5.4.1:

$$p = 2 \int_{0}^{1} \frac{dt}{\sqrt{1-t^2}} = \int_{-1}^{0} \frac{dt}{\sqrt{1-t^2}} + \int_{0}^{1} \frac{dt}{\sqrt{1-t^2}} = \int_{-1}^{1} \frac{dt}{\sqrt{1-t^2}}.$$

Of course, from calculus we know that $p = \pi$ but that will be irrelevant for our purposes. All that matters is that $p > 0$.

We wish now to define a function which will turn out to be the inverse cosine function. Motivated by calculus, we feel the inverse sine of x can be given as $\int_{0}^{x} 1/\sqrt{1-t^2}\, dt$. Since $\cos^{-1}(x) = \pi/2 - \sin^{-1}(x)$, we will define a function $I(x)$ as in Equation (4-1).

Lemma 5.4.2. *The function* $I(x)$ *given by*

$$I(x) = \frac{p}{2} - \int_{0}^{x} \frac{dt}{\sqrt{1-t^2}} \quad \text{for } -1 \le x \le 1 \tag{4-1}$$

is a bijection from $[-1,1]$ *to* $[0,p]$.

PROOF. By the Fundamental Theorem of Calculus, $I(x)$ is differentiable for $-1 < x < 1$ and, in fact,

$$I'(x) = -\frac{1}{\sqrt{1-x^2}}. \tag{4-2}$$

Since a differentiable function is continuous, we know that $I(x)$ is continuous for $-1 < x < 1$. Since the improper integrals that define $I(1)$ and $I(-1)$ converge, $I(x)$ is actually continuous for $-1 \le x \le 1$.

Equation (4-2) shows that $I'(x) < 0$ for $-1 < x < 1$ so that $I(x)$ is a strictly decreasing function. In particular, $I(x)$ is injective.

Since I is decreasing, its largest value occurs at the left endpoint and is

$$I(-1) = \frac{p}{2} - \int_{0}^{-1} \frac{dt}{\sqrt{1-t^2}} = \frac{p}{2} + \int_{-1}^{0} \frac{dt}{\sqrt{1-t^2}} = \frac{p}{2} + \frac{p}{2} = p.$$

Thus $I(x) \le p$ for all x in $[-1,1]$. On the other hand, the smallest value of $I(x)$ occurs at $x = 1$ and is

$$I(1) = \frac{p}{2} - \int_0^1 \frac{dt}{\sqrt{1-t^2}} = \frac{p}{2} - \frac{p}{2} = 0$$

so that $I(x) \geq 0$ for all x in $[-1,1]$. We have therefore shown that $I:[-1,1] \rightarrow [0,p]$ and is a continuous decreasing function. I is surjective because it is continuous and sends endpoints of $[-1,1]$ to endpoints of $[0,p]$. (This is the Intermediate Value Theorem.) Thus I is bijective. $\square$

Since $I:[-1,1] \rightarrow [0,p]$ is a bijection, it has an inverse, which we shall call c. Of course, we expect that $c(\theta)$ is really $\cos(\theta)$.

Definition. The **cosine function** $c:[0,p] \rightarrow [-1,1]$ is the inverse of I: $[-1,1] \rightarrow [0,p]$.
The **sine function** $s:[0,p] \rightarrow [0,1]$ is the function given by

$$s(\theta) = \sqrt{1 - c^2(\theta)} \quad \text{where } c^2(\theta) = (c(\theta))^2.$$

Lemma 5.4.3. *$c(\theta)$ and $s(\theta)$ are both differentiable for $0 < \theta < p$.*

Proof. From calculus we know that $c'(\theta)$ exists because $I'(x) \neq 0$. In fact, using Equation (4-2), we see that

$$c'(\theta) = \frac{1}{I'(c(\theta))} = -\sqrt{1 - c^2(\theta)} = -s(\theta) \quad \text{for } 0 < \theta < p. \quad (4\text{-}3)$$

On the other hand, we have

$$s'(\theta) = \frac{1}{2\sqrt{1-c^2(\theta)}}(-2c(\theta) \cdot c'(\theta)) = c(\theta) \quad \text{for } 0 < \theta < p. \quad (4\text{-}4)$$

Furthermore,

$$c''(\theta) = -c(\theta) \quad \text{and} \quad s''(\theta) = -s(\theta) \quad \text{for } 0 < \theta < p. \quad (4\text{-}5)$$

$\square$

Note that Equations (4-3), (4-4), and (4-5) are the familiar differential equations for the sine and cosine functions.
In order to prove the angle addition axiom holds for m_E, we shall need to prove the addition law for the cosine function. This in turn depends on the uniqueness of the solution of the initial value problem for the special case of differential Equation (4-5).

Lemma 5.4.4. *Let a and b be real numbers with $0 < a < b$. Then the only solution of the initial value problem*

$$\boxed{\begin{array}{l} \textit{Solve: } y''(\theta) = -y(\theta) \quad \textit{for } 0 < \theta < b \\ \quad \textit{with } y(a) = 0 \quad \textit{and} \quad y'(a) = 0 \end{array}} \quad (4\text{-}6)$$

is given by $y(\theta) = 0$ for all $0 < \theta < b$.

PROOF. Note that $y(\theta) \equiv 0$ is a solution of Problem (4-6). Let $z(\theta)$ be any solution of Problem (4-6). We want to show that $z(\theta) = 0$ for $0 < \theta < b$. Let $w(\theta) = z'(\theta)$ so that $(z'(\theta), w'(\theta)) = (w(\theta), -z(\theta))$ since $w'(\theta) = z''(\theta) = -z(\theta)$.

We may integrate a vector valued function such as $(w(\theta), -z(\theta))$ by integrating each component separately. Alternatively, we can define such an integral as the limit of a sum. Either way, we have

$$(z(\theta), w(\theta)) = \int_a^\theta (z'(t), w'(t))\, dt = \int_a^\theta (w(t), -z(t))\, dt. \qquad (4\text{-}7)$$

In the proof of Proposition 3.1.6, we saw that $\|X + Y\| \le \|X\| + \|Y\|$. By induction we have

$$\left\|\sum_{i=1}^n X_i\right\| \le \sum_{i=1}^n \|X_i\|$$

where $X_i \in \mathbb{R}^2$. By taking the appropriate limits this yields the inequality

$$\left\|\int_a^\theta X(t)\, dt\right\| \le \int_a^\theta \|X(t)\|\, dt \quad \text{for } a \le \theta < b \qquad (4\text{-}8)$$

where $X(t) = (w(t), -z(t))$. If we let

$$u(\theta) = \|(z(\theta), w(\theta))\| = \sqrt{z^2(\theta) + w^2(\theta)} \qquad (4\text{-}9)$$

then we can combine Equations (4-7), (4-8), and (4-9) as

$$u(\theta) = \|(z(\theta), w(\theta))\|$$

$$= \left\|\int_a^\theta (w(t), -z(t))\, dt\right\| \le \int_a^\theta \|(w(t), -z(t))\|\, dt$$

$$= \int_a^\theta u(t)\, dt.$$

Thus

$$u(\theta) \le \int_a^\theta u(t)\, dt \quad \text{where } u(\theta) \ge 0 \text{ and } u(a) = 0. \qquad (4\text{-}10)$$

We will show that $u(\theta) \equiv 0$ for $0 < \theta < b$. This implies that $z(\theta) \equiv 0$ (and $w(\theta) \equiv 0$).

By Inequality (4-10) we have

$$u(\theta) - \int_a^\theta u(t)\, dt \le 0$$

and so for any θ,

$$e^{-\theta} u(\theta) - e^{-\theta} \int_a^\theta u(t)\, dt \le 0.$$

But the expression on the left in this last inequality is the derivative of $e^{-\theta} \int_a^\theta u(t)\, dt$. Thus $e^{-\theta} \int_a^\theta u(t)\, dt$ is a decreasing function whose value at a is 0. Hence

$$e^{-\theta} \int_a^\theta u(t)\,dt \le 0 \quad \text{for } a \le \theta < b.$$

Thus $\int_a^\theta u(t)\,dt \le 0$ for $a \le \theta < b$. But since $u(\theta)$ is non-negative, $0 \le u(\theta) \le \int_a^\theta u(t)\,dt \le 0$ so that $u(\theta) = 0$ for $a \le \theta < b$.

To handle the case $0 < \theta \le a$ consider $-(z(\theta), w(\theta)) = \int_\theta^a (w(t), -z(t))\,dt$ and show that $e^\theta \int_\theta^a u(t)\,dt$ is increasing. This yields $u(\theta) = 0$ for $0 < \theta \le a$. Hence $0 = u(\theta) = \|(z(\theta), w(\theta))\|$ for $0 < \theta < b$ so that $z(\theta) = 0$ for $0 < \theta < b$. $\square$

Lemma 5.4.5. *Suppose that $0 < b \le p$ and that $f : [0, b] \to \mathbb{R}$ is continuous with $f''(\theta) = -f(\theta)$ for $0 < \theta < b$. Then there exist unique real numbers A and B such that*

$$f(\theta) = Ac(\theta) + Bs(\theta) \quad \text{for } 0 \le \theta \le b. \tag{4-11}$$

PROOF. Consider the initial value problem

$$\boxed{\begin{aligned} &\textit{Solve: } y''(\theta) = -y(\theta) \quad \text{for } 0 < \theta < b, \\ &\qquad \text{with } y(a) = f(a) \quad \text{and} \quad y'(a) = f'(a). \end{aligned}} \tag{4-12}$$

If $y(\theta)$ is a solution of Problem (4-12) then $y(\theta) - f(\theta)$ is a solution of Problem (4-6) so that $y(\theta) = f(\theta)$ for $0 < \theta < b$.

On the other hand, if $y(\theta) = Ac(\theta) + Bs(\theta)$, then $y''(\theta) = -y(\theta)$ so that $y(\theta)$ is a solution of Problem (4-12) if and only if A and B satisfy

$$\begin{cases} Ac(a) + Bs(a) = f(a) \\ -As(a) + Bc(a) = f'(a). \end{cases} \tag{4-13}$$

Since $c(a) \cdot c(a) - (-s(a)) \cdot s(a) = c^2(a) + s^2(a) = 1 \ne 0$, Equations (4-13) have a unique solution, namely

$$A = f(a) \cdot c(a) - f'(a) \cdot s(a), \qquad B = f'(a) \cdot c(a) + f(a) \cdot s(a).$$

For these unique values of A and B

$$f(\theta) = y(\theta) = Ac(\theta) + Bs(\theta) \quad \text{for } 0 < \theta < b.$$

Since both $f(\theta)$ and $Ac(\theta) + Bs(\theta)$ are continuous on $[0, b]$, $f(\theta) = Ac(\theta) + Bs(\theta)$ for $0 \le \theta \le b$. $\square$

What we have really done so far is prove the existence and uniqueness of solutions to the differential equation $f'' = -f$ with a given set of initial conditions. We now use this result to prove the addition law for cosine.

Lemma 5.4.6. *If θ, φ, and $\theta + \varphi$ are all in $[0, p]$ then*

$$c(\theta + \varphi) = c(\theta)c(\varphi) - s(\theta)s(\varphi). \tag{4-14}$$

PROOF. If $\varphi = 0$ or if $\varphi = p$ (so that $\theta = 0$) the result follows easily. (See Problem A1.) Now we assume that φ is fixed with $0 < \varphi < p$ and set $f(\theta) = c(\theta + \varphi)$. By Lemma 5.4.5 with $b = p - \varphi$ we have

$$c(\theta + \varphi) = f(\theta) = Ac(\theta) + Bs(\theta).$$

Now $f'(0)$ exists because $f(\theta)$ is differentiable for $-\varphi < \theta < p - \varphi$. Thus $A = f(0) = c(\varphi)$ and $B = f'(0) = c'(\varphi) = -s(\varphi)$. Hence

$$c(\theta + \varphi) = c(\varphi)c(\theta) - s(\varphi)s(\theta). \qquad \square$$

This completes our development of the cosine and inverse cosine functions. We are now ready to prove that the Euclidean Plane has an angle measure. Recall that our motivation is the fact that if θ is the measure of the angle between the vectors $(A - B)$ and $(C - B)$ then

$$\langle A - B, C - B \rangle = \|A - B\|\,\|C - B\|\cos\theta.$$

Definition. If A, B, C are noncollinear points in the Euclidean Plane, then the **Euclidean angle measure** of $\angle ABC$ is

$$m(\angle ABC) = I\left(\frac{\langle A - B, C - B \rangle}{\|A - B\|\,\|C - B\|}\right).$$

We should note that since $A \neq B$ and $B \neq C$, $\|A - B\|$ and $\|C - B\|$ are nonzero. Furthermore, the Cauchy-Schwarz Inequality (Proposition 3.1.5) says that $-1 \leq \langle A - B, C - B \rangle/\|A - B\|\,\|C - B\| \leq 1$ so that the definition makes sense. (We have dropped the subscript E to ease the notation a bit. Also, m is radian measure and not degree measure.)

Proposition 5.4.7. *For all angles* $\angle ABC$ *in* $\mathscr{E}$

$$0 < m(\angle ABC) < p.$$

PROOF. Since $0 \leq I(x) \leq p$ for all x between -1 and 1, we need only show that 0 and p cannot occur as the measure of an angle. Since A, B, and C are not collinear, $A - B \neq t(C - B)$ for any t and $C - B \neq (0,0)$. Thus by Proposition 3.1.5

$$-1 < \frac{\langle A - B, C - B \rangle}{\|A - B\|\,\|C - B\|} < 1.$$

Since $I(-1) = p$, $I(1) = 0$, and I is injective,

$$0 < m(\angle ABC) = I\left(\frac{\langle A - B, C - B \rangle}{\|A - B\|\,\|C - B\|}\right) < p. \qquad \square$$

Proposition 5.4.8 (Angle Construction). *In the Euclidean Plane let* $\overline{BA}$ *be a ray in the edge of the half plane* H_1 *and let* r *be a real number with* $0 < r < p$. *Then there exists a unique ray* $\overline{BC}$ *with* $C \in H_1$ *and* $m(\angle ABC) = r$.

PROOF. Let $X = (A - B)/\|A - B\|$. Let W be either $X^\perp$ or $-X^\perp$ where the sign is chosen so that $H_1 = \{P \mid \langle P - B, W \rangle > 0\}$. Set

$$C = B + c(r)X + s(r)W \quad \text{and} \quad A' = B + X.$$

Then $\angle ABC = \angle A'BC$ and

$$m(\angle A'BC) = I\left(\frac{\langle X, c(r)X + s(r)W \rangle}{\|X\| \|c(r)X + s(r)W\|}\right)$$

$$= I(c(r)) = r$$

because $\|X\| = \|W\| = \|c(r)X + s(r)W\| = 1$ by Problem A2 and $\langle X, W \rangle = 0$. Since $\langle C - B, W \rangle = s(r) > 0$, $C \in H_1$. Thus we have the existence of a ray with the desired property. See Figure 5-22 which illustrates the case where $W = -X^{\perp}$.

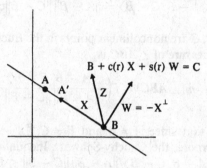

Figure 5-22

To show uniqueness of the ray suppose that $D \in H_1$ with $m(\angle ABD) = r$. We must show that $\overrightarrow{BD} = \overrightarrow{BC}$.

Choose $D' \in \overrightarrow{BD}$ with $\|B - D'\| = 1$. Then $\angle ABD = \angle A'BD'$. Let $Z = D' - B$. We claim that $Z = \langle Z, X \rangle X + \langle Z, W \rangle W$. Since $\langle W, W \rangle = 1$, $\langle Z - \langle Z, W \rangle W, W \rangle = \langle Z, W \rangle - \langle Z, W \rangle = 0$. Thus, since $W = \pm X^{\perp}$, Proposition 4.2.1 implies that $Z - \langle Z, W \rangle W = tX$ for some $t \in \mathbb{R}$. We find t by taking the scalar product with X:

$$t = t\langle X, X \rangle = \langle tX, X \rangle = \langle Z - \langle A, W \rangle W, X \rangle = \langle Z, X \rangle.$$

Thus $Z = \langle Z, W \rangle W + \langle Z, X \rangle X$ as claimed.

Since $\|X\| = \|Z\| = 1$, we have

$$r = m(\angle A'BD') = I\left(\frac{\langle X, Z \rangle}{\|X\| \|Z\|}\right) = I(\langle X, Z \rangle).$$

Thus $c(r) = \langle X, Z \rangle$. Since $\|Z\| = 1$, we have

$$\langle Z, W \rangle = \pm\sqrt{1 - \langle Z, X \rangle^2} = \pm\sqrt{1 - c^2(r)} = \pm s(r).$$

Since $D' \in H_1$, $\langle Z, W \rangle > 0$ and so $\langle Z, W \rangle = +s(r)$. Thus $Z = c(r)X + s(r)W$ and $D' = C$. Hence $\overrightarrow{BD} = \overrightarrow{BD'} = \overrightarrow{BC}$ and there is a unique ray $\overrightarrow{BC}$ with $C \in H_1$ and $m(\angle ABC) = r$. $\qquad\square$

In order to verify the Angle Addition Axiom it is necessary to prove first two results which you would normally expect to be consequences of Angle Addition.

Proposition 5.4.9. *In the Euclidean Plane, if* $D \in \text{int}(\angle ABC)$ *then* $m(\angle ABD) < m(\angle ABC)$.

Proof. Let $X = A - B$, $Y = C - B$, and $Z = D - B$. By replacing A, C, and D by other points on the appropriate rays we can make $\|X\| = \|Y\| = \|Z\| = 1$. See Figure 5-23.

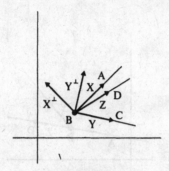

Figure 5-23

Since $D \in \text{int}(\angle ABC)$, $\langle Z, X^{\perp} \rangle$ and $\langle Y, X^{\perp} \rangle$ have the same sign. Choose W to be either $X^{\perp}$ or $-X^{\perp}$ in a manner so that $\langle Z, W \rangle > 0$. As in the proof of Proposition 5.4.8 we have

$$Y = c(r)X + s(r)W \quad \text{and} \quad Z = c(\rho)X + s(\rho)W$$

where $r = m(\angle ABC)$ and $\rho = m(\angle ABD)$.

Now $Y^{\perp} = \pm(s(r)X - c(r)W)$. Since $D \in \text{int}(\angle ABC)$, $\langle X, Y^{\perp} \rangle$ and $\langle Z, Y^{\perp} \rangle$ have the same sign. Thus

$$\langle X, s(r)X - c(r)W \rangle = s(r)$$

and

$$\langle c(\rho)X + s(\rho)W, s(r)X - c(r)W \rangle = c(\rho)s(r) - s(\rho)c(r)$$

must both be positive. Hence

$$s(\rho)c(r) < c(\rho)s(r) \quad \text{and} \quad \frac{c(r)}{s(r)} < \frac{c(\rho)}{s(\rho)}.$$

By Problem A3, $f(\theta) = c(\theta)/s(\theta)$ is a strictly decreasing function. Thus $\rho < r$ and $m(\angle ABD) < m(\angle ABC)$. $\square$

Proposition 5.4.10. *In the Euclidean Plane, if* $\angle ABC$ *and* $\angle CBD$ *form a linear pair then* $m(\angle ABC) + m(\angle CBD) = p$.

PROOF. Let $X = A - B$, $Y = C - B$, and $Z = D - B$ and assume that $\|X\| = \|Y\| = \|Z\| = 1$ as before. Note that since B is between A and D we have $Z = -X$. See Figure 5-24. By Problem A4, $I(-x) = p - I(x)$. Thus

$$
\begin{aligned}
m(\angle ABC) + m(\angle CBD) &= I(\langle X, Y \rangle) + I(\langle Y, Z \rangle) \\
&= I(\langle X, Y \rangle) + I(-\langle Y, X \rangle) \\
&= I(\langle X, Y \rangle) + p - I(\langle X, Y \rangle) \\
&= p. \qquad \qquad \qquad \square
\end{aligned}
$$

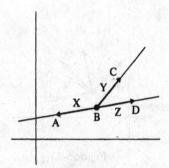

Figure 5-24

Proposition 5.4.11 (Angle Addition). *In the Euclidean Plane if $D \in \text{int}(\angle ABC)$ then $m(\angle ABD) + m(\angle DBC) = m(\angle ABC)$.*

PROOF. Choose E so that B is between A and E. Then $C \in \text{int}(\angle DBE)$. By Propositions 5.4.9 and 5.4.10

$$
m(\angle DBC) < m(\angle DBE) = p - m(\angle ABD)
$$

so that $m(\angle ABD) + m(\angle DBC) < p$. Hence we can apply Lemma 5.4.6 with $\theta = m(\angle ABD)$ and $\varphi = m(\angle DBC)$. See Figure 5-25.

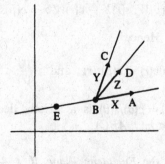

Figure 5-25

As before $X = A - B, Y = C - B, Z = D - B$ and $\|X\| = \|Y\| = \|Z\| = 1$. Note that $m(\angle ABC) = I(\langle X, Y\rangle), m(\angle ABD) = I(\langle X, Z\rangle)$, and $m(\angle DBC) = I(\langle Z, Y\rangle)$. Hence

$$
\begin{aligned}
c(m(\angle ABD) + m(\angle DBC)) &= c(I(\langle X, Z\rangle) + I(\langle Z, Y\rangle)) \\
&= c(I(\langle X, Z\rangle))c(I(\langle Z, Y\rangle)) \\
&\quad - s(I(\langle X, Z\rangle))s(I(\langle Z, Y\rangle)) \\
&= \langle X, Z\rangle\langle Z, Y\rangle - \sqrt{1 - \langle X, Z\rangle^2}\sqrt{1 - \langle Z, Y\rangle^2}.
\end{aligned}
$$

By Problem A5, $\langle X, Z^{\perp}\rangle = \pm\sqrt{1 - \langle X, Z\rangle^2}$ and $\langle Y, Z^{\perp}\rangle = \pm\sqrt{1 - \langle Y, Z\rangle^2}$.

Since $D \in \text{int}(\angle ABC), \langle X, Z^{\perp}\rangle$ and $\langle Y, Z^{\perp}\rangle$ must have opposite signs so that

$$
\langle X, Z^{\perp}\rangle\langle Y, Z^{\perp}\rangle = -\sqrt{1 - \langle X, Z\rangle^2}\sqrt{1 - \langle Y, Z\rangle^2}.
$$

Hence

$$
\begin{aligned}
c(m(\angle ABD) + m(\angle DBC)) &= \langle X, Z\rangle\langle Y, Z\rangle + \langle X, Z^{\perp}\rangle\langle Y, Z^{\perp}\rangle \\
&= \langle(\langle X, Z\rangle Z + \langle X, Z^{\perp}\rangle Z^{\perp}), Y\rangle \\
&= \langle X, Y\rangle \\
&= c(m(\angle ABC)).
\end{aligned}
$$

Since the function $c(\theta)$ is injective, we have

$$
m(\angle ABD) + m(\angle DBC) = m(\angle ABC). \qquad \square
$$

Propositions 5.4.7, 5.4.8, and 5.4.11 show that m is an angle measure based on p. By Problem A4 of Section 5.1, the function $m_E = (180/p)m$ is an angle measure based on 180. It is the measure we actually use in the Euclidean Plane. The function $I(x)$ is really $\cos^{-1}(x)$ (in radians). $c(\theta)$ is $\cos(\theta)$ and $s(\theta)$ is $\sin(\theta)$. It is possible to build up all of the trigonometric functions from what we have here and have them defined for all θ, not just $0 \le \theta \le \pi$. You should feel free now to compute

$$
m_E(\angle ABC) = \cos^{-1}\left(\frac{\langle A - B, C - B\rangle}{\|A - B\|\|C - B\|}\right)
$$

using $\cos^{-1}(x)$ in terms of degrees.

Example 5.4.12. Let $A = (2, 1)$, $B = (3, -2)$ and $C = (-1, 3)$. What is $m_E(\angle ABC)$?

SOLUTION. $A - B = (-1, 3), C - B = (-4, 5)$

$$
\begin{aligned}
m_E(\angle ABC) &= \cos^{-1}\frac{\langle(-1, 3), (-4, 5)\rangle}{\sqrt{10} \cdot \sqrt{41}} \\
&= \cos^{-1}\left(\frac{19}{\sqrt{410}}\right) \\
&\doteq 20.225. \qquad \square
\end{aligned}
$$

We now turn our attention to the Poincaré Plane. Recall the following definitions from Section 5.1.

Definition. If $\overrightarrow{BA}$ is a ray in the Poincaré Plane, where $B = (x_B, y_B)$ and $A = (x_A, y_A)$, then the **Euclidean tangent** to $\overrightarrow{BA}$ at B is

$$T_{BA} = \begin{cases} (0, y_A - y_B) & \text{if } \overleftrightarrow{AB} \text{ is type I} \\ (y_B, c - x_B) & \text{if } \overleftrightarrow{AB} \text{ is } {}_cL_r,\ x_B < x_A \\ -(y_B, c - x_B) & \text{if } \overleftrightarrow{AB} \text{ is } {}_cL_r,\ x_B > x_A. \end{cases}$$

The **Euclidean tangent ray** to $\overrightarrow{BA}$ is the Euclidean ray $\overrightarrow{BA'}$ where $A' = B + T_{BA}$. The **Poincaré measure** of $\angle ABC$ in $\mathbb{H}$ is

$$m_H(\angle ABC) = m_E(\angle A'BC') = \cos^{-1} \frac{\langle T_{BA}, T_{BC} \rangle}{\|T_{BA}\| \|T_{BC}\|}.$$

We must show that m_H is an angle masure. Because it is defined in terms of m_E we expect that the basic results about Poincaré angle measure should follow fairly easily from similar statements about Euclidean angle measure.

Proposition 5.4.13. *For every hyperbolic angle* $\angle ABC$, $0 < m_H(\angle ABC) < 180$.

PROOF. This is immediate since $0 < m_E(\angle A'BC') < 180$. $\square$

The key step for the rest of this section is the next proposition. It tells us that for each possible tangent direction there is a unique Poincaré ray.

Proposition 5.4.14. *Let* $B = (x_B, y_B) \in \mathbb{H}$ *and let* $T = (t_1, t_2) \neq (0, 0)$. *Then there exists a unique ray* $\overrightarrow{BA}$ *in* $\mathbb{H}$ *with* $T_{BA} = \lambda T$ *for some* $\lambda > 0$.

PROOF.

Case 1. $t_1 = 0$. The ray should be of type I. Let λ be any positive number with $y_B + \lambda t_2 > 0$. This is possible since $y_B > 0$. Let $A = (x_B, y_B + \lambda t_2) \in \mathbb{H}$. Then $T_{BA} = (0, \lambda t_2) = \lambda T$. This gives existence.

On the other hand, if $C = (x_C, y_C) \in \mathbb{H}$ with $T_{BC} = \mu T$ for some $\mu > 0$ then the first component of T_{BC} must be zero so that $\overleftrightarrow{BC}$ is a type I line. Thus $\overleftrightarrow{BC} = \overleftrightarrow{BA}$. Finally, since $y_A - y_B$ and $y_C - y_B$ have the same sign as t_2, B is not between A and C. Hence, $\overrightarrow{BC} = \overrightarrow{BA}$.

Case 2. $t_1 \neq 0$. Let $\lambda = y_B/|t_1|$, $c = x_B + (y_B t_2/t_1)$, and $r = \lambda \|T\|$. Then

$$(x_B - c)^2 + (y_B)^2 = \left(\frac{y_B}{t_1}\right)^2 (t_2)^2 + (y_B)^2$$
$$= \lambda^2 (t_2)^2 + \lambda^2 (t_1)^2$$
$$= \lambda^2 \|T\|^2 = r^2.$$

Thus $B \in {}_cL_r$. Choose $A \in {}_cL_r$ with $x_A > x_B$ if $t_1 > 0$ and $x_A < x_B$ if $t_1 < 0$. Then $c - x_B = y_B t_2 / t_1$ so that

$$T_{BA} = \begin{cases} (y_B, c - x_B) = (\lambda t_1, \lambda t_2) = \lambda T & \text{if } t_1 > 0 \\ -(y_B, c - x_B) = -(-\lambda t_1, -\lambda t_2) = \lambda T & \text{if } t_1 < 0. \end{cases}$$

Hence $T_{BA} = \lambda T$ and we have existence.

On the other hand, suppose that $C \in \mathbb{H}$ and $T_{BC} = \mu T$ for some $\mu > 0$. Then $\overleftrightarrow{BC}$ is a type II line ${}_dL_s$ since $t_1 \neq 0$, and $T_{BC} = \pm(y_B, d - x_B) = \mu(t_1, t_2)$. Thus $\mu = \pm y_B / t_1$, where the $\pm$ sign must be the sign of t_1 since $\mu > 0$. Thus $\mu = y_B / |t_1| = \lambda$. Hence

$$\pm(d - x_B) = \mu t_2$$

or

$$d = x_B \pm \mu t_2 = x_B + (y_B t_2 / t_1) = c$$

$$s = \sqrt{(x_B - d)^2 + (y_B)^2} = \sqrt{(x_B - c)^2 + (y_B)^2} = \lambda \|T\| = r.$$

Thus A and C belong to ${}_cL_r$. We need only show that B is not between A and C to have $\overrightarrow{BC} = \overrightarrow{BA}$.

$$T_{BC} = \pm(y_B, d - x_B) = \mu T = \lambda T = T_{BA}$$

where the $\pm$ sign is the sign of t_1. But by the definition of T_{BC} and T_{BA}, the $\pm$ sign is the sign of $x_C - x_B$ and of $x_A - x_B$. Hence $x_C - x_B$ and $x_A - x_B$ have the same sign so that B is not between A and C. Thus there is a unique ray $\overrightarrow{BA}$ with $T_{BA} = \lambda T$ for some $\lambda > 0$. $\qquad\square$

Before we go any further, let us adopt some informal terminology regarding the sides of a line. If l is a vertical Euclidean line or a type I Poincaré line, it is clear what "left side" and "right side" mean. Similarly, if l is a non-vertical Euclidean line or a type II Poincaré line, then "top side" and "bottom side" have intuitive meaning. This terminology could be made formal if needed. Given a Poincaré ray $\overrightarrow{BA}$, there is a tangent Euclidean ray $\overrightarrow{BA'}$ where $A = B + T_{BA}$. Note that $\overrightarrow{BA}$ and $\overrightarrow{BA'}$ are either both vertical or both not vertical. Then given a side of one ray, there is a corresponding side of the other ray (left, right, top, or bottom). The proof of the next result is left to Problem A6. See Figure 5-26.

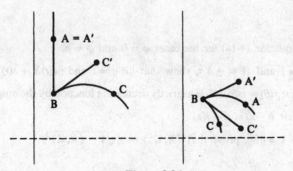

Figure 5-26

Proposition 5.4.15. *Let $\overrightarrow{BA}$ be a Poincaré ray and let $\overrightarrow{BA}'$ be the tangent Euclidean ray. Then the side of the Poincaré line $\overleftrightarrow{BA}$ that contains C corresponds to the side of the Euclidean line $\overleftrightarrow{BA}'$ that contains $C' = B + T_{BC}$. (See Figure 5-26.)*

Proposition 5.4.16 (Angle Construction). *Let $\overrightarrow{BA}$ be a ray in $\mathbb{H}$ which lies in the edge of the half plane H_1 and suppose that $0 < \theta < 180$. Then there is a unique ray $\overrightarrow{BC}$ in $\mathbb{H}$ with $C \in H_1$ and $m_H(\angle ABC) = \theta$.*

PROOF. Let $A' = B + T_{BA}$. There are exactly two Euclidean rays, $\overrightarrow{BC}'$ and $\overrightarrow{BD}'$, with $m_E(\angle A'BC') = \theta = m_E(\angle A'BD')$. C' and D' lie on opposite sides of the Euclidean line $\overleftrightarrow{BA}'$. By Proposition 3.14 there are unique Poincaré rays $\overrightarrow{BC}$ and $\overrightarrow{BD}$ which have $\overrightarrow{BC}'$ and $\overrightarrow{BD}'$ as tangents. By Proposition 5.4.15, C and D must lie on opposite sides of the Poincaré line $\overleftrightarrow{BA}$. Hence exactly one of them lies in H_1. Assume it is C. Then $m_H(\angle ABC) = m_E(\angle A'BC') = \theta$, and we have existence.

On the other hand, if $F \in H_1$ with $m_H(\angle ABF) = \theta$, then $m_E(\angle A'BF') = \theta$. Thus $\overrightarrow{BF}' = \overrightarrow{BC}'$ and $\overrightarrow{BF} = \overrightarrow{BC}$. $\qquad\square$

Proposition 5.4.17 (Angle Addition). *In the Poincaré Plane if $D \in \text{int}(\angle ABC)$ then $m_H(\angle ABD) + m_H(\angle DBC) = m_H(\angle ABC)$.*

PROOF. Let $A' = B + T_{BA}$, $C' = B + T_{BC}$, and $D' = B + T_{BD}$. By Proposition 5.4.14, $D' \in \text{int}(\angle A'BC')$, where $\angle A'BC'$ is the Euclidean angle. Then

$$m_H(\angle ABD) + m_H(\angle DBC) = m_E(\angle A'BD') + m_E(\angle D'BC')$$
$$= m_E(\angle A'BC')$$
$$= m_H(\angle ABC). \qquad\square$$

Propositions 5.4.13, 5.4.16, 5.4.17 prove that m_H is an angle measure on $\mathbb{H}$.

PROBLEM SET 5.4

Part A.

1. Verify Equation (4-14) for the cases $\varphi = 0$ and $\varphi = p$.

2. If $\|X\| = 1$ and $W = \pm X^\perp$, show that $\|W\| = 1$ and $\|c(r)X + s(r)W\| = 1$.

3. Prove that $f(\theta) = c(\theta)/s(\theta)$ is a strictly decreasing function by showing that $f'(\theta) < 0$.

4. Prove that $I(-x) = p - I(x)$.

5. If $\|X\| = 1 = \|Z\|$, prove that $\langle X, Z^\perp \rangle = \pm\sqrt{1 - \langle X, Z \rangle^2}$.

6. Prove Proposition 5.4.15.

7. Let θ be a number with $0 < \theta < p$. Let A, B, $C \in \mathbb{R}^2$ with $A = (1,0)$, $B = (0,0)$ and $C = (c(\theta), s(\theta))$. Prove that $m(\angle ABC) = \theta$. This means that $c(\theta)$ and $s(\theta)$ are the cosine and sine functions defined in trigonometry. See Figure 5-27.

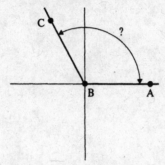

Figure 5-27

CHAPTER 6
Neutral Geometry

6.1 The Side-Angle-Side Axiom

One of the most fundamental problems in mathematics is finding the appropriate notion of equivalence for each particular area of mathematics. In geometry the appropriate notion of equivalence is that of "congruence." We have already discussed congruence for segments and angles. In this chapter we will define and work with congruence between triangles.

Before we take up the study of triangle congruences, it is appropriate to discuss briefly the general notion of congruence of geometric figures. Intuitively, two figures are congruent if one can be "picked up and laid down exactly on the other" so that the two coincide. Euclid used this "method of superposition" but only sparingly. Roughly, it means that the figure is not distorted as it is moved. Whatever the first statement in quotes means exactly, it should include the fact that there is a bijection between the two figures and that both the corresponding sides and the corresponding angles are congruent. We shall make these ideas precise below. The general notion of superposition is made concrete with the idea of an isometry in Chapter 11.

We shall now formally define what is meant by congruent triangles. This will consist of six conditions to verify in order to show that two triangles are congruent. We will then introduce a new axiom (Side-Angle-Side, SAS) which allows us to verify only three conditions to show that two triangles are congruent. In later sections we shall develop various results that follow from SAS.

As in any case in which we add an axiom, we must determine whether our standard examples satisfy the axiom. We will see that although the Taxicab Plane does not, both the Poincaré and Euclidean Planes do satisfy the new axiom.

> **Convention.** In $\triangle ABC$, if there is no confusion, we will denote $\angle ABC$ by $\angle B$, etc., so that
>
> $$\angle A = \angle CAB, \qquad \angle B = \angle ABC, \quad \text{and} \quad \angle C = \angle BCA.$$

Definition. Let $\triangle ABC$ and $\triangle DEF$ be two triangles in a protractor geometry and let $f:\{A, B, C\} \rightarrow \{D, E, F\}$ be a bijection between the vertices of the triangles. f is a **congruence** if

$$\overline{AB} \simeq \overline{f(A)f(B)}, \qquad \overline{BC} \simeq \overline{f(B)f(C)}, \qquad \overline{CA} \simeq \overline{f(C)f(A)}$$

and

$$\angle A \simeq \angle f(A), \qquad \angle B \simeq \angle f(B), \qquad \angle C \simeq \angle f(C).$$

Two triangles, $\triangle ABC$ and $\triangle DEF$, are **congruent** if there is a congruence $f:\{A, B, C\} \rightarrow \{D, E, F\}$. If the congruence is given by $f(A) = D$, $f(B) = E$, and $f(C) = F$, then we write $\triangle ABC \simeq \triangle DEF$.

A congruence is pictured in Figure 6-1. In this case $f(A) = E$, $f(B) = D$, and $f(C) = F$. Thus $\triangle ABC \simeq \triangle EDF$. Note that the notation $\simeq$ for congruent triangles includes the particular bijection. Thus, it is *incorrect* to write $\triangle ABC \simeq \triangle DEF$ in Figure 6-1, *even though* $\triangle DEF = \triangle EDF$. Given a bijection between the vertices of two triangles there is induced a bijection between sides and between angles. Therefore, a congruence is a bijection for which corresponding sides are congruent and for which corresponding angles are congruent. As an aid in visualization, it is useful to mark corresponding sides with the same number of slash marks when they are known to be congruent. Similarly, if they are known to be congruent, we mark corresponding angles with the same Greek letter which gives their measure.

The fundamental question of this section is: How much do we need to know about a triangle so that it is determined up to congruence? More precisely, if we are given $\triangle ABC$ and $\triangle DEF$ for which some sides of $\triangle ABC$ are congruent to the corresponding sides of $\triangle DEF$ and some angles of $\triangle ABC$ are congruent to the corresponding angles of $\triangle DEF$, is $\triangle ABC$ congruent to $\triangle DEF$? As the question is worded, the answer is no. Certainly if $\overline{AB} \simeq \overline{DE}$ and $\angle A \simeq \angle D$ it need not be true that $\triangle ABC \simeq \triangle DEF$. (There

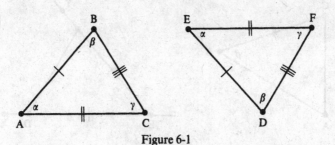

Figure 6-1

are examples in every protractor geometry.) What happens in the case of two sides and the included angle? That is, if $\overline{AB} \simeq \overline{DE}$, $\overline{AC} \simeq \overline{DF}$ and $\angle A \simeq \angle D$, must $\triangle ABC$ be congruent to $\triangle DEF$? In the Euclidean Plane the answer is yes, but in an arbitrary protractor geometry the answer need not be yes. (An example in the Taxicab Plane is given in Example 6.1.1.) Thus if we want our protractor geometries to have this property we must add it to our list of axioms. Because the question involves two sides and the angle between them, the axiom is called Side-Angle-Side or SAS. To see what might "go wrong" in the general case we will consider the following situation.

Suppose that we are given $\triangle ABC$ and a ray $\overrightarrow{EX}$ which lies on the edge of a half plane H_1. Then we can construct the following by the Segment Construction Theorem (Theorem 3.3.6) and the Angle Construction Theorem (Theorem 5.3.10)

(a) A unique point $D \in \overrightarrow{EX}$ with $\overline{BA} \simeq \overline{ED}$
(b) A unique ray $\overrightarrow{EY}$ with $Y \in H_1$ and $\angle ABC \simeq \angle XEY$
(c) A unique point $F \in \overrightarrow{EY}$ with $\overline{BC} \simeq \overline{EF}$.

See Figure 6-2. Is $\triangle ABC \simeq \triangle DEF$? Intuitively it should be (and it will be if SAS is satisfied). However, since we know nothing about the rulers for $\overline{DF}$ and $\overline{AC}$, we have no way of showing that $\overline{AC} \simeq \overline{DF}$. In fact, Example 6.1.1 will show that $\overline{AC}$ need not be congruent to $\overline{DF}$.

The philosophical problem is as follows. To get a protractor geometry, we put two different structures on an incidence geometry. One was the notion of rulers and the other was that of protractors. There was no assumption in the axioms which said that the rulers and protractors must "get along." That is, no relation was assumed between the rulers and the angle measure. In fact, rulers for one line need not be related to rulers for another line and protractors at one point need not be related to protractors at another point. Because of this we should not expect $\triangle ABC$ to be congruent to $\triangle DEF$ in the above construction.

Example 6.1.1. In the Taxicab Plane let $A = (1, 1)$, $B = (0, 0)$, $C = (-1, 1)$, $E = (0, 0)$, $X = (3, 0)$, and let H_1 be the half plane above the x-axis. Carry out the construction outlined above and check to see whether or not $\triangle ABC$ is congruent to $\triangle DEF$.

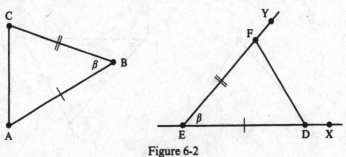

Figure 6-2

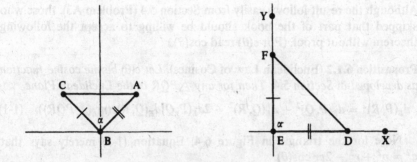

Figure 6-3

SOLUTION. $d_T(B, A) = 1 + 1 = 2$ so that $D = (2, 0)$. $m_E(\angle ABC) = 90$ so that we may take $Y = (0, 3)$. $d_T(B, C) = 1 + 1 = 2$ so that $F = (0, 2)$. Since Taxicab angle measure is the same as Euclidean angle measure, we have

$$m_E(\angle BCA) = 45 = m_E(\angle EFD) \quad \text{so that} \quad \angle BCA \simeq \angle EFD$$
$$m_E(\angle CAB) = 45 = m_E(\angle FDE) \quad \text{so that} \quad \angle CAB \simeq \angle FDE.$$

On the other hand,

$$d_T(A, C) = 2 \quad \text{and} \quad d_T(D, F) = 2 + 2 = 4 \quad \text{so that} \quad \overline{AC} \not\simeq \overline{DF}.$$

Hence $\triangle ABC$ is not congruent to $\triangle DEF$. See Figure 6-3. $\qquad\square$

Example 6.1.1 shows that we need to add another axiom to our protractor geometry in order to have our intuition about triangle congruence be valid. The axiom we will add can be remembered informally as: *If two sides and the included angle of a triangle are congruent to two sides and the included angle of another triangle, then the two triangles are congruent.*

Definition. A protractor geometry satisfies the **Side-Angle-Side Axiom** (SAS) if whenever $\triangle ABC$ and $\triangle DEF$ are two triangles with $\overline{AB} \simeq \overline{DE}$, $\angle B \simeq \angle E$, and $\overline{BC} \simeq \overline{EF}$, then $\triangle ABC \simeq \triangle DEF$.

Definition. A **neutral geometry** (or **absolute geometry**) is a protractor geometry which satisfies SAS.

The traditional term "absolute geometry" is somewhat misleading because it connotes some finality or uniqueness of the resulting object of study. We have chosen to use the term "neutral" geometry which was introduced by Prenowitz and Jordan [1965]. This term indicates we are taking a neutral course relative to a choice of parallel axioms. See Section 7.3.

We have seen that the Taxicab Plane is not a neutral geometry. In Problem B11 you will show that the Moulton Plane is not a neutral geometry either. However, our two basic models are neutral geometries. To show that the Euclidean Plane is a neutral geometry requires the familiar law of cosines.

Although the result follows easily from Section 5.4 (Problem A3), those who skipped that part of the book should be willing to accept the following theorem without proof. (For $c(\theta)$ read $\cos(\theta)$.)

Proposition 6.1.2 (Euclidean Law of Cosines). *Let $c(\theta)$ be the cosine function as developed in Section 5.4. Then for any $\triangle PQR$ in the Euclidean Plane*

$$d_E(P,R)^2 = d_E(P,Q)^2 + d_E(Q,R)^2 - 2d_E(P,Q)d_E(Q,R)c(m_E(\angle PQR)). \quad (1\text{-}1)$$

Note for the triangle in Figure 6-4, Equation (1-1) merely says that $q^2 = p^2 + r^2 - 2pr \cos(\theta)$.

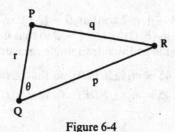

Figure 6-4

Proposition 6.1.3. The Euclidean Plane $\mathscr{E}$ satisfies SAS.

PROOF. Let $\triangle ABC$ and $\triangle DEF$ be given with $\overline{AB} \simeq \overline{DE}$, $\angle B \simeq \angle E$, and $\overline{BC} \simeq \overline{EF}$. Then by Proposition 6.1.2

$$\begin{aligned}
(AC)^2 &= (AB)^2 + (BC)^2 - 2(AB)(BC)c(m_E(\angle B)) \\
&= (DE)^2 + (EF)^2 - 2(DE)(EF)c(m_E(\angle E)) \\
&= (DF)^2.
\end{aligned}$$

Thus $AC = DF$ so that $\overline{AC} \simeq \overline{DF}$. Now solve Equation (1-1) for $c(m_E(\angle PQR))$:

$$c(m_E(\angle PQR)) = \frac{(PQ)^2 + (QR)^2 - (PR)^2}{2(PQ)(QR)}.$$

Hence as a special case

$$\begin{aligned}
c(m_E(\angle BAC)) &= \frac{(BA)^2 + (AC)^2 - (BC)^2}{2(BA)(AC)} \\
&= \frac{(ED)^2 + (DF)^2 - (EF)^2}{2(ED)(DF)} \\
&= c(m_E(\angle EDF)).
\end{aligned}$$

Since the function $c(\theta)$ is injective (for $0 < \theta < 180$)

$$m_E(\angle BAC) = m_E(\angle EDF) \quad \text{and} \quad \angle A \simeq \angle D.$$

Similarly $\angle C \simeq \angle F$ so that $\triangle ABC \simeq \triangle DEF$.   □

To prove that the Poincaré Plane satisfies SAS is much harder. Although a proof can be given with the material developed thus far, we assume that the next theorem is true for now. We will present a proof in Chapter 11 when we study isometries.

Proposition 6.1.4. *The Poincaré Plane $\mathcal{H}$ is a neutral geometry.*

Definition. A triangle in a protractor geometry is **isosceles** if (at least) two sides are congruent. Otherwise, the triangle is **scalene**. The triangle is **equilateral** if all three sides are congruent. If $\triangle ABC$ is isosceles with $\overline{AB} \simeq \overline{BC}$, then the **base angles** of $\triangle ABC$ are $\angle A$ and $\angle C$.

Our first application of SAS is the following theorem on isosceles triangles. The Latin name (literally "the bridge of asses") refers to the complicated figure Euclid used in his proof, which looked like a bridge, and to the fact that only someone as dull as an ass would fail to understand it. (See Heath's translation of Euclid for a further discussion of the name.) The proof which follows is due to Pappus (4th Century AD).

Theorem 6.1.5 (*Pons Asinorum*). *In a neutral geometry, the base angles of an isosceles triangle are congruent.*

PROOF. The proof proceeds by showing that $\triangle ABC$ is congruent to itself! Let $\triangle ABC$ be isosceles with $\overline{AB} \simeq \overline{CB}$. The congruence will be given by $f(A) = C$, $f(B) = B$, $f(C) = A$. This has the effect of flipping the triangle over along an axis through B perpendicular to $\overleftrightarrow{AC}$. See Figure 6-5.

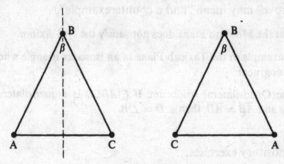

Figure 6-5

Since $\overline{AB} \simeq \overline{CB}$, $\angle ABC \simeq \angle CBA$, and $\overline{CB} \simeq \overline{AB}$, $\triangle ABC \simeq \triangle CBA$ by SAS. But this means that $\angle BAC \simeq \angle BCA$ so that the base angles are congruent. □

PROBLEM SET 6.1

Part A.

1. Prove that congruence is an equivalence relation on the set of all triangles in a protractor geometry.

2. In $\mathscr{H}$ if $A = (0, 1)$, $B = (0, 2)$, $C = (0, 4)$, and $D = (1, \sqrt{3})$, then show that $\triangle ABD \simeq \triangle CBD$ without using Proposition 6.1.4.

3. Prove Proposition 6.1.2, assuming only the results of Section 5.4.

4. Let $\triangle ABC$ be an isosceles triangle in a neutral geometry with $\overline{AB} \simeq \overline{CA}$. Let M be the midpoint of $\overline{BC}$. Prove that $\overline{AM} \perp \overline{BC}$.

5. Prove that in a neutral geometry every equilateral triangle is **equiangular**; that is, all its angles are congruent.

6. Use the Euclidean law of cosines (Proposition 6.1.2) to show that if $\triangle ABC$ is a triangle in the Euclidean Plane which has a right angle at C then $(AB)^2 = (AC)^2 + (BC)^2$.

7. Let $\triangle ABC$ be a triangle in the Euclidean Plane with $\angle C$ a right angle. If $m_E(\angle B) = \theta$ prove that $c(\theta) = BC/AB$ and $s(\theta) = AC/AB$. (Hint: Use Proposition 6.1.2 and Problem A6.)

8. Let $\square ABCD$ be a quadrilateral in a neutral geometry with $\overline{CD} \simeq \overline{CB}$. If $\overline{CA}$ is the bisector of $\angle DCB$ prove that $\overline{AB} \simeq \overline{AD}$.

9. Let $\square ABCD$ be a quadrilateral in a neutral geometry and assume that there is a point $M \in \overline{BD} \cap \overline{AC}$. If M is the midpoint of both $\overline{BD}$ and $\overline{AC}$ prove that $\overline{AB} \simeq \overline{CD}$.

10. Suppose there are points A, B, C, D, E in a neutral geometry with $A—D—B$ and $A—E—C$ and A, B, C not collinear. If $\overline{AD} \simeq \overline{AE}$ and $\overline{DB} \simeq \overline{EC}$ prove that $\angle EBC \simeq \angle DCB$.

Part B. "Prove" may mean "find a counterexample".

11. Show that the Moulton Plane does not satisfy the SAS Axiom.

12. Give an example in the Taxicab Plane of an isosceles triangle whose base angles are not congruent.

13. Prove the Quadrilateral Asinorum: If $\square ABCD$ is a quadrilateral in a neutral geometry and $\overline{AB} \simeq \overline{AD}$, then $\angle D \simeq \angle B$.

Part C. Expository exercises.

14. Write an essay discussing the statement "a neutral geometry is the type of geometry we dealt with in high school."

15. Now that you have received an introduction to the formal basis of geometry, read about informal geometry. (Hoffer [1981] is a good reference.) Compare and contrast these two approaches in an essay. Which of the two is more appropriate for middle or high school students? Is there a single correct answer?

6.2 Basic Triangle Congruence Theorems

The SAS Axiom tells us that if three certain parts of one triangle are congruent to the three corresponding parts of another triangle then the triangles are congruent. Of course, each triangle has six measurable parts—three sides and three angles—so that there are other possible choices for what three parts to compare. In this section we shall prove congruence theorems which involve other choices of angles or sides. The first of these results is informally remembered as: *If two angles and the included side of one triangle are congruent to two angles and the included side of another triangle, then the two triangles are congruent.*

Definition. A protractor geometry satisfies the **Angle-Side-Angle Axiom** (ASA) if whenever $\triangle ABC$ and $\triangle DEF$ are two triangles with $\angle A \simeq \angle D$, $\overline{AB} \simeq \overline{DE}$, and $\angle B \simeq \angle E$, then $\triangle ABC \simeq \triangle DEF$.

Theorem 6.2.1. *A neutral geometry satisfies ASA.*

PROOF. Let $\triangle ABC$ and $\triangle DEF$ be two triangles with $\angle A \simeq \angle D$, $\overline{AB} \simeq \overline{DE}$, and $\angle B \simeq \angle E$. See Figure 6-6. By the Segment Construction Theorem there is a unique point $G \in \overrightarrow{DF}$ with $\overline{DG} \simeq \overline{AC}$. We will show that $\triangle ABC \simeq \triangle DEG$ and that $G = F$ so that $\triangle ABC \simeq \triangle DEF$.

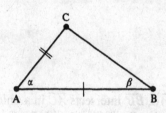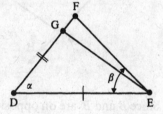

Figure 6-6

Since $\overline{AB} \simeq \overline{DE}$, $\angle A = \angle BAC \simeq \angle EDG = \angle D$, and $\overline{AC} \simeq \overline{DG}$, SAS implies that $\triangle BAC \simeq \triangle EDG$. Hence $\angle ABC \simeq \angle DEG$. But $\angle ABC \simeq \angle DEF$ by hypothesis so that $\angle DEF \simeq \angle DEG$. Since $G \in \overrightarrow{DF}$, F and G are on the same side of $\overleftrightarrow{DE}$. By the Angle Construction Theorem $\overrightarrow{EF} = \overrightarrow{EG}$. Hence

$$\{F\} = \overrightarrow{EF} \cap \overrightarrow{DF} = \overrightarrow{EG} \cap \overrightarrow{DF} = \{G\}$$

so that $F = G$. Thus $\triangle BAC \simeq \triangle EDF$; i.e., $\triangle ABC \simeq \triangle DEF$. $\qquad\square$

The next result is left to Problem A1.

Theorem 6.2.2 (Converse of *Pons Asinorum*). *In a neutral geometry, given $\triangle ABC$ with $\angle A \simeq \angle C$, then $\overline{AB} \simeq \overline{CB}$ and the triangle is isosceles.*

Stated informally, the next axiom tells us: *If the three sides of one triangle are congruent to the three sides of another triangle, then the triangles are congruent.*

Definition. A protractor geometry satisfies the **Side-Side-Side Axiom** (SSS) if whenever $\triangle ABC$ and $\triangle DEF$ are two triangles with $\overline{AB} \simeq \overline{DE}$, $\overline{BC} \simeq \overline{EF}$, and $\overline{CA} \simeq \overline{FD}$, then $\triangle ABC \simeq \triangle DEF$.

Theorem 6.2.3. *A neutral geometry satisfies SSS.*

PROOF. Let $\triangle ABC$ and $\triangle DEF$ be two triangles with $\overline{AB} \simeq \overline{DE}$, $\overline{BC} \simeq \overline{EF}$, and $\overline{CA} \simeq \overline{FD}$. The key to the proof is to make a copy of $\triangle DEF$ on the underside of $\triangle ABC$ using SAS. See Figure 6-7. By the Angle Construction Theorem, there is a unique ray $\overrightarrow{AH}$ with H on the opposite side of $\overleftrightarrow{AC}$ as B such that $\angle CAH \simeq \angle FDE$. By the Segment Construction Theorem, there is a unique point $B' \in \overrightarrow{AH}$ with $\overline{AB'} \simeq \overline{DE}$.

Since $\overline{CA} \simeq \overline{FD}$, $\angle CAB' \simeq \angle FDE$, and $\overline{B'A} \simeq \overline{ED}$, SAS implies that $\triangle CAB' \simeq \triangle FDE$. (This is what we meant by copying $\triangle DEF$ on the underside of $\triangle ABC$.) To complete the proof we show that $\triangle ABC \simeq \triangle AB'C$.

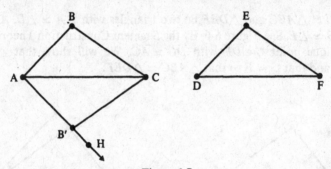

Figure 6-7

Since B and B' are on opposite sides of $\overleftrightarrow{AC}$, $\overline{BB'}$ intersects $\overleftrightarrow{AC}$ in a unique point G. There are five possibilities: (i) $G—A—C$, (ii) $G = A$, (iii) $A—G—C$, (iv) $G = C$, or (v) $A—C—G$. The first three cases are shown in Figure 6-8. Case (i) and (v) are really the same, as are cases (ii) and (iv). We complete the proof for case (i) and leave the other two as Problem A3.

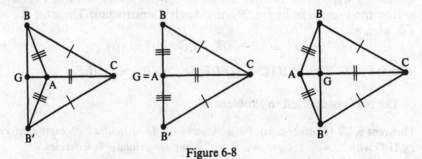

Figure 6-8

Assume G—A—C so that B, A, and B' are not collinear. $\triangle BAB'$ is isosceles since $\overline{BA} \simeq \overline{ED} \simeq \overline{B'A}$. Thus $\angle ABB' \simeq \angle AB'B$. Similarly $\triangle BCB'$ is isosceles and $\angle CBB' \simeq \angle CB'B$. Since G—A—C, $A \in \text{int}(\angle CBG) = \text{int}(\angle CBB')$ by Theorem 4.4.6. Similarly, $A \in \text{int}(\angle CB'B)$.

By the Angle Subtraction Theorem, $\angle CBA \simeq \angle CB'A$. Since $\overline{BA} \simeq \overline{ED} \simeq \overline{B'A}$ and $\overline{BC} \simeq \overline{EF} \simeq \overline{B'C}$, $\triangle ABC \simeq \triangle AB'C$ by SAS. Hence $\triangle ABC \simeq \triangle AB'C \simeq \triangle DEF$. $\square$

In Chapter 4 we showed that PSA and PP are equivalent axioms: if a metric geometry satisfies one of them then it also satisfies the other. A similar situation is true for SAS and ASA. We already know that SAS implies ASA (Theorem 6.2.1). The next result, whose proof is left to Problem A4, gives the converse. We can ask whether SSS is also equivalent to SAS. This situation is more complicated, as we'll see in Section 6.6.

Theorem 6.2.4. *If a protractor geometry satisfies ASA then it also satisfies SAS and is thus a neutral geometry.*

Recall that Theorem 5.3.5 said that if $B \in l$ then there is a unique line through B perpendicular to l. Now we would like to study the case where $B \notin l$. The proof of Theorem 6.2.3 suggests a method for proving the existence of a line through B perpendicular to l.

Theorem 6.2.5. *In a neutral geometry, given a line l and a point $B \notin l$, then there exists at least one line through B perpendicular to l.*

PROOF. Let A and C be two distinct points on l. By the Angle Construction Theorem there is a unique ray $\overrightarrow{AH}$ with H on the opposite side of $l = \overleftrightarrow{AC}$ as B and $\angle CAH \simeq \angle CAB$. See Figure 6-9. By the Segment Construction Theorem there is a unique point $B' \in \overrightarrow{AH}$ with $\overline{AB'} \simeq \overline{AB}$. Since B and B' are on opposite sides of l, $\overline{BB'}$ intersects l at a unique point G.

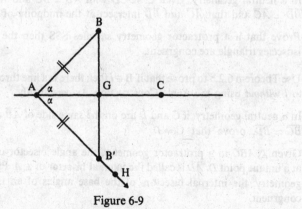

Figure 6-9

If $G \neq A$ then $\triangle BAG \simeq \triangle B'AG$ by SAS so that $\angle AGB \simeq \angle AGB'$. Thus $\angle AGB$ is a right angle by Problem A15 of Section 5.3. Hence $\overline{BB'} \perp l$.

If $G = A$ then $\angle BAC$ and $\angle B'AC$ form a linear pair of congruent angles so that $\overline{BB'} \perp l$.   □

Note that we did not claim that the perpendicular through B was unique. This is true, but we shall need to prove the Exterior Angle Theorem before we can show uniqueness. However, it is important to note that no additional axioms are needed in order to prove uniqueness—it will follow from the axioms of a neutral geometry. If the geometry is not a neutral geometry, then perpendiculars need not exist, and even when they do, they need not be unique. See Problem B23 of Section 5.3 for an example in the Moulton Plane.

PROBLEM SET 6.2

Part A.

1. Prove Theorem 6.2.2.

2. Prove that in a neutral geometry every equiangular triangle is also equilateral.

3. Complete the proof of Theorem 6.2.3 for cases (ii) and (iii).

4. Prove Theorem 6.2.4.

5. In a neutral geometry, given $\triangle ABC$ with $\overline{AB} \simeq \overline{BC}$, A—D—E—C, and $\angle ABD \simeq \angle CBE$, prove that $\overline{DB} \simeq \overline{EB}$.

6. In a neutral geometry, given $\triangle ABC$ with A—D—E—C, $\overline{AD} \simeq \overline{EC}$, and $\angle CAB \simeq \angle ACB$, prove that $\angle ABE \simeq \angle CBD$.

7. In a neutral geometry, given $\square ABCD$ with $\overline{AB} \simeq \overline{CD}$ and $\overline{AD} \simeq \overline{BC}$, prove that $\angle A \simeq \angle C$ and $\angle B \simeq \angle D$.

8. In a neutral geometry, given $\triangle ABC$ with A—D—B, A—E—C, $\angle ABE \simeq \angle ACD$, $\angle BDC \simeq \angle BEC$, and $\overline{BE} \simeq \overline{CD}$, prove that $\triangle ABC$ is isosceles.

9. In a neutral geometry, given $\square ABCD$ with $\overline{AB} \simeq \overline{BC}$ and $\overline{AD} \simeq \overline{CD}$, prove that $\overline{BD} \perp \overline{AC}$ and that $\overline{AC}$ and $\overline{BD}$ intersect at the midpoint of $\overline{AC}$.

10. Prove that if a protractor geometry satisfies SSS then the base angles of any isosceles triangle are congruent.

11. Use Theorem 6.2.5 to prove that if $B \in l$ then there is a line through B perpendicular to l *without* using the Angle Construction *Axiom* directly.

12. In a neutral geometry, if C and D are on the same side of $\overline{AB}$ and if $\overline{AC} \simeq \overline{AD}$ and $\overline{BC} \simeq \overline{BD}$, prove that $C = D$.

13. Given $\triangle ABC$ in a protractor geometry, the angle bisector of $\angle A$ intersects $\overline{BC}$ at a unique point D. $\overline{AD}$ is called the **internal bisector** of $\angle A$. Prove that in a neutral geometry, the internal bisectors of the base angles of an isosceles triangle are congruent.

Part B. "Prove" may mean "find a counterexample".

14. State and prove the SSA Congruence Theorem.

15. Prove that the Taxicab Plane does not satisfy SSS. (Hint: Problem A10, or find non-congruent equilateral triangles of side 2.)

6.3 The Exterior Angle Theorem and Its Consequences

This section is primarily concerned with theorems in a neutral geometry whose conclusions involve inequalities which compare the corresponding parts of two triangles. Of particular interest is the "common sense" result that if two sides of a triangle are not congruent then neither are their opposite angles. In fact, the larger side is opposite the larger angle. In order to prove this result, the notion of an exterior angle is needed. Our basic tool is then the Exterior Angle Theorem (Theorem 6.3.3) which is used to prove not only the result above but also the SAA Congruence Theorem, the Triangle Inequality (which relates the lengths of the three sides of a triangle) and the Open Mouth Theorem (which says that the larger the measure of an angle, the more it "opens").

Definition. In a metric geometry, the line segment $\overline{AB}$ is **less than** (or **smaller than**) the line segment $\overline{CD}$ (written $\overline{AB} < \overline{CD}$) if $AB < CD$. $\overline{AB}$ is **greater than** (or **larger than**) $\overline{CD}$ if $AB > CD$. The symbol $\overline{AB} \leq \overline{CD}$ means that either $\overline{AB} < \overline{CD}$ or $\overline{AB} \simeq \overline{CD}$.

Definition. In a protractor geometry, the angle $\angle ABC$ is **less than** (or **smaller than**) the angle $\angle DEF$ (written $\angle ABC < \angle DEF$) if $m(\angle ABC) < m(\angle DEF)$. $\angle ABC$ is **greater than** (or **larger than**) $\angle DEF$ if $\angle DEF < \angle ABC$. The symbol $\angle ABC \leq \angle DEF$ means that either $\angle ABC < \angle DEF$ or $\angle ABC \simeq \angle DEF$.

It is possible to give an alternative but equivalent description of less than without referring to length or angle measure. The theorems below use congruence and betweenness to do this. Their proofs are left as Problems A1 and A2.

Theorem 6.3.1. *In a metric geometry, $\overline{AB} < \overline{CD}$ if and only if there is a point $G \in \text{int}(\overline{CD})$ with $\overline{AB} \simeq \overline{CG}$.*

Theorem 6.3.2. *In a protractor geometry, $\angle ABC < \angle DEF$ if and only if there is a point $G \in \text{int}(\angle DEF)$ with $\angle ABC \simeq \angle DEG$.*

Definition. Given $\triangle ABC$ in a protractor geometry, if $A-C-D$ then $\angle BCD$ is an **exterior angle** of $\triangle ABC$. $\angle A$ and $\angle B$ are the **remote interior angles** of the exterior angle $\angle BCD$. (See Figure 6-10.)

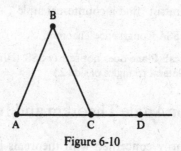

Figure 6-10

The introduction of the notion of an exterior angle and the next theorem are the keys to the results of this section. Note that at each vertex of a triangle there are two exterior angles. These are congruent by Problem A3.

Theorem 6.3.3 (Exterior Angle Theorem). *In a neutral geometry, any exterior angle of $\triangle ABC$ is greater than either of its remote interior angles.*

PROOF. Let $\triangle ABC$ be given and assume that A—C—D. We will prove that $\angle BCD > \angle ABC$ and then argue that $\angle BCD > \angle BAC$ also.

Let M be the midpoint of $\overline{BC}$ and let E be the point on $\overrightarrow{AM}$ with A—M—E and $\overline{ME} \simeq \overline{MA}$. See Figure 6-11. Since $\angle AMB$ and $\angle EMC$ are a vertical pair, they are congruent. Since $\overline{BM} \simeq \overline{MC}$, we have $\triangle AMB \simeq \triangle EMC$ by SAS. Hence

$$\angle ABC = \angle ABM \simeq \angle ECM = \angle ECB.$$

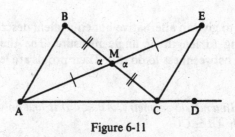

Figure 6-11

However, since A—M—E, we have $E \in \text{int}(\angle BCD)$ by Problem A6 of Section 4.4. Thus by Theorem 6.3.2, $\angle ABC \simeq \angle ECB < \angle BCD$ or $\angle BCD > \angle ABC$.

To show that $\angle BCD > \angle BAC$, choose D' so that B—C—D'. See Figure 6-12. By the first part of the proof, $\angle ACD' > \angle BAC$. By Problem A3, $\angle ACD' \simeq \angle BCD$. Hence $\angle BCD > \angle BAC$. $\qquad\qquad\square$

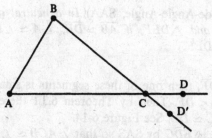

Figure 6-12

Corollary 6.3.4. *In a neutral geometry, there is exactly one line through a given point P perpendicular to a given line l.*

PROOF. If $P \in l$, then the result follows from Theorem 5.3.5. Thus we consider only the case where $P \notin l$. We already know that there is such a line by Theorem 6.2.5. Now suppose that there are two distinct lines l' and l'' through P both of which are perpendicular to l. See Figure 6-13.

Let $\{A\} = l' \cap l$ and $\{C\} = l'' \cap l$. Since l' and l'' are distinct, they cannot have two points in common so that $A \neq C$. Choose D with $A-C-D$. Then the right angle $\angle DCP$ is an exterior angle of $\triangle APC$ and is thus greater than $\angle CAP$ by the Exterior Angle Theorem. Because $\angle CAP$ is a right angle and two right angles are always congruent, this is a contradiction.

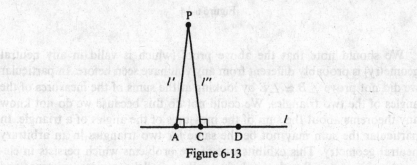

Figure 6-13

Hence there cannot be two distinct lines through P which are perpendicular to l. □

The Exterior Angle Theorem can be used to prove another triangle congruence theorem. This new theorem will be useful in our study of right triangles in the next section. Informally, it says: *If two angles and a side of one triangle are congruent to the corresponding two angles and side of another triangle, then the triangles are congruent.* This result when coupled with the ASA Theorem shows that it does not matter which pair of angles and which side we choose, as long as we use the corresponding angles and side from the other triangle.

Theorem 6.3.5 (Side-Angle-Angle, SAA). *In a neutral geometry, given two triangles* $\triangle ABC$ *and* $\triangle DEF$, *if* $\overline{AB} \simeq \overline{DE}$, $\angle A \simeq \angle D$, *and* $\angle C \simeq \angle F$, *then* $\triangle ABC \simeq \triangle DEF$.

PROOF. If $\overline{AC} \not\simeq \overline{DF}$, then one of these segments is smaller than the other. Suppose that $\overline{AC} < \overline{DF}$. Then by Theorem 6.3.1 there is a point G with $D{-}G{-}F$ and $\overline{AC} \simeq \overline{DG}$. See Figure 6-14.

Now $\triangle BAC \simeq \triangle EDG$ by SAS so that $\angle ACB \simeq \angle DGE$. Since $\angle DGE$ is an exterior angle of $\triangle GEF$, $\angle DGE > \angle DFE$ by the Exterior Angle Theorem. However, $\angle ACB \simeq \angle DFE$ by hypothesis, so that

$$\angle ACB \simeq \angle DGE > \angle DFE \simeq \angle ACB$$

which is impossible. Hence it cannot be that $\overline{AC} < \overline{DF}$. Similarly we cannot have $\overline{DF} < \overline{AC}$. Thus $\overline{AC} \simeq \overline{DF}$ and $\triangle BAC \simeq \triangle EDF$ by SAS. □

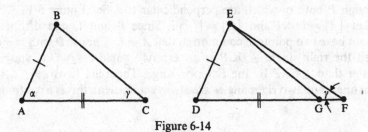

Figure 6-14

We should note that the above proof (which is valid in any neutral geometry) is probably different from any you have seen before. In particular we did not prove $\angle B \simeq \angle E$ by looking at the sums of the measures of the angles of the two triangles. We could not do this because we do not know any theorems about the sum of the measures of the angles of a triangle. In particular the sum may not be the same for two triangles in an arbitrary neutral geometry. This exhibits one of the problems which persists in elementary geometry. People regularly use the parallel postulate (existence and uniqueness of parallels) or its consequences (such as the sum of the angle measures of a triangle is 180) to prove certain theorems which really don't need that postulate. More succinctly, they are proving theorems which are valid in an arbitrary neutral geometry in a Euclidean fashion. (Of course, these authors only claim to have proved these theorems for Euclidean geometry because their proofs are not valid in a general neutral geometry.)

Theorem 6.3.6. *In a neutral geometry, if two sides of a triangle are not congruent, neither are the opposite angles. Furthermore, the larger angle is opposite the longer side.*

PROOF. In $\triangle ABC$ assume that $\overline{AB} > \overline{AC}$. We want to show that $\angle C > \angle B$. Now there exists a unique point D with $A{-}C{-}D$ and $\overline{AD} \simeq \overline{AB}$ (Why?).

See Figure 6-15. Since A—C—D, $C \in \text{int}(\angle ABD)$ and $\angle ABC < \angle ABD$. However, $\triangle BAD$ is isosceles with $\overline{AB} \simeq \overline{AD}$ so that $\angle ABD \simeq \angle ADB$. By the Exterior Angle Theorem for $\triangle BCD$, $\angle ADB < \angle ACB$. Thus

$$\angle ABC < \angle ABD \simeq \angle ADB < \angle ACB$$

so that $\angle B < \angle C$. $\square$

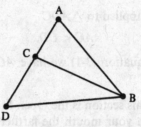

Figure 6-15

The converse of Theorem 6.3.6 is also true. Its proof is left to Problem A5.

Theorem 6.3.7. *In a neutral geometry, if two angles of a triangle are not congruent, neither are the opposite sides. Furthermore, the longer side is opposite the larger angle.*

The Triangle Inequality which we present next can be proved in a variety of manners depending on the context. We have already seen a proof in the Euclidean plane which used vector concepts (Proposition 3.1.6). Like Theorem 6.3.6, it is really a theorem in neutral geometry.

Theorem 6.3.8 (Triangle Inequality). *In a neutral geometry the length of one side of a triangle is strictly less than the sum of the lengths of the other two sides.*

PROOF. We must show that for $\triangle ABC$

$$AC < AB + BC.$$

This will be done by grafting an "extra section" of length AB onto $\overline{BC}$. See Figure 6-16.

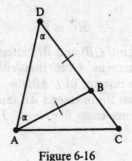

Figure 6-16

Let $D \in \overline{CB}$ so that $C{-}B{-}D$ and $\overline{BD} \simeq \overline{AB}$. Then

$$CD = CB + BD = BC + AB. \qquad (3\text{-}1)$$

By Theorem 4.4.6, $B \in \text{int}(\angle DAC)$ so that $\angle DAB < \angle DAC$. Since $\triangle DBA$ is isosceles, $\angle DAB \simeq \angle ADB$ so that

$$\angle ADB < \angle DAC.$$

Thus by Theorem 6.3.6 applied to $\triangle ADC$

$$AC < CD.$$

Combining this with Equation (3-1) we have $AC < BC + AB$ as required.
$\square$

The last theorem of this section is the Open Mouth Theorem, which says that the wider you open your mouth the farther apart your lips are. (It is called the Hinge Theorem by some authors.) Whereas the previous theorems in this section dealt with inequalities in a single triangle, the open Mouth Theorem gives an inequality relating two triangles.

Theorem 6.3.9 (Open Mouth Theorem). *In a neutral geometry, given two triangles $\triangle ABC$ and $\triangle DEF$ with $\overline{AB} \simeq \overline{DE}$ and $\overline{BC} \simeq \overline{EF}$, if $\angle B > \angle E$ then $\overline{AC} > \overline{DF}$.*

PROOF. We first construct a copy of $\triangle DEF$ along the side $\overline{BC}$ of $\triangle ABC$. There is a unique point H on the same side of $\overleftrightarrow{BC}$ as A with $\angle HBC \simeq \angle DEF$ and $\overline{BH} \simeq \overline{ED}$ (Why?). Then $\triangle DEF \simeq \triangle HBC$ by SAS. See Figure 6-17.

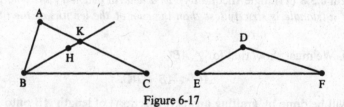

Figure 6-17

Hence

$$\overline{HC} \simeq \overline{DF}. \qquad (3\text{-}2)$$

Since $\angle E < \angle B$, $H \in \text{int}(\angle B)$ and $\overrightarrow{BH}$ intersects $\overline{AC}$ at a unique point K by the Crossbar Theorem. Note that either $B{-}H{-}K$, $H = K$, or $B{-}K{-}H$. Let $\overrightarrow{BL}$ be the bisector of $\angle ABH = \angle ABK$. Using the Crossbar Theorem again, we see that $\overrightarrow{BL}$ intersects $\overline{AK}$ (and $\overline{AC}$) at a unique point M. See Figure 6-18 for the three cases $B{-}H{-}K$, $H = K$, and $B{-}K{-}H$.

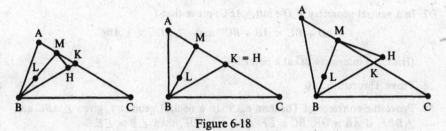

Figure 6-18

Now $\triangle ABM \simeq \triangle HBM$ by SAS so that

$$\overline{AM} \simeq \overline{HM}. \tag{3-3}$$

If $H \neq K$ then C, H, and M are not collinear, and the Triangle Inequality implies

$$HC < HM + MC.$$

This is also true if $K = H$ since C—H—M implies $HC < MC < HM + MC$. Since $\overline{AM} \simeq \overline{HM}$ (Congruence (3-3)) we have

$$HC < HM + MC = AM + MC = AC$$

where the last equality comes from the fact that A—M—C. Finally since $\overline{HC} \simeq \overline{DF}$ (Congruence (3-2)) we see that

$$DF < AC \quad \text{or} \quad \overline{AC} > \overline{DF}. \qquad \square$$

Although the following result is left as an exercise, it will prove to be extremely useful.

Theorem 6.3.10. *In a neutral geometry, a line segment joining a vertex of a triangle to a point on the opposite side is shorter than the longer of the remaining two sides. More precisely, given $\triangle ABC$ with $\overline{AB} \leq \overline{CB}$, if A—D—C then $\overline{DB} < \overline{CB}$.*

PROBLEM SET 6.3

Part A.

1. Prove Theorem 6.3.1.

2. Prove Theorem 6.3.2.

3. In a protractor geometry prove the two exterior angles of $\triangle ABC$ at the vertex C are congruent.

4. In a neutral geometry prove that the base angles of an isosceles triangle are acute.

5. Prove Theorem 6.3.7.

*6. Prove the General Triangle Inequality for a neutral geometry: If A, B and C are distinct points in a neutral geometry, then $AC \leq AB + BC$. Furthermore, equality holds if and only if A—B—C.

7. In a neutral geometry, if $D \in \text{int}(\triangle ABC)$ prove that

$$AD + DC < AB + BC \quad \text{and} \quad \angle ADC > \angle ABC.$$

(Hint: $\overrightarrow{AD}$ intersects $\overline{BC}$ at a point E.)

8. Prove Theorem 6.3.10.

*9. Prove the converse of Theorem 6.3.9: In a neutral geometry, given $\triangle ABC$ and $\triangle DEF$, if $\overline{AB} \simeq \overline{DE}$, $\overline{BC} \simeq \overline{EF}$, and $\overline{AC} > \overline{DF}$, then $\angle B > \angle E$.

10. In a neutral geometry prove that a triangle with an obtuse angle must have two acute angles.

Part B. "Prove" may mean "find a counterexample".

11. Let m be any angle measure on $\mathbb{R}^2$. Prove that $\{\mathbb{R}^2, \mathcal{L}_E, d_T, m\}$ does not satisfy SAS. (Thus an angle measure can never be found for the Taxicab Plane so that the resulting object is a neutral geometry.)

12. In a neutral geometry, given $\triangle ABC$ such that the internal bisectors of $\angle A$ and $\angle C$ are congruent, prove that $\triangle ABC$ is isosceles. (Hint: Assume $\angle A < \angle C$ and consider Figure 6-19 where $\overline{AQ} \simeq \overline{CP}$. How are $\angle AQD$ and $\angle CPE$ related?)

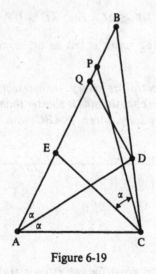

Figure 6-19

13. Replace the word "neutral" in the hypothesis of Theorem 6.3.6 with the word "protractor". Is the conclusion still valid?

6.4 Right Triangles

In this section we shall prove some basic results about right triangles in a neutral geometry. Besides the standard congruence theorems, the most important result will be the often quoted theorem that the shortest distance from a point to a line is given by the perpendicular distance.

A word of caution is needed here. The first thing that many of us think about when we hear the phrase "right triangle" is the classical Pythagorean Theorem. This theorem (which states that the square of the length of the hypotenuse is the sum of the squares of the lengths of the other two sides) is very much a Euclidean theorem. That is, it is true in the Euclidean Plane but *not* in all neutral geometries (see Problem A6). Thus in each proof of this section which deals with a general neutral geometry we must avoid the use of the Pythagorean Theorem.

Definition. If an angle of $\triangle ABC$ is a right angle, then $\triangle ABC$ is a **right triangle**. A side opposite a right angle in a right triangle is called a **hypotenuse**.

If $\triangle ABC$ has a right angle at C we insert a small box at $\angle C$ to indicate the right angle. See Figure 6-20.

Definition. $\overline{AB}$ is **the longest side** of $\triangle ABC$ if $\overline{AB} > \overline{AC}$ and $\overline{AB} > \overline{BC}$. $\overline{AB}$ is **a longest side** of $\triangle ABC$ if $\overline{AB} \geq \overline{AC}$ and $\overline{AB} \geq \overline{BC}$.

A triangle will always have "a" longest side, but there may not be "the" longest side. (Can you think of examples?)

Theorem 6.4.1. *In a neutral geometry, there is only one right angle and one hypotenuse for each right triangle. The remaining angles are acute, and the hypotenuse is the longest side of the triangle.*

PROOF. Let $\triangle ABC$ be a right triangle with $\angle C$ a right angle. Let D be such that $D-C-B$ (see Figure 6-20). $\angle DCA$ is a right angle and by the Exterior Angle Theorem, $\angle B < \angle DCA$ and $\angle A < \angle DCA$. Thus both $\angle A$ and $\angle B$ are acute and there is only one right angle (and hence one hypotenuse). Finally, $\overline{BC} < \overline{AB}$ and $\overline{AC} < \overline{AB}$ by Theorem 6.3.7. Hence the hypotenuse $\overline{AB}$ is the longest side.   □

Definition. If $\triangle ABC$ is a right triangle with right angle at C then the **legs** of $\triangle ABC$ are $\overline{AC}$ and $\overline{BC}$.

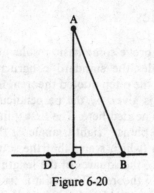

Figure 6-20

Theorem 6.4.1 says that the hypotenuse is longer than either leg. This result really needs the strength of the SAS Axiom. We saw in Problem A12 of Section 5.2 that there are triangles in the Moulton Plane with two right angles. (The situation is even worse on the Riemann Sphere. There it is possible to have a reasonable angle measure and a triangle with three right angles.)

We shall now prove that the shortest distance between a point and a line in a neutral geometry is the perpendicular distance.

Theorem 6.4.2 (Perpendicular Distance Theorem). *In a neutral geometry, if l is a line, $Q \in l$, and $P \notin l$ then*

(i) *if $\overleftrightarrow{PQ} \perp l$ then*

$$PQ \leq PR \qquad \text{for all } R \in l \qquad (4\text{-}1)$$

(ii) *if $PQ \leq PR$ for all $R \in l$ then $\overleftrightarrow{PQ} \perp l$.*

PROOF. If $\overleftrightarrow{PQ} \perp l$ and $R \in l$, then either $Q = R$ (so that $PQ = PR$) or else $Q \neq R$ (and so P, Q, R are not collinear). In the latter case $\triangle PQR$ has a right angle at Q. (See Figure 6-21.) By Theorem 6.4.1, $PQ < PR$ since $\overline{PR}$ is the hypotenuse of $\triangle PQR$. Hence $PQ \leq PR$ for all $R \in l$ if $\overleftrightarrow{PQ} \perp l$.

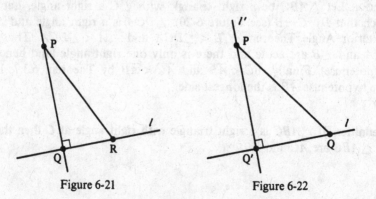

Figure 6-21   Figure 6-22

Conversely, suppose that $PQ \le PR$ for all $R \in l$. We must show that $\overrightarrow{PQ} \perp l$. Let l' be the unique line through P which is perpendicular to l (Corollary 6.3.4). See Figure 6-22. Let $l \cap l' = \{Q'\}$. We need to show that $Q = Q'$. If $Q \ne Q'$, then $\triangle PQ'Q$ is a right triangle and $PQ' < PQ$. But $PQ \le PR$ for all $R \in l$ by hypothesis so that, in particular, $PQ \le PQ'$. This contradiction of $PQ' < PQ$ shows that $Q = Q'$. Hence $\overrightarrow{PQ} = l'$ is perpendicular to l.   ☐

It is illuminating to recast Theorem 6.4.2 in terms of the concept of the distance from a point to a line.

Definition. Let l be a line and P a point in a neutral geometry. If $P \notin l$, let Q be the unique point of l such that $\overrightarrow{PQ} \perp l$. The **distance from** P **to** l is

$$d(P, l) = \begin{cases} d(P, Q) & \text{if } P \notin l \\ 0 & \text{if } P \in l. \end{cases}$$

In terms of this definition, Theorem 6.4.2 reads:

Theorem 6.4.2'. *For any line l in a neutral geometry and $P \notin l$*

$$d(P, l) \le d(P, R) \quad \text{for all } R \in l.$$

Furthermore, $d(P, l) = d(P, R)$ if and only if $\overrightarrow{PR} \perp l$.

Definition. If l is the unique perpendicular to $\overrightarrow{AB}$ through the vertex C of $\triangle ABC$ and if $l \cap \overrightarrow{AB} = \{D\}$, then $\overline{CD}$ is the **altitude** from C. D is the **foot** of the altitude (or of the perpendicular) from C.

Note that in general the foot of the altitude from C need not lie on $\overline{AB}$. However, it seems clear (and we now shall prove) that if $\overline{AB}$ is a longest side then the foot will actually belong to $\overline{AB}$. See Figure 6-23.

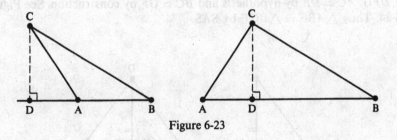

Figure 6-23

Theorem 6.4.3. *In a neutral geometry, if $\overline{AB}$ is a longest side of $\triangle ABC$ and if D is the foot of the altitude from C, then A—D—B.*

PROOF. Either $D—A—B$, $D = A$, $A—D—B$, $D = B$, or $A—B—D$. The first and last cases are essentially the same, as are the second and fourth. We shall show that neither the first nor the second case can occur. This will imply that the only possibility is $A—D—B$.

Now $\overline{CB}$ is the hypotenuse of the right triangle $\triangle CBD$ so that $\overline{DB} < \overline{CB}$. If $D—A—B$ then $\overline{AB} < \overline{DB}$ so that

$$\overline{AB} < \overline{DB} < \overline{CB}$$

which contradicts the fact that $\overline{AB}$ is a longest side. If $D = A$ then $\overline{AB} = \overline{DB}$ so that

$$\overline{AB} = \overline{DB} < \overline{CB}$$

which is again a contradiction. Hence $A—D—B$.   □

For a right triangle in a neutral geometry, any two sides suffice to determine the triangle up to congruence. If the two sides are the legs this follows from SAS. If one of the sides is the hypotenuse then SAS cannot be used. Instead we need the next result which says that if the hypotenuse and a leg of one right triangle are congruent to the hypotenuse and leg of a second right triangle then the two triangles are congruent.

Theorem 6.4.4 (Hypotenuse-Leg, HL). *In a neutral geometry if $\triangle ABC$ and $\triangle DEF$ are right triangles with right angles at C and F, and if $\overline{AB} \simeq \overline{DE}$ and $\overline{AC} \simeq \overline{DF}$, then $\triangle ABC \simeq \triangle DEF$.*

PROOF. As was the case in other triangle congruence theorems, we shall construct an intermediate triangle which is congruent to both $\triangle ABC$ and $\triangle DEF$.

Let G be the unique point such that $E—F—G$ and $\overline{FG} \simeq \overline{BC}$. Since E, F, G are collinear and $\angle DFE$ is a right angle, so is $\angle DFG$. Hence $\angle ACB \simeq \angle DFG$. $\overline{AC} \simeq \overline{DF}$ by hypothesis and $\overline{BC} \simeq \overline{GF}$ by construction. See Figure 6-24. Thus $\triangle ABC \simeq \triangle DGF$ by SAS.

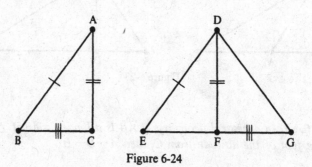

Figure 6-24

Since $\triangle ABC \simeq \triangle DGF$, $\overline{AB} \simeq \overline{DG}$. But $\overline{AB} \simeq \overline{DE}$ by hypothesis so that $\overline{DE} \simeq \overline{DG}$ and $\triangle EDG$ is isosceles. Thus $\angle DEF = \angle DEG \simeq \angle DGE = \angle DGF$ and $\triangle DEF \simeq \triangle DGF$ by SAA. Hence $\triangle ABC \simeq \triangle DEF$. $\qquad\square$

When you saw this theorem in high school the proof may have used the Pythagorean Theorem to show that $\overline{BC} \simeq \overline{EF}$ so that SSS could be applied. The proof given above is valid in any neutral geometry, not just those models where the Pythagorean Theorem holds. As was stated at the beginning of this section, to prove something in neutral geometry, the Pythagorean Theorem cannot be used. Similarly, we should not use the "fact" that the sum of the angle measures of a triangle is 180 to prove the next theorem.

Theorem 6.4.5 (Hypotenuse-Angle, HA). *In a neutral geometry, let $\triangle ABC$ and $\triangle DEF$ be right triangles with right angles at C and F. If $\overline{AB} \simeq \overline{DE}$ and $\angle A \simeq \angle D$, then $\triangle ABC \simeq \triangle DEF$.*

Definition. The **perpendicular bisector** of the segment $\overline{AB}$ in a neutral geometry is the (unique) line l through the midpoint M of $\overline{AB}$ and which is perpendicular to $\overline{AB}$.

The next result contains a useful description of the perpendicular bisector.

Theorem 6.4.6. *In a neutral geometry the perpendicular bisector l of the segment $\overline{AB}$ is the set*
$$\mathscr{B} = \{P \in \mathscr{S} \,|\, AP = BP\},$$

PROOF. We first show $\mathscr{B} \subset l$. Let $P \in \mathscr{B}$. We must show that $P \in l$. If $P \in \overline{AB}$, then $AP = BP$ implies $A{-}P{-}B$ and $P = M$ is the midpoint of $\overline{AB}$. (See Problems A11 and A12 of Section 3.3.) Hence $P \in l$. If $P \notin \overline{AB}$ let l' be the unique perpendicular to $\overleftrightarrow{AB}$ through P. See Figure 6-25. Let $l' \cap \overleftrightarrow{AB} = \{N\}$. Then N is not A or B (otherwise $\triangle APB$ would be a right triangle with one of $\overline{PA}$ and $\overline{PB}$ the hypotenuse and the other a leg. This contradicts $PA = PB$.). Hence we have two triangles $\triangle PNA$ and $\triangle PNB$ which are congruent

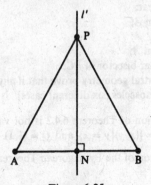

Figure 6-25

by HL. Hence $\overline{AN} \simeq \overline{NB}$ and $N = M$ is the midpoint. Thus $l' = l$ and $P \in l$. Hence $\mathcal{B} \subset l$.

We now show that $l \subset \mathcal{B}$. We assume $P \in l$ and show $P \in \mathcal{B}$. If $P \in \overleftrightarrow{AB}$ then $P = M$ and $P \in \mathcal{B}$. If $P \notin \overleftrightarrow{AB}$ then $\angle PMA \simeq \angle PMB$ since both are right angles. Thus $\triangle PMA \simeq \triangle PMB$ by SAS and $\overline{PA} \simeq \overline{PB}$. Hence $P \in \mathcal{B}$ and $l \subset \mathcal{B}$. This means that $l = \mathcal{B}$.   □

Theorem 6.4.7. *In a neutral geometry, if $\overline{BD}$ is the bisector of $\angle ABC$ and if E and F are the feet of the perpendiculars from D to $\overrightarrow{BA}$ and $\overrightarrow{BC}$ then $\overline{DE} \simeq \overline{DF}$.*

PROOF. Problem A10. See Figure 6-26.   □

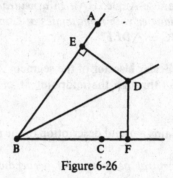

Figure 6-26

PROBLEM SET 6.4

Part A.

1. Prove Theorem 6.4.5.

2. In a neutral geometry, if D is the foot of the altitude of $\triangle ABC$ from C and A—B—D, then prove $\overline{CA} > \overline{CB}$.

3. If M is the midpoint of $\overline{BC}$ then $\overline{AM}$ is called a **median** of $\triangle ABC$.
   a. Prove that in a neutral geometry if $\triangle ABC$ is isosceles with base $\overline{BC}$ then the following are collinear:
      (i) the median from A;
      (ii) the bisector of $\angle A$;
      (iii) the altitude from A;
      (iv) the perpendicular bisector of $\overline{BC}$.
   b. Conversely, in a neutral geometry prove that if *any* two of (i)–(iv) are collinear then the triangle is isosceles (six different cases).

4. Show that the conclusion of Theorem 6.4.2 is not valid in the Taxicab Plane by taking $P = (-1, 1)$, $l = \{(x, y) | y = x\}$ and $Q = (1, 1)$.

5. Show that the conclusion of the Pythagorean Theorem is not valid in the Taxicab Plane.

6. Show that the conclusion of the Pythagorean Theorem is not valid in the Poincaré Plane by considering $\triangle ABC$ with $A = (2, 1)$, $B = (0, \sqrt{5})$, and $C = (0, 1)$. Thus

the Pythagorean Theorem does not hold in every neutral geometry.

7. Show that the hypotenuse need not be the longest side of a right triangle in a protractor geometry by examining $\triangle ABC$ of Example 6.1.1.

8. In a neutral geometry, if $\triangle ABC$ is a right triangle with right angle at C and if
   $B-D-C$, then prove $\triangle BDA$ is obtuse.

9. In a neutral geometry, if $\triangle ABC$ and $\triangle DEF$ are triangles with $\angle B \simeq \angle E$, $\overline{BC} \simeq$
   $\overline{EF}$, $\overline{CA} \simeq \overline{FD}$, and $\angle A$ and $\angle D$ are either both acute or both obtuse, then prove
   that $\triangle ABC \simeq \triangle DEF$.

10. Prove Theorem 6.4.7.

11. In a neutral geometry prove that the bisector of $\angle ABC$ is $\mathscr{B} = \{B\} \cup \{X \in$
    $\text{int}(\angle ABC) | d(X, \overrightarrow{BC}) = d(X, \overrightarrow{BA})\}$.

12. In a neutral geometry, let $\overrightarrow{BD}$ and $\overrightarrow{CE}$ be the bisectors of $\angle B$ and $\angle C$ of $\triangle ABC$.
    Prove that $\overrightarrow{BD} \cap \overrightarrow{CE}$ contains a single point F. Prove that $\overrightarrow{AF}$ is the bisector of $\angle A$.
    (Hint: Use Problem A11.)

13. In a neutral geometry, let l_1, l_2, and l_3 be the perpendicular bisectors of the three
    sides of $\triangle ABC$. If $D \in l_1 \cap l_2$, prove that $D \in l_3$.

Part B. "Prove" may mean "find a counterexample".

PROBLEM SET 6.4B   Prove or give a counterexample.

14. In a neutral geometry, if $A-D-B$ and $\overline{CD}$ is an altitude of $\triangle ABC$, then prove
    that $\overline{AB}$ is a longest side of $\triangle ABC$.

15. Show that the conclusion of Theorem 6.4.6 is false in the Taxicab Plane by
    considering $A = (1, 0)$ and $B = (0, 1)$.

16. Find the error or errors in the following alleged "proof" that in a neutral geometry
    any triangle is isosceles.

    Let M be the midpoint of $\overline{AC}$ and let l be the perpendicular to $\overline{AC}$ at M. Let $\overrightarrow{BQ}$
    be the angle bisector of $\angle ABC$ and let $D \in l \cap \overrightarrow{BQ}$. If E is the foot of the
    perpendicular from D to $\overrightarrow{BC}$ and if F is the foot of the perpendicular from D to
    $\overrightarrow{BA}$, then $\overline{FD} \simeq \overline{ED}$ by Theorem 6.4.7. $\overline{AD} \simeq \overline{CD}$ by Theorem 6.4.6. Hence
    $\triangle AFD \simeq \triangle CED$ by HL and $\overline{AF} \simeq \overline{CE}$. Since $\triangle BDF \simeq \triangle BDE$ (by HA), $\overline{BF} \simeq$
    $\overline{BE}$. Hence $BA = BF + FA = BE + EC = BC$ and $\overline{BA} \simeq \overline{BC}$. See Figure 6-27.

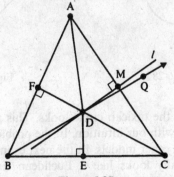

Figure 6-27

6.5 Circles and Their Tangent Lines

In this section we will define the concepts of a circle and of a tangent to a circle. We shall see that in a metric geometry some rather strange examples of circles exist. In the latter part of the section we shall see that the familiar properties of tangents do indeed hold in a neutral geometry. In particular, we will show that there is a unique tangent at each point on a circle in a neutral geometry.

Definition. If C is a point in a metric geometry $\{\mathscr{S}, \mathscr{L}, d\}$ and if $r > 0$, then

$$\mathscr{C} = \mathscr{C}_r(C) = \{P \in \mathscr{S} \mid PC = r\}$$

is a **circle** with **center** C and **radius** r. If A and B are distinct points of $\mathscr{C}$ then $\overline{AB}$ is a **chord** of $\mathscr{C}$. If the center C is a point on the chord $\overline{AB}$, then $\overline{AB}$ is a **diameter** of $\mathscr{C}$. For any $Q \in \mathscr{C}$, $\overline{CQ}$ is called a **radius segment** of $\mathscr{C}$.

Example 6.5.1. Find and sketch the circle of radius 1 with center $(0,0)$ in the Euclidean Plane and in the Taxicab Plane.

SOLUTION. In the Euclidean Plane we have

$$\mathscr{C}_1((0,0)) = \{(x,y) \mid x^2 + y^2 = 1\}.$$

This is sketched in Figure 6-28.
   In the Taxicab Plane we have

$$\mathscr{C}_1((0,0)) = \{(x,y) \mid |x| + |y| = 1\}.$$

This set consists of four line segments, with slope ± 1. See Figure 6-29. $\square$

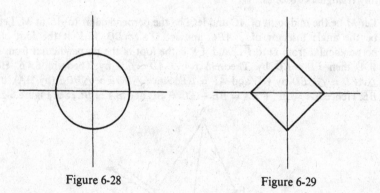

Figure 6-28   Figure 6-29

Note how strange the taxicab circle looks. This is one more indication we should be careful with our intuition. In the problems you will find some examples of circles in other models. In the next example we shall see that a Poincaré circle actually looks like a Euclidean circle, a result which is unexpected!

Example 6.5.2. Show that $\mathscr{A} = \{(x, y) \in \mathbb{H} \mid x^2 + (y - 5)^2 = 16\}$ is the Poincaré circle $\mathscr{C}$ with center $(0, 3)$ and radius $\ln 3$.

SOLUTION. Let $(a, b) \in \mathscr{A}$. If $a = 0$ then $b = 1$ or 9. Clearly, $d((0, 1), (0, 3)) = \ln 3 = d((0, 3), (0, 9))$ so that $(a, b) \in \mathscr{C}$ in this case.

Now assume that $(a, b) \in \mathscr{A}$ and $a \neq 0$. We must find c and r so that both $(a, b) \in {}_cL_r$ and $(0, 3) \in {}_cL_r$ in order to compute $d((a, b), (0, 3))$. By Equations (1-5) and (1-6) of Section 2.1 we have

$$c = \frac{b^2 - 3^2 + a^2}{2a} \quad \text{and} \quad r = \sqrt{c^2 + 9}. \tag{5-1}$$

Since $(a, b) \in \mathscr{A}$ we have $a^2 + b^2 - 10b + 25 = 16$ so that

$$a^2 + b^2 - 9 = 16 - 25 + 10b - 9 = 10b - 18.$$

Substituting this last result into Equations (5-1) we obtain

$$c = \frac{5b - 9}{a}$$

$$r = \sqrt{c^2 + 9} = \sqrt{\frac{25b^2 - 90b + 81}{a^2} + 9}$$

$$= \sqrt{\frac{16b^2 + 9(a^2 + (b - 5)^2 - 16)}{a^2}}$$

$$= \sqrt{\frac{16b^2}{a^2}} = \frac{4b}{|a|}.$$

Now $d((a, b), (0, 3)) = |\ln((a - c + r)/b \cdot 3/(-c + r))|$ and

$$\frac{a - c + r}{b} \cdot \frac{3}{-c + r} = \frac{a - \dfrac{5b - 9}{a} + \dfrac{4b}{|a|}}{b} \cdot \frac{3}{-\dfrac{5b - 9}{a} + \dfrac{4b}{|a|}}$$

$$= \frac{(a^2 - 5b + 9 \pm 4b)3}{b(-5b + 9 \pm 4b)} \quad \text{where } \pm \text{ denotes the sign of } a$$

$$= \frac{(16 - b^2 + 10b - 25 - 5b + 9 \pm 4b)3}{b(-5b + 9 \pm 4b)}$$

$$= \frac{(-b^2 + 5b \pm 4b)3}{b(-5b + 9 \pm 4b)}$$

$$= \begin{cases} \dfrac{(-b^2 + 9b)3}{b(9 - b)} = 3 & \text{if } a > 0 \\[4mm] \dfrac{(-b^2 + b)3}{b(9 - 9b)} = \dfrac{1}{3} & \text{if } a < 0. \end{cases}$$

Hence $d((a, b), (0, 3)) = |\ln 3| = |\ln \tfrac{1}{3}|$. Thus $(a, b) \in \mathscr{C}$ and $\mathscr{A} \subset \mathscr{C}$.

Finally, if $(a, b) \in \mathscr{C}$, let l be the Poincaré line through $(0, 3)$ and (a, b). Then l is part of either a Euclidean line or a Euclidean circle depending on whether l is of type I or type II. Either way, from our previous knowledge of Euclidean geometry (which will be carefully demonstrated in the next section) $l \cap \mathscr{A}$ has exactly two points. Likewise $l \cap \mathscr{C}$ has two points (by Segment Construction). Since $\mathscr{A} \subset \mathscr{C}$, $l \cap \mathscr{A} \subset l \cap \mathscr{C}$. Since both sets have two points, $l \cap \mathscr{A} = l \cap \mathscr{C}$ and $(a, b) \in l \cap \mathscr{C} = l \cap \mathscr{A}$ so that $(a, b) \in \mathscr{A}$. Thus $\mathscr{C} \subset \mathscr{A}$ and $\mathscr{C} = \mathscr{A}$. See Figure 6-30.   □

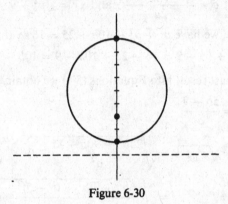

Figure 6-30

Our first result tells us that in a neutral geometry the center and radius of a circle are determined by any three points on the circle. The nice thing about this theorem is its constructive nature. The proof will proceed by starting with the three points on the circle and actually constructing the center of the circle as the intersection of perpendicular bisectors of any two of the chords as in Figure 6-31. (Note, we do not claim that *any* three points determine a circle.)

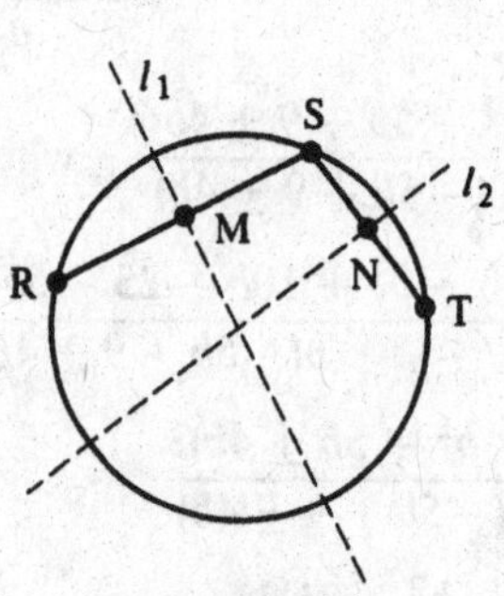

Figure 6-31

Theorem 6.5.3. *In a neutral geometry, let* $\mathscr{C}_1 = \mathscr{C}_r(C)$ *and* $\mathscr{C}_2 = \mathscr{C}_s(D)$. *If* $\mathscr{C}_1 \cap \mathscr{C}_2$ *contains at least three points, then* $C = D$ *and* $r = s$. *Thus, three points of a circle in a neutral geometry uniquely determine that circle.*

PROOF. Let R, S and T be three distinct points in $\mathscr{C}_1 \cap \mathscr{C}_2$. Let l_1 be the perpendicular bisector of $\overline{RS}$ and let l_2 be the perpendicular bisector of $\overline{ST}$. (See Figure 6-31.) By Theorem 6.4.6, $C \in l_1$ since $RC = SC = r$. Similarly, $C \in l_2$ so that $C \in l_1 \cap l_2$. By using Theorem 6.4.6 again (and the fact that $R, S, T \in \mathscr{C}_2$) we see that $D \in l_1 \cap l_2$. Thus, either $C = D$ or else l_1 and l_2 have two distinct points (C and D) in common and so are equal. We now show that the last case cannot happen.

Assume for the moment that $l_1 = l_2$. There are two possibilities for $\{R, S, T\}$. If $\{R, S, T\}$ is a collinear set and $l_1 = l_2$ then the midpoint of $\overline{RS}$ must be equal to the midpoint of $\overline{ST}$. However, the only way this can happen is if $R = T$, which is a contradiction.

On the other hand, if $l_1 = l_2$ and $\{R, S, T\}$ is non-collinear, let M and N be the midpoints of $\overline{RS}$ and $\overline{ST}$. Since $l_1 = l_2 = \overleftrightarrow{MN}$ (by assumption), $\triangle MNS$ has a right angle at M and another at N. However, this contradicts the fact that a right triangle has exactly one right angle (Theorem 6.4.1). Thus $l_1 \neq l_2$ and so $C = D$.

Since $r = RC = RD = s$, the radius of the circle is determined once the center and any point on the circle are known.   □

Contained in the proof of the above result is the following fact.

Corollary 6.5.4. *For any circle in a neutral geometry, the perpendicular bisector of any chord contains the center.*

Definition. Let $\mathscr{C}$ be the circle with center C and radius r. The **interior** of $\mathscr{C}$ is the set

$$\text{int}(\mathscr{C}) = \{P \in \mathscr{S} \mid CP < r\}.$$

The **exterior** of $\mathscr{C}$ is the set

$$\text{ext}(\mathscr{C}) = \{P \in \mathscr{S} \mid CP > r\}.$$

Theorem 6.5.5. *If $\mathscr{C}$ is a circle in a neutral geometry then $\text{int}(\mathscr{C})$ is convex.*

PROOF. Let $\mathscr{C} = \mathscr{C}_r(C)$. Suppose that A and B belong to $\text{int}(\mathscr{C})$ so that

$$AC < r \quad \text{and} \quad BC < r. \tag{5-2}$$

Let $D \in \overline{AB}$ with A—D—B. We must show that $D \in \text{int}(\mathscr{C})$. There are two cases to consider.

First suppose that $C \in \overleftrightarrow{AB}$. See Figure 6-32. Let f be a ruler for $\overleftrightarrow{AB}$ with origin at C. The points of $\overleftrightarrow{AB} \cap \text{int}(\mathscr{C})$ have coordinates between $-r$ and r. Now the coordinate of D is between that of A and B. By Inequalities (5-2), the coordinates of A and B are between $-r$ and r. Hence, so is the coordinate of D and $D \in \overleftrightarrow{AB} \cap \text{int}(\mathscr{C})$. Thus $D \in \text{int}(\mathscr{C})$.

For the second case, suppose that $C \notin \overleftrightarrow{AB}$ so that we have a triangle, $\triangle ABC$. See Figure 6-33. We must show that $D \in \text{int}(\mathscr{C})$. By Theorem 6.3.10

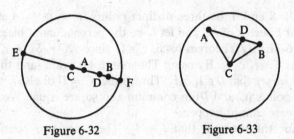

Figure 6-32   Figure 6-33

either $\overline{CD} < \overline{CA}$ or $\overline{CD} < \overline{CB}$. Since both $\overline{CA}$ and $\overline{CB}$ have length less than r, so does $\overline{CD}$. Thus $D \in \text{int}(\mathscr{C})$ and $\overline{AB} \subset \text{int}(\mathscr{C})$. Hence $\text{int}(\mathscr{C})$ is convex. $\square$

According to Problem A4, in a neutral geometry the intersection of a line and a circle consists of either zero, one, or two points. Since the first case is less interesting, we only name the latter two.

Definition. In a metric geometry, a line l is a **tangent** to the circle $\mathscr{C}$ if $l \cap \mathscr{C}$ contains exactly one point (which is called the **point of tangency**). l is called a **secant** of the circle $\mathscr{C}$ if $l \cap \mathscr{C}$ has exactly two points.

In a general metric geometry there are strange situations which can occur with respect to tangents (see Problems A5 and A6). However, in the context of a neutral geometry the situation is more as expected. In Corollary 6.5.7 we shall show that in a neutral geometry every point of a circle is the point of tangency of a unique tangent line. In Theorem 6.5.10, we shall see that in a neutral geometry if P is in the exterior of $\mathscr{C}$ then there are exactly two lines through P which are tangent to $\mathscr{C}$.

Theorem 6.5.6. *In a neutral geometry, let $\mathscr{C}$ be a circle with center C and let $Q \in \mathscr{C}$. If t is a line through Q, then t is tangent to $\mathscr{C}$ if and only if t is perpendicular to the radius segment $\overline{CQ}$.*

PROOF. First assume that t is perpendicular to $\overline{CQ}$ and that P is any point on t with $P \neq Q$. $\triangle PQC$ is a right triangle with hypotenuse $\overline{PC}$. See Figure 6-34.

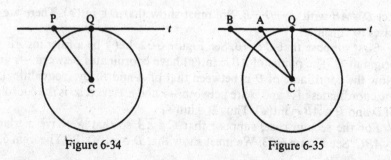

Figure 6-34   Figure 6-35

Thus $\overline{PC} > \overline{QC}$, which means $P \notin \mathscr{C}$ (in fact $P \in \mathrm{ext}(\mathscr{C})$). Hence t cannot intersect $\mathscr{C}$ at a second point and so t is tangent to $\mathscr{C}$.

Conversely, suppose that t is tangent to $\mathscr{C}$ at Q. Let A be the foot of the perpendicular from C to t. If $A \neq Q$ there exists a unique point $B \in t$ with Q—A—B and $\overline{QA} \simeq \overline{AB}$. See Figure 6-35. Then $\triangle CAB \simeq \triangle CAQ$ by SAS so that $\overline{CB} \simeq \overline{CQ}$. Thus $B \in \mathscr{C}$ and t intersects $\mathscr{C}$ at two distinct points, which contradicts the hypothesis that t is tangent. Hence $A = Q$ and t is perpendicular to $\overline{CA} = \overline{CQ}$. $\square$

Corollary 6.5.7 (Existence and Uniqueness of Tangents). *In a neutral geometry, if $\mathscr{C}$ is a circle and $Q \in \mathscr{C}$ then there is a unique line t which is tangent to $\mathscr{C}$ and whose point of tangency is Q.*

PROOF. Let C be the center of $\mathscr{C}$. Since there is a unique perpendicular to $\overline{CQ}$ at the point Q (Corollary 6.3.4), the result follows from Theorem 6.5.6.
                                                                             $\square$

In the more general setting of a metric geometry, a given point Q may not be the point of tangency for any line. There are also instances in which there are many lines which are tangent to $\mathscr{C}$ at Q. These pathologies are explored in the problems at the end of this section.

The next result tells us that under certain circumstances a line must intersect a circle. This will help us show that a line which intersects the interior of a circle is a secant and that from a given external point there are two tangent lines.

In order to prove this result we will use a technique, which may be new to the reader, called a "continuity argument." Recall that

$$[a, b] = \{x \in \mathbb{R} \mid a \leq x \leq b\}$$

and the following ideas from calculus.

Definition. $h : \mathbb{R} \to \mathbb{R}$ is **continuous** at $t_0 \in \mathbb{R}$ if for every $\varepsilon > 0$ there is a $\delta > 0$ such that

$$|h(t) - h(t_0)| < \varepsilon \quad \text{if } |t - t_0| < \delta.$$

(Thus if t is "near" t_0 then $h(t)$ is "near" $h(t_0)$)

Intermediate Value Theorem. *If $h : [a, b] \to \mathbb{R}$ is continuous at every $t_0 \in [a, b]$ and if y is a number between $h(a)$ and $h(b)$ then there is a point $s \in [a, b]$ with $h(s) = y$.*

The Intermediate Value Theorem says that a continuous function takes on all of the values between its values at the two endpoints. To successfully apply a continuity argument, one must come up with an appropriate function, prove that it is continuous, and apply the Intermediate Value Theorem. This technique is used in the next theorem. We used this method to show that $\mathscr{H}$ satisfies PSA in Chapter 4.

Theorem 6.5.8. *Let r be a positive real number and let A, B, C be points in a neutral geometry such that $AC < r$ and $\overline{AB} \perp \overline{AC}$. Then there is a point $D \in \overline{AB}$ with $CD = r$.*

PROOF. Let E be a point on $\overrightarrow{AB}$ with $d(A, E) = r$ as in Figure 6-36. Since the hypotenuse of $\triangle ACE$ is $\overline{CE}$ and $AC < r$

$$d(C, E) > r \quad \text{and} \quad d(A, C) < r. \tag{5-3}$$

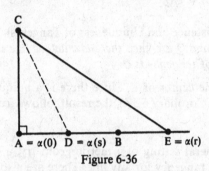

Figure 6-36

An appropriate function would seem to be $g : \overline{AE} \to R$ given by

$$g(P) = d(P, C).$$

However, in order to apply a continuity argument we need to use the real numbers as domain. Fortunately we can transfer g back to $\mathbb{R}$ by the use of a ruler.

Let $f : \overrightarrow{AB} \to \mathbb{R}$ be the ruler with origin at A and with E positive. Let $\alpha : \mathbb{R} \to \overrightarrow{AB}$ be the inverse of f. Then $\alpha(0) = A$ and $\alpha(r) = E$. Let $h(t) = g(\alpha(t)) = d(\alpha(t), C)$ for $t \in [0, r]$.

We shall first show that h is continuous at any point $t_0 \in [0, r]$. Let $\varepsilon > 0$ be given. We must find a $\delta > 0$ such that if $|t - t_0| < \delta$ then $|h(t) - h(t_0)| < \varepsilon$. (We will end up with $\delta = \varepsilon$.)

First note that the Triangle Inequality

$$d(C, \alpha(t_0)) + d(\alpha(t), \alpha(t_0)) \geq d(C, \alpha(t))$$

implies

$$d(\alpha(t), \alpha(t_0)) \geq d(C, \alpha(t)) - d(C, \alpha(t_0)). \tag{5-4}$$

Likewise $d(C, \alpha(t)) + d(\alpha(t), \alpha(t_0)) \geq d(C, \alpha(t_0))$ implies

$$d(C, \alpha(t)) - d(C, \alpha(t_0)) \geq -d(\alpha(t), \alpha(t_0)). \tag{5-5}$$

We may combine Inequalities (5-4) and (5-5) to obtain

$$|d(C, \alpha(t)) - d(C, \alpha(t_0))| \leq d(\alpha(t), \alpha(t_0)).$$

Because f is a ruler, $d(\alpha(t), \alpha(t_0)) = |f(\alpha(t)) - f(\alpha(t_0))| = |t - t_0|$ and

$$|h(t) - h(t_0)| \leq |t - t_0|. \tag{5-6}$$

We now let $\delta = \varepsilon$. If $|t - t_0| < \delta$ then by Inequality (5-6) $|h(t) - h(t_0)| < \varepsilon$ and h is a continuous function.

From Inequality (5-3) we have

$$h(0) = d(C, A) < r \quad \text{and} \quad h(r) = d(C, E) > r.$$

Hence by the Intermediate Value Theorem with $y = r$ we conclude there is an $s \in [0, r]$ with $h(s) = r$. We may thus let $D = \alpha(s)$ so that $D \in \overrightarrow{AB}$ and

$$CD = d(C, D) = d(C, \alpha(s)) = h(s) = r. \qquad \square$$

Note that this theorem says that $\mathscr{C}_r(C)$ intersects $\overrightarrow{AB}$ at a point D if $\overrightarrow{AC} \perp \overrightarrow{AB}$ and $d(A, C) < r$. It is possible to show that the point D found above is unique by using Theorem 6.3.10. However, this fact also follows from the next result.

Theorem 6.5.9 (Line-Circle Theorem). *In a neutral geometry, if a line l intersects the interior of a circle $\mathscr{C}$, then l is a secant.*

PROOF. Suppose that $\mathscr{C} = \mathscr{C}_r(C)$. If $C \in l$ then there are exactly two points on l whose distance from C is r by the Segment Construction Axiom. Hence if $C \in l$, l is a secant.

Now suppose that $C \notin l$. By hypothesis there is a point $E \in l \cap \text{int}(\mathscr{C})$, so that $EC < r$. Let A be the foot of the perpendicular from C to l. See Figure 6-37. If $A = E$ then $CA = CE < r$. If $A \neq E$ then $\triangle CAE$ has a right angle at A so that $CA < CE < r$. Hence $CA < r$ in either case.

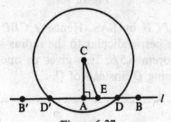

Figure 6-37

Choose $B, B' \in l$ with $B{-}A{-}B'$. By Theorem 6.5.8 there are points $D \in \overrightarrow{AB}$ and $D' \in \overrightarrow{AB'}$ with $CD = r$ and $CD' = r$. Thus $D \in \mathscr{C}$ and $D' \in \mathscr{C}$. Hence $l \cap \mathscr{C}$ has at least two points. By Problem A4, $l \cap \mathscr{C}$ has exactly two points and so l is a secant. $\qquad \square$

A standard proof of the Line-Circle Theorem uses the Pythagorean Theorem (Chapter 16 of Moise [1990]). Such a proof is only valid for the Euclidean Plane. The proof given above is valid in any neutral geometry. Thus the Line-Circle Theorem is really a neutral (rather than Euclidean) theorem.

The next theorem discusses how many tangent lines can be drawn to a given circle $\mathscr{C}$ from an external point.

Theorem 6.5.10 (External Tangent Theorem). *In a neutral geometry, if $\mathscr{C}$ is a circle and $P \in \text{ext}(\mathscr{C})$, then there are exactly two lines through P tangent to $\mathscr{C}$.*

PROOF. We shall prove the existence of two tangent lines. In Problem A9 you will show there are no more. Since $P \in \text{ext}(\mathscr{C})$, $CP > r$ and there exists a unique point A with $C{-}A{-}P$ and $CA = r$. Let l be the perpendicular to $\overleftrightarrow{CP}$ at A. $CA = r < CP$ so that A is interior to the circle $\mathscr{C}'$ with center C and radius CP. By Theorem 6.5.9 l intersects $\mathscr{C}'$ at points Q and Q'. Since $CQ = CP > r$, there exists a unique point B such that $C{-}B{-}Q$ and $CB = r$. Note $B \in \mathscr{C}$. See Figure 6-38.

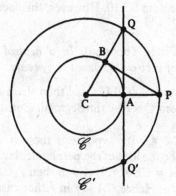

Figure 6-38

Now $\triangle QCA \simeq \triangle PCB$ by SAS. Hence $\angle CBP$ is a right angle since $\angle CAQ$ is. Thus $\overleftrightarrow{BP}$ is perpendicular to the radius segment $\overline{CB}$ and is thus tangent to $\mathscr{C}$ by Theorem 6.5.6. This gives us one tangent through P. A second is found by using Q' instead of Q. $\qquad\square$

PROBLEM SET 6.5

Part A.

1. Consider $\{\mathbb{R}^2, \mathscr{L}_E\}$ with the max distance d_S (Problem B18 of Section 2.2). Sketch the circle $\mathscr{C}_1((0, 0))$.

2. In the Poincaré Plane show that $\{(x, y)|x^2 + (y - 5)^2 = 9\}$ is the circle with center $(0, 4)$ and radius $\ln 2$.

3. If $\overline{AB}$ is a chord of a circle in a neutral geometry but is not a diameter, prove that the line through the midpoint of $\overline{AB}$ and the center of the circle is perpendicular to $\overline{AB}$.

4. Prove that a line in a neutral geometry intersects a circle at most twice. (Hint: Theorem 6.3.10.)

5. In the Taxicab Plane prove that for the circle $\mathscr{C} = \mathscr{C}_1((0, 0))$:
   a. There are exactly four points at which a tangent to $\mathscr{C}$ exists.
   b. At each point in part (a) there are infinitely many tangent lines.

6. For the circle of Problem A1, how many points have tangent lines?

7. Use Theorem 6.3.10 to prove that the point D of Theorem 6.5.8 is unique.

8. Prove that for the max plane (Problem A1) there are two distinct circles with at least 3 common points. What is the maximum number of common points which distinct circles may have?

9. Complete the proof of Theorem 6.5.10.

10. In a neutral geometry, if $\mathscr{C}$ is a circle with $A \in \text{int}(\mathscr{C})$ and $B \in \text{ext}(\mathscr{C})$, prove that $\overline{AB} \cap \mathscr{C} \neq \varnothing$.

11. Suppose that the perpendicular bisectors of $\overline{AB}$ and $\overline{BC}$ intersect at P in a neutral geometry. Prove there is a circle $\mathscr{C}$ with $A, B, C \in \mathscr{C}$.

Part B. "Prove" may mean "find a counterexample".

12. In a neutral geometry, prove that the union of a circle and its interior is convex.

13. In a neutral geometry, prove that the union of a circle and its exterior is convex.

14. In a neutral geometry, if $\overline{AB}$ is a chord in a circle of radius r, prove that $AB \leq 2r$. Furthermore, prove that equality holds if and only if $\overline{AB}$ is a diameter.

15. Let $\overline{AB}$ and $\overline{DE}$ be chords of a circle with center C in a neutral geometry. Prove that $\overline{AB} \simeq \overline{DE}$ if and only if $d(C, \overleftrightarrow{AB}) = d(C, \overleftrightarrow{DE})$.

16. Prove that if $\mathscr{C}$ is a cricle in a neutral geometry, l is a line, and if $l \cap \mathscr{C} \neq \varnothing$ while $l \cap \text{int}(\mathscr{C}) = \varnothing$, then l is tangent to $\mathscr{C}$.

17. In the Moulton Plane
   a. Show that $\{(x, y) | x^2 + y^2 = 1\}$ and $\{(x, y) | (x - 2)^2 + y^2 = 1\}$ are circles.
   b. Show that $\{(x, y) | (x + 1)^2 + y^2 = 4\}$ is not a circle.
   c. Carefully sketch the circle $\mathscr{C}_5((-1, 0))$.

18. For the model of Problem B20 of Section 4.1, carefully sketch the circles $\mathscr{C}_{1/2}((0, 2))$ and $\mathscr{C}_{7/2}((0, 2))$.

19. We say that a line l is a **subtangent** of the circle $\mathscr{C}$ if $l \cap \mathscr{C} \neq \varnothing$ but $l \cap \text{int}(\mathscr{C}) = \varnothing$. Note that a tangent is a subtangent. Show that for the Taxicab Plane
   a. At every point of $\mathscr{C}$ there is a subtangent.
   b. Through every point $P \in \text{ext}(\mathscr{C})$ there are exactly two subtangents.

20. Prove the results of Problem B19 do not generalize to an arbitrary metric geometry by considering Problem B18.

21. In a neutral geometry, prove that a diameter of a circle $\mathscr{C}$ bisects a chord of $\mathscr{C}$ if and only if the diameter is perpendicular to the chord.

22. In the Missing Strip Plane (see Section 4.3) sketch the circle $\mathscr{C}_1((1, 0))$.

23. Find an example of a circle $\mathscr{C}$ and a point $Q \in \text{ext}(\mathscr{C})$ such that there are more than two lines through Q tangent to $\mathscr{C}$.

24. In a neutral geometry, if $A, B \in \mathscr{C}_r(C)$, prove that $AB \leq 2r$. If "neutral" is omitted from the hypothesis, is the result still true?

25. Find an example of a circle $\mathscr{C}$ and a point $P \in \text{int}(\mathscr{C})$ and a line l containing P such that l is tangent to $\mathscr{C}$.

26. Find three non-collinear points in $\mathscr{H}$ which do not all lie on the same circle.

6.6 The Two Circle Theorem

By Theorem 6.5.3 we know that two distinct circles in a neutral geometry intersect in at most two points. The main point of this section is to give a condition for when two circles intersect in exactly two points. This result, called the Two Circle Theorem, will follow directly from a converse of the Triangle Inequality.

Our first result is called the Sloping Ladder Theorem. It tells us that if a ladder leans against a wall and the bottom is pulled out from the wall, then the top slides down.

Theorem 6.6.1 (Sloping Ladder Theorem). *In a neutral geometry with right triangles $\triangle ABC$ and $\triangle DEF$ whose right angles are at C and F, if $\overline{AB} \simeq \overline{DE}$ and $\overline{AC} > \overline{DF}$, then $\overline{BC} < \overline{EF}$.*

PROOF. Let $G \in \overrightarrow{CA}$ so that $\overline{CG} \simeq \overline{DF}$. Let $H \in \overrightarrow{CB}$ so that $\overline{CH} \simeq \overline{FE}$. Then $\triangle GCH \simeq \triangle DFE$ by SAS and so $\overline{GH} \simeq \overline{DE} \simeq \overline{AB}$. Since $\overline{EF} \simeq \overline{CH}$, if we can show that $C—B—H$, we are finished. We shall do this by showing that $H = B$ and $C—H—B$ both lead to a contradiction.

Assume $H = B$. Then $\triangle ABC \simeq \triangle GHC$ by HL and $\overline{AC} \simeq \overline{GC} \simeq \overline{DF}$ which contradicts the hypothesis $\overline{AC} > \overline{DF}$. Hence $H \neq B$. Assume $C—H—B$ as in Figure 6-39. Then $\overline{GH} < \overline{GB} < \overline{AB}$, which contradicts $\overline{GH} \simeq \overline{DE} \simeq \overline{AB}$. Hence we must have $C—B—H$ and $\overline{EF} \simeq \overline{HC} > \overline{BC}$ since $HC = HB + BC$. $\square$

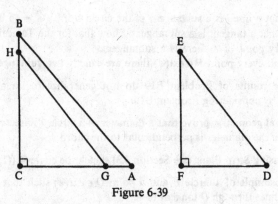

Figure 6-39

The proof of the next result is left as Problem A1.

Theorem 6.6.2. *Let $\overline{AB}$ and $\overline{DE}$ be two chords of the circle $\mathscr{C} = \mathscr{C}_r(C)$ in a neutral geometry. If $\overline{AB}$ and $\overline{DE}$ are both perpendicular to a diameter of $\mathscr{C}$ at points P and Q with $C\text{---}P\text{---}Q$, then $DQ < AP < r$.*

We shall next prove the converse of the Triangle Inequality. Recall that theorem stated that in a neutral geometry the sum of the lengths of any two sides of a triangle is greater than the length of the third.

Theorem 6.6.3 (Triangle Construction Theorem). *Let $\{\mathscr{S}, \mathscr{L}, d, m\}$ be a neutral geometry and let a, b, c be three positive numbers such that the sum of any two is greater than the third. Then there is a triangle in $\mathscr{S}$ whose sides have length a, b, and c.*

PROOF. We may rename the numbers in such a way that c is largest. Let A and B be two points with $AB = c$ and let f be a coordinate system for $\overleftrightarrow{AB}$ with $f(A) = 0$ and $f(B) = c$. Let $\mathscr{C}_1 = \mathscr{C}_b(A)$ and $\mathscr{C}_2 = \mathscr{C}_a(B)$. We will show that $\mathscr{C}_1 \cap \mathscr{C}_2 \neq \varnothing$. If there is a point $C \in \mathscr{C}_1 \cap \mathscr{C}_2$ then since $C \in \mathscr{C}_1 = \mathscr{C}_b(A)$, $AC = b$. Since $C \in \mathscr{C}_2 = \mathscr{C}_a(B)$ then $BC = a$. Thus $\triangle ABC$ will be exactly what we need. (See Figure 6-40.) We proceed by a continuity argument.

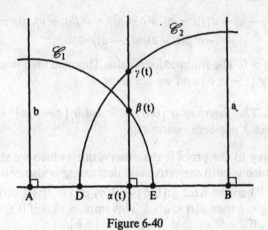

Figure 6-40

Now $\mathscr{C}_1$ intersects $\overleftrightarrow{AB}$ at a point E with $f(E) = b$ while $\mathscr{C}_2$ intersects $\overleftrightarrow{AB}$ at a point D with $f(D) = c - a \geq 0$. Thus

$$0 \leq c - a < b \leq c.$$

We shall let α be the inverse of f and choose a half plane H_1 determined by $\overleftrightarrow{AB}$. Note that

$$\alpha(0) = A, \qquad \alpha(c - a) = D, \qquad \alpha(b) = E \quad \text{and} \quad \alpha(c) = B.$$

For each $t \in [0, b]$ let $\beta(t)$ be defined as the point in H_1 which is on both $\mathscr{C}_1$ and the line through $\alpha(t)$ which is perpendicular to $\overleftrightarrow{AB}$. See Figure 6-40.

For each $t \in [c - a, c]$ let $\gamma(t)$ be defined as the point in H_1 which is on both $\mathscr{C}_2$ and the line through $\alpha(t)$ which is perpendicular to $\overleftrightarrow{AB}$.

Define

$$k(t) = d(\gamma(t), \alpha(t)) - d(\beta(t), \alpha(t)).$$

The key observation is that *if* there is a number s for which $k(s) = 0$ then

$$d(\gamma(s), \alpha(s)) = d(\beta(s), \alpha(s)).$$

Since $\gamma(s), \alpha(s), \beta(s)$ are collinear and $\gamma(s)$ and $\beta(s)$ are on the same side of $\overleftrightarrow{AB}$ then $\gamma(s) = \beta(s)$. Thus $\gamma(s) = \beta(s) \in \mathscr{C}_1 \cap \mathscr{C}_2$ and the proof would be finished.

We will prove in Lemma 6.6.4 below that both $g : [0, b] \to \mathbb{R}$ given by

$$g(t) = d(\gamma(t), \alpha(t))$$

and $h : [c - a, c] \to \mathbb{R}$ given by

$$h(t) = d(\beta(t), \alpha(t))$$

are continuous. Thus

$$k(t) = g(t) - h(t)$$

is continuous on $[c - a, b] = [0, b] \cap [c - a, c]$. However $\gamma(c - a) = D = \alpha(c - a)$ so that

$$\begin{aligned} k(c - a) &= d(\gamma(c - a), \alpha(c - a)) - d(\beta(c - a), \alpha(c - a)) \\ &= -d(\beta(c - a), \alpha(c - a)) < 0. \end{aligned}$$

Similarly $k(b) > 0$. The Intermediate Value Theorem then implies that $k(s) = 0$ for some $s \in [c - a, b]$ and we are done. $\qquad\qquad\qquad\square$

Lemma 6.6.4. *The functions $g : [0, b] \to \mathbb{R}$ and $h : [c - a, c] \to \mathbb{R}$ in the proof of Theorem 6.6.3 are both continuous.*

PROOF. The key to the proof is the observation (which we shall prove) that a function which is both surjective and decreasing is also continuous.

Step 1. By Theorem 6.6.2 $g : [0, b] \to \mathbb{R}$ by $g(t) = d(\gamma(t), \alpha(t))$ is strictly decreasing: if $t_0 < t$ then $g(t) < g(t_0)$. This implies that if $0 \leq t \leq b$ then $0 = g(b) \leq g(t) \leq g(0) = b$ so that $\operatorname{image}(g) \subset [0, b]$.

Step 2. We claim $\operatorname{image}(g) = [0, b]$. If $0 < r < b$ let P be the point in H_1 which is on the perpendicular to $\overleftrightarrow{AB}$ at A so that $AP = r$. Then by Theorem 6.5.9 there is a point $Q \in \overleftrightarrow{AB}$ with $PQ = b$. (See Figure 6-41.) Let R be the point in H_1 where $\mathscr{C}_1$ intersects the perpendicular to $\overleftrightarrow{AB}$ at Q. Then $\triangle PAQ \simeq \triangle RQA$ by HL so that $RQ = PA = r$. Thus if $f(Q) = s$, then $Q = \alpha(s)$ and $R = \gamma(s)$ so that $g(s) = r$. Hence $r \in \operatorname{image}(g)$ and $[0, b] \subset \operatorname{image}(g)$. Thus $\operatorname{image}(g) = [0, b]$.

Step 3. We show that g is continuous at t_0 if $0 < t_0 < b$. Let $\varepsilon > 0$. Then there are numbers r_1 and r_2 in $[0, b]$ with

$$g(t_0) - \varepsilon < r_1 < g(t_0) < r_2 < g(t_0) + \varepsilon.$$

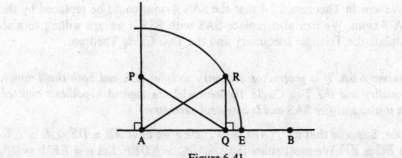

Figure 6-41

By Step 2 there are numbers $s_1 > t_0$ and $s_2 < t_0$ in $[0, b]$ with $g(s_i) = r_i$. If δ is the smaller of the two numbers $s_1 - t_0$ and $t_0 - s_2$ then whenever $|t - t_0| < \delta$ we have

$$s_2 - t_0 < t - t_0 < s_1 - t_0$$

or

$$s_2 < t < s_1$$

and since g is decreasing

$$g(t_0) - \varepsilon < r_1 = g(s_1) < g(t) < g(s_2) = r_2 < g(t_0) + \varepsilon$$

so that

$$|g(t) - g(t_0)| < \varepsilon.$$

Hence g is continuous at t_0. A slight modification of the proof shows that g is continuous at 0 and b also.

The proof that h is continuous is similar except in this case $h : [c - a, c] \to [0, a]$ is surjective and strictly increasing. $\square$

By this point you probably have noticed that we have used continuity (and the Intermediate Value Theorem) quite a bit. This should not be too surprising. We mentioned earlier that the ruler postulate was very powerful. By parametrizing lines we are quite naturally led to real valued functions. The Plane Separation Axiom is really a kind of continuity axiom—note that in two of the examples we gave where PSA did not hold (Missing Strip Plane and the geometry of Problem B20 in Section 4.1), the rulers look like they have "holes" in them.

The Two Circle Theorem can now be proven directly from Theorem 6.6.3. This is left as Problem A2.

Theorem 6.6.5 (Two Circle Theorem). *In a neutral geometry, if $\mathscr{C}_1 = \mathscr{C}_b(A)$, $\mathscr{C}_2 = \mathscr{C}_a(B)$, $AB = c$, and if each of a, b, c is less than the sum of the other two, then $\mathscr{C}_1$ and $\mathscr{C}_2$ intersect in exactly two points, and these points are on opposite sides of $\overleftrightarrow{AB}$.*

In Problem A3, you will give another necessary and sufficient condition for two circles to intersect in two points.

We saw in Theorem 6.2.4 that the SAS Axiom could be replaced by the ASA Axiom. We can also replace SAS with SSS if we are willing to also postulate the Triangle Inequality and the Two Circle Theorem.

Theorem 6.6.6. *If a protractor geometry satisfies SSS and both the Triangle Inequality and the Two Circle Theorem with the neutral hypothesis omitted, then it also satisfies SAS and is a neutral geometry.*

PROOF. Suppose that for $\triangle ABC$ and $\triangle DEF$ we have $\overline{AB} \simeq \overline{DE}$, $\angle B \simeq \angle E$, and $\overline{BC} \simeq \overline{EF}$. We must show that $\triangle ABC \simeq \triangle DEF$. Let $a = EF$, $b = DF$, and $c = DE$.

The Triangle Inequality guarantees that each of a, b, c is less than the sum of the other two. Hence $\mathscr{C}_c(B)$ intersects $\mathscr{C}_b(C)$ in a unique point G on the same side of $\overleftrightarrow{BC}$ as A by the Two Circle Theorem. See Figure 6-42. Then $\triangle GBC \simeq \triangle DEF$ by SSS.

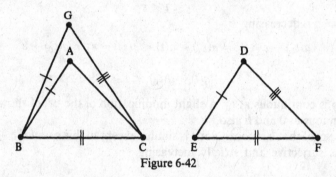

Figure 6-42

Hence $\angle GBC \simeq \angle DEF \simeq \angle ABC$. Since A and G are on the same side of $\overleftrightarrow{BC}$, $\overrightarrow{BA} = \overrightarrow{BG}$ by the Angle Construction Theorem. Then $\overline{GB} \simeq \overline{DE} \simeq \overline{AB}$ implies $G = A$. Hence $\triangle ABC = \triangle GBC \simeq \triangle DEF$.   □

Theorem 6.6.6 says that a protractor geometry with SSS, the Triangle Inequality and the Two Circle Theorem is a neutral geometry. It may be that not both the Triangle Inequality and the Two Circle Theorem are needed. (Of course, they are needed in our proof.) We do not know of any protractor geometry which satisfies SSS for which either the Two Circle Theorem or the Triangle Inequality fail.

PROBLEM SET 6.6

Part A.

1. Prove Theorem 6.6.2.

2. Prove Theorem 6.6.5.

3. Prove that in a neutral geometry, two circles $\mathscr{C}_1$ and $\mathscr{C}_2$ intersect in exactly two points if and only if $\mathscr{C}_1 \cap \text{int}(\mathscr{C}_2) \neq \varnothing$ and $\mathscr{C}_1 \cap \text{ext}(\mathscr{C}_2) \neq \varnothing$.

4. Prove that in a neutral geometry a circle of radius r has a chord of length c if and only if $0 < c \leq 2r$.

5. Let $a = 2$, $b = 3$ and $c = 4$ and construct a triangle $\triangle ABC$ in the Euclidean Plane with $AB = 4$, $AC = 3$ and $BC = 2$. (Hint: Imitate the proof of Theorem 6.6.3. Start with $A = (0, 0)$ and $B = (4, 0)$.)

6. Let $a = 2$, $b = 3$, and $c = 4$ and carry through the steps of Theorem 6.3 to construct a triangle $\triangle ABC$ in the Taxicab Plane with $AB = 4$, $AC = 3$ and $BC = 2$. (Remark: This is not a neutral geometry so we don't know, *a priori*, that the construction will work.)

7. In a neutral geometry prove that for any $s > 0$ there is an equilateral triangle each of whose sides has length s.

Part B. "Prove" may mean "find a counterexample".

8. Prove that in a neutral geometry, if two circles $\mathscr{C}_1$ and $\mathscr{C}_2$ have exactly one point Q in common then they have a common tangent line. (That is, there is a line l which is tangent to both $\mathscr{C}_1$ and $\mathscr{C}_2$ at Q.)

9. Let $\mathbb{Q}^2 = \{(x, y) \in \mathbb{R}^2 \mid x \text{ and } y \text{ are both rational}\}$. $\{\mathbb{Q}^2, \mathscr{L}_E^*, d_E\}$ is an incidence geometry formed by using Euclidean lines with rational slope intersected with $\mathbb{Q}^2$ and Euclidean distance.
   a. Prove that $\{\mathbb{Q}^2, \mathscr{L}_E^*, d_E\}$ is an incidence geometry. (Hint: Problem B21 of Problem Set 2.1.) It is *not* a metric geometry.
   b. Show that the conclusions of Theorems 6.6.3 and 6.6.5 are false for this geometry. (This points out dramatically the need for rulers in these theorems. There are many, many holes in $\mathbb{Q}^2$ and so any kind of continuity argument is out.)

10. Let $a = \ln 2$, $b = \ln 3$, and $c = \ln 4$ and carry out the steps of Theorem 6.6.3 to construct a triangle in $\mathscr{H}$ with $AB = \ln 4$, $AC = \ln 3$, and $BC = \ln 2$. Start with $B = (0, 1)$ and $C = (0, 2)$. Find the exact coordinates of A.

Part C. Expository exercises.

11. We now have at least three ways of defining a neutral geometry by imposing additional axioms on a protractor geometry: (1) SAS (our selection), (2) ASA, or (3) SSS, the Triangle Inequality, and the Two Circle Theorem. Discuss the relative merits of each of these choices. Would you choose (2) or (3)? Why?

6.7 The Synthetic Approach

As we remarked in Section 2.2, there are two main approaches to developing geometry: the metric approach of Birkhoff which we are using (and which is also followed in Martin [1975], Moise [1990], and Prenowitz-Jordan

[1965]) and the axiomatic or synthetic approach first firmly established by Hilbert (and which may be found in Greenberg [1980] and Borsuk-Szmielew [1960]). In this section we shall present a brief overview of how the synthetic approach would be used to obtain the results we have developed so far. We shall not present any proofs.

In Chapter 1 we said the choice of axioms is guided by three principles: correspondence to an intuitive picture, richness of the theory (i.e., many interesting theorems can be proven), and *consistency*. Consistency means that the axiom system will not lead to a contradiction. Many authors want two other properties for their axiom system: *minimality* and *categoricalness*.

Minimality really consists of two notions. The first is that the fewest undefined terms and axioms are used. For example, it is more economical to assume SAS than it is to assume SSS, the Triangle Inequality and the Two Circle Theorem. The other notion included in minimality is the *independence* of the axiom system. This means that no axiom can be proved from any of the others. There is no need to assume both SAS and ASA, for example.

Categoricalness means that there are sufficiently many axioms so that all models of the axiom system are equivalent under some natural sense of equivalence. For example, if we add to our current axiom system the Euclidean parallel postulate then we will have a categorical axiom system for Euclidean geometry—the Euclidean Plane that we have developed is, up to an isometry, the only model. We prove this result in Theorem 11.1.20.

Many axiom systems for geometry have been proposed since Euclid. If you are interested in such foundational questions, see Chapter 15 of Martin [1975], Chapter 3 of Greenberg [1980], or Borsuk-Szmielew [1960]. The axioms we give below are essentially those of Hilbert but have been modified somewhat.

In the synthetic approach, congruence axioms replace distance and angular measure. The set of axioms for neutral geometry are as follows.

A neutral geometry consists of a set $\mathscr{S}$ whose elements are called points, a collection $\mathscr{L}$ of subsets of $\mathscr{S}$ called lines, a ternary relation ( )—( )—( ) ("between"), and a binary relation $\simeq$ satisfying the following axioms:

Incidence:
(1) If A, B are distinct points in $\mathscr{S}$ then there is a unique line $l \in \mathscr{L}$ with $A, B \in l$.
(2) Every line has at least two points.
(3) There is a set of three noncollinear points.

Betweenness:
(4) If A—B—C then A, B, C are distinct collinear points and C—B—A.
(5) If A, B, C are distinct collinear points then either A—B—C or B—C—A or C—A—B.
(6) If $A \neq C$ then there are points B and D such that A—B—C and A—C—D.

Separation:
- (7) For each line l there are two subsets H_1 and H_2 of $\mathscr{S}$ such that
  - (a) $H_1 \cap H_2 = \varnothing$, $\mathscr{S} - l = H_1 \cup H_2$;
  - (b) H_1 and H_2 are convex;
  - (c) If $A \in H_1$ and $B \in H_2$ then $\overline{AB} \cap l \neq \varnothing$.

Congruence:
- (8) $\simeq$ is an equivalence relation on the set of segments.
- (9) Given $\overline{PQ}$ and $\overrightarrow{AB}$ there exists a unique point $C \in \overrightarrow{AB}$ such that $\overline{AC} \simeq \overline{PQ}$.
- (10) If A—B—C, D—E—F, $\overline{AB} \simeq \overline{DE}$, and $\overline{BC} \simeq \overline{EF}$, then $\overline{AC} \simeq \overline{DF}$.
- (11) $\simeq$ is an equivalence relation on the set of angles.
- (12) Given $\angle ABC$ and a ray $\overrightarrow{ED}$ which lies in the edge of the half plane H, then there is a unique ray $\overrightarrow{EF}$ with $F \in H$ and $\angle DEF \simeq \angle ABC$.
- (13) If $\overline{AB} \simeq \overline{DE}$, $\angle ABC \simeq \angle DEF$, $\overline{BC} \simeq \overline{EF}$, then $\triangle ABC \simeq \triangle DEF$. (Triangle congruence as before.)

Note that we never talk about angle measure (only angle congruence) or the distance between points (only segment congruence). Thus in a synthetic approach, we cannot talk about an angle whose measure is 35 or 90 or about points being 2 units apart.

However, with these axioms it is possible to recover all the work we have done up through Theorem 6.5.7 except those results that explicitly deal with distance or angle measure. The order that results are derived changes somewhat. For example, existence of midpoints comes after SAS. Some definitions and statements of theorems must be changed to avoid mention of distance or angle measure. For example, in the synthetic approach, $\angle ABC$ is called a right angle if there is a point D with A—B—D and $\angle ABC \simeq \angle CBD$. One shows that if an angle is congruent to a right angle then it is a right angle, and that all right angles are congruent.

As a second example, inequality of segments and angles can be defined by congruence as in Theorems 6.3.1 and 6.3.2. The Triangle Inequality is stated in synthetic language as: For any $\triangle ABC$ there is a point D such that A—B—D, $\overline{BD} \simeq \overline{AC}$, and $\overline{BC} < \overline{AD}$.

The remaining theorems in Sections 6.5 and 6.6 relied heavily on the ruler postulate. Indeed, many of the proofs used a continuity agreement. In the synthetic approach two axioms are added to take the place of the continuity part of the ruler axiom. These are axioms (14) and (15) below. (Actually axiom (14) suffices to derive the results of Sections 6.5 and 6.6.) These two continuity axioms can be replaced by a single axiom due to Dedekind which is very topological in nature.

Continuity:
- (14) If $\mathscr{C}_1$ and $\mathscr{C}_2$ are two circles with $\mathscr{C}_1 \cap \operatorname{int}(\mathscr{C}_2) \neq \varnothing$ and $\mathscr{C}_1 \cap \operatorname{ext}(\mathscr{C}_2) \neq \varnothing$ then $\mathscr{C}_1 \cap \mathscr{C}_2$ contains exactly two points.
- (15) Given $\overline{AB}$ and $\overline{CD}$ there are points $P_0, P_1, \ldots, P_n \in \overrightarrow{AB}$ with $P_0 = A$, P_0—P_1—P_2—$\cdots$—P_n, $\overline{P_{i-1}P_i} \simeq \overline{CD}$ for all i and $B \in \overline{P_0 P_n}$.

It can be shown (see Borsuk-Szmielew) that if (14) and (15) are replaced by Dedekind's axiom and a suitable choice of parallel axioms is made (see Chapter 7) then the system of axioms is categorical so that there is essentially only one model of Euclidean geometry and one of Hyperbolic geometry, namely our models $\mathscr{E}$ and $\mathscr{H}$.

We have followed Birkhoff's metric approach because we feel that the metric axioms of rulers and protractors are more intuitive, natural, and easier to follow. Of course, as either the metric or the synthetic approach yield the same body of theorems, *chacun à son gout*. A synthetic development can be found in Chapter 8 of Moise [1990] or Chapter 3 of Greenberg [1980].

PROBLEM SET 6.7

Part C. Expository exercises.

1. Starting with the references in this section, write an essay that describes the various axiom systems that have been proposed for geometry.

CHAPTER 7
The Theory of Parallels

7.1 The Existence of Parallel Lines

The concept of parallel lines has led to both the most fruitful and the most frustrating developments in plane geometry. Euclid (c. 330–275 B.C.E.) defined two segments to be parallel if no matter how far they are extended in both directions, they never meet. Note that he was interested in *segments* rather than *lines*. This follows the general preference at that time for finite objects. The idea of *never* meeting is, however, infinite in nature. How then does one determine if two lines are parallel?

By a stroke of genius Euclid adopted as his Fifth Postulate:

> If a line falling on two straight lines makes the interior angles on one side less than two right angles, then the two lines, if extended indefinitely, intersect on that side on which the interior angles are less than two right angles.

From this he deduced the important result that if l is a line and $P \notin l$ then there is a unique line through P parallel to l.

From Euclid's time to the mid-nineteenth century, geometers were disturbed by the Fifth Postulate. During that time the prevailing viewpoint was that postulates were "self-evident truths" and this postulate, because of its infinite nature, was not self-evident enough to be accepted without proof. In the 5th century Proclus argued that it was conceivable that two lines could approach each other asymptotically the way a hyperbola approaches an asymptote. It was generally felt (or hoped) that the Fifth Postulate need not be a separate axiom but instead could be derived from the other axioms and their consequences. In modern terminology it was felt that the Fifth Postulate was not "independent" of the remaining axioms. Of course we now know that the Fifth Postulate is independent thanks to the example of the Poincaré Plane.

The history of the parallel postulate is fascinating. In fact, many mathematicians attempted to prove the Fifth Postulate, and some thought they had succeeded. The list of those presenting fallacious proofs includes Ptolemy (2nd century A.D.), Proclus (410–485), Nasîr-Eddîn (1201–1274), John Wallis (1616–1703), Giordano Vitale (1633–1711), Gerolamo Saccheri (1667–1733), Johann Lambert (1728–1777), John Playfair (1748–1819), and Charles Dodgson (Lewis Carroll) (1832–1897). All failed, usually because at some point in their arguments they made assumptions (equidistant lines, similarity) that were equivalent to the desired result, or else they argued about infinite area or nonexistent points. For an extended discussion of these and other "proofs," see the first two chapters of Bonola [1955] or various sections in Martin [1975] and Greenberg [1980]. See also Heath's translation of Euclid [1956].

We shall see in this chapter that the axioms which we have adopted so far (and which are a refinement of those of Euclid) are sufficient only for proving the *existence* of parallel lines, but not the uniqueness. This phenomenon was probably first noticed by Carl Frederich Gauss (1777–1855) at the beginning of the 19th century. He never published his work in the subject, but in various letters he hinted that he had come very close to discovering what we call hyperbolic geometry. The first published accounts were given independently by Nicholai Lobachevsky (1792–1856) in 1829 and by Janos Bolyai (1802–1860) in 1832. They asserted the consistency of a neutral geometry in which Euclid's Fifth Postulate did not hold and developed much of the resulting theory of such a geometry. Translations of their work may be found in Bonola [1955]. The first proofs of the consistency of this new non-Euclidean geometry were given by Eugenio Beltrami (1835–1900) and Felix Klein (1849–1925). These proofs involved developing models. Later models were given by Poincaré, including our model $\mathscr{H}$. (The Klein and Poincaré models will be discussed briefly in Chapter 11.)

We have seen several instances where we have a choice of equivalent axioms—PSA or PP, SAS or ASA. As we complete our axiom system in this chapter, we again have a choice of axioms, but with an important difference. The two choices for a parallel axiom will not be equivalent. In fact, the All or None Theorem (Theorem 7.3.10) will tell us that exactly one of the two choices holds in any particular model of neutral geometry.

To begin our discussion of the theory of parallels, we need the concept of a transversal. Once this has been established, we can define the notions of alternate interior angles and corresponding angles (both of which are quite familiar from high school geometry) and then obtain a sufficient condition for two lines to be parallel (Theorem 7.1.2).

Definition. Given three distinct lines l, l_1, and l_2, we say that l is a **transversal** of l_1 and l_2 if l intersects both l_1 and l_2, but in different points.

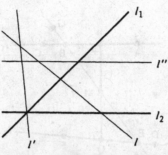

Figure 7-1

In Figure 7-1, l is a transversal of l_1 and l_2. l' is not a transversal because it does not intersect l_1 and l_2 in distinct points. l'' is not a transversal of l_1 and l_2 because it does not intersect l_2 at all.

Definition. Assume that the line $\overleftrightarrow{GH}$ is transversal to $\overleftrightarrow{AC}$ and $\overleftrightarrow{DF}$ in a metric geometry and that $\overleftrightarrow{AC} \cap \overleftrightarrow{GH} = \{B\}$ and $\overleftrightarrow{DF} \cap \overleftrightarrow{GH} = \{E\}$. If the points A, B, C, D, E, F, G and H are situated in such a way that

(i) $A—B—C$, $D—E—F$, and $G—B—E—H$, and
(ii) A and D are on the same side of $\overleftrightarrow{GH}$

then $\angle ABE$ and $\angle FEB$ are a pair of **alternate interior angles** and $\angle ABG$ and $\angle DEB$ are a pair of **corresponding angles**. (See Figure 7-2.)

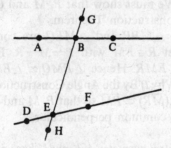

Figure 7-2

Note that in Figure 7-2 $\angle DEB$ and $\angle CBE$ are also alternate interior angles. Furthermore, $\angle CBG$ and $\angle FEB$ are corresponding angles, as are $\angle ABE$ and $\angle DEH$, as well as $\angle CBE$ and $\angle FEH$.

The following result gives a sufficient condition for two lines to have a common perpendicular. This means that if two lines satisfy the condition then they must have a common perpendicular. However, examples can be found in $\mathcal{H}$ to show that the converse is not true.

Theorem 7.1.1. Let l_1 and l_2 be two lines in a neutral geometry. If there is a transversal l of l_1 and l_2 with a pair of alternate interior angles congruent then there is a line l' which is perpendicular to both l_1 and l_2.

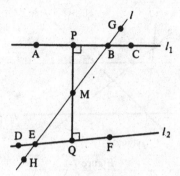

Figure 7-3

PROOF. Let $l_1 = \overleftrightarrow{AC}$, $l_2 = \overleftrightarrow{DF}$, and $l = \overleftrightarrow{GH}$ where $l_1 \cap l = \{B\}$, $l_2 \cap l = \{E\}$, $A—B—C$, $D—E—F$, $G—B—E—H$, and A and D are on the same side of $\overleftrightarrow{GH}$ as in Figure 7-3. If the alternate interior angles are right angles then $\overleftrightarrow{GH}$ gives the desired line l'. Otherwise, one of the two pairs of alternate interior angles consists of a pair of acute angles. We need only investigate that case.

Assume that $\angle ABE \simeq \angle FEB$ is acute as in Figure 7-3. Let M be the midpoint of $\overline{EB}$ and let P be the foot of the perpendicular from M to $\overleftrightarrow{AC}$. Since $\angle ABE = \angle ABM$ is acute, A and P lie on the same side of $\overleftrightarrow{GH}$. (See Problem A1.) Likewise if Q is the foot of the perpendicular from M to $\overleftrightarrow{DF}$ then Q and F are on the same side of $\overleftrightarrow{GH}$. Hence P and Q are on opposite sides of $\overleftrightarrow{GH}$. (Why?) We must show that P, M and Q are collinear. This is done by the Angle Construction Theorem.

The right triangles $\triangle MBP$ and $\triangle MEQ$ are congruent by HA. Thus $\angle BMP \simeq \angle EMQ$. Let $R \in \overrightarrow{PM}$ with $P—M—R$. By the Vertical Angle Theorem $\angle BMP \simeq \angle EMR$. Hence $\angle EMQ \simeq \angle EMR$. Q and R are on the same side of $\overleftrightarrow{GH}$. (Why?) By the Angle Construction Theorem $\angle EMQ = \angle EMR$. Hence $Q \in \text{int}(\overrightarrow{MR}) \subset \overrightarrow{PM}$ so that P, M and Q are collinear. Hence $\overleftrightarrow{PQ} = l'$ is the desired common perpendicular. $\square$

Theorem 7.1.2. *In a neutral geometry, if l_1 and l_2 have a common perpendicular, then l_1 is parallel to l_2. In particular, if there is a transversal to l_1 and l_2 with alternate interior angles congruent, then $l_1 \| l_2$.*

PROOF. Suppose that l is perpendicular to l_1 at P and to l_2 at Q. If $l_1 = l_2$ the first part is trivial. Hence we assume that $l_1 \neq l_2$ and proceed with a proof by contradiction.

Suppose $l_1 \cap l_2$ contains a point R. Then $P \neq R$, $Q \neq R$, and P, Q, R are not collinear. See Figure 7-4. But then $\triangle PQR$ has two right angles, which is impossible. Thus $l_1 \cap l_2 = \varnothing$ and $l_1 \| l_2$. The "in particular" statement follows from Theorem 7.1.1. $\square$

By Theorem 7.1.2, if l_1 and l_2 have a common perpendicular then $l_1 \| l_2$. Is the converse true: If $l_1 \| l_2$, do l_1 and l_2 have a common perpendicular?

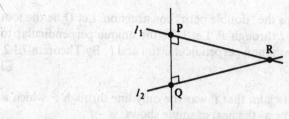

Figure 7-4

Many fallacious "proofs" of the Fifth Postulate assume the answer is yes. The next example shows that the answer is not always. Another example is supplied in Problem A6.

Example 7.1.3. In the Poincaré Plane let $l = {_0}L$ and $l' = {_1}L_1$. See Figure 7-5. Show that $l \| l'$ but that there is no line perpendicular to both l and l'.

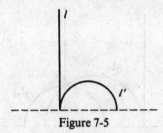

Figure 7-5

SOLUTION. First we note that $l \cap l' = \emptyset$. After all, if $(x, y) \in l \cap l'$ then $x = 0$ and $(x - 1)^2 + y^2 = 1$. But this would imply that $y = 0$ which is not true for a point $(x, y) \in \mathbb{H}$. Thus $l \| l'$.

By Problem B19 of Section 5.3, the only lines perpendicular to l take the form $_0L_r$ for $r > 0$. The line $_0L_r$ intersects $_1L_1$ only if $r < 2$, and in that case the point of intersection is $(r^2/2, \sqrt{r^2 - (r^4/4)}) = B$.

A tangent vector to $_0L_r$ at B is $(-\sqrt{r^2 - (r^4/4)}, r^2/2)$ while a tangent vector to $_1L_1$ at B is $(-\sqrt{r^2 - (r^4/4)}, (r^2/2) - 1)$. $_0L_r$ is perpendicular to $_1L_1$ if and only if

$$\left\langle \left(-\sqrt{r^2 - \frac{r^4}{4}}, \frac{r^2}{2}\right), \left(-\sqrt{r^2 - \frac{r^4}{4}}, \frac{r^2}{2} - 1\right) \right\rangle = 0.$$

But the left hand side of this equation is

$$r^2 - \frac{r^4}{4} + \frac{r^4}{4} - \frac{r^2}{2} = \frac{r^2}{2} \neq 0.$$

Hence no line is perpendicular to both l and l'. □

Theorem 7.1.4. *In a neutral geometry, let l be a line and $P \notin l$. Then there is a line l' through P which is parallel to l.*

PROOF. We shall perform the "double perp" construction. Let Q be the foot of the perpendicular to l through P. Let l' be the unique perpendicular to $\overleftrightarrow{PQ}$ at P. Then $\overleftrightarrow{PQ}$ is a common perpendicular to l and l'. By Theorem 7.1.2, $l \| l'$. $\square$

Note that we did not claim that l' was the only line through P which is parallel to l. It may not be as the next example shows.

Example 7.1.5. Show that in the Poincaré Plane there is more than one line through $P = (3, 4)$ which is parallel to $_{-5}L$.

SOLUTION. $_3L$ and $_3L_4$ are both parallel to $_{-5}L$. See Figure 7-6. In fact, there are an infinite number of lines through $(3, 4)$ parallel to $_{-5}L$: $_cL_r$ is parallel to $_{-5}L$ if $0 < c$ and $r = \sqrt{(c - 3)^2 + 16}$. Note also that even $_0L_5$ is parallel to $_{-5}L$. This is because $(-5, 0) \notin \mathbb{H}$.

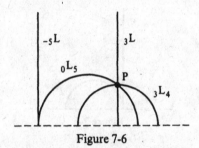

Figure 7-6

We shall see later that $_3L$ and $_0L_5$ are somewhat special. In a sense all other parallels are between these two lines. Furthermore, neither of these two lines has a common perpendicular with $_{-5}L$, but each of the other parallels does. This will be shown in the next chapter when we classify parallel lines. $\square$

In order to prove the uniqueness of parallels, Euclid introduced his Fifth Postulate. There are many equivalent formulations of Euclid's Fifth Postulate, some of which will be discussed in Chapter 9. (Martin [1975] lists 26!) The formulation we state below is Euclid's original version. We will show that this is equivalent to the uniqueness of parallels (which is usually called *Playfair's Postulate*). The definition which follows is merely a mathematically precise form of the quote from Euclid at the beginning of this section.

Definition. A protractor geometry satisfies **Euclid's Fifth Postulate** (EFP) if whenever $\overleftrightarrow{BC}$ is a transversal of $\overleftrightarrow{DC}$ and $\overleftrightarrow{AB}$ with

(i) A and D on the same side of $\overleftrightarrow{BC}$
(ii) $m(\angle ABC) + m(\angle BCD) < 180$

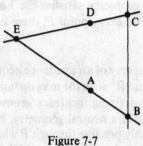

Figure 7-7

then $\overleftrightarrow{AB}$ and $\overleftrightarrow{CD}$ intersect at a point E on the same side of $\overleftrightarrow{BC}$ as A and D. (See Figure 7-7.)

Theorem 7.1.6. *If l is a line and $P \notin l$ in a neutral geometry which satisfies EFP, then there exists a unique line l' through P which is parallel to l.*

PROOF. Let l' be the line of the "double perp" construction of Theorem 7.1.4 so that Q is the foot of the perpendicular from P to l. Suppose that $\overleftrightarrow{AB}$ is another line through P with $A—P—B$. See Figure 7-8.

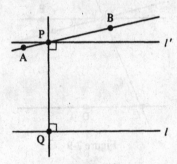

Figure 7-8

If $\overleftrightarrow{AB} \neq l'$ then one of the angles $\angle APQ$ or $\angle BPQ$ is acute. We may assume that $\angle APQ$ is acute. Since the angles at Q are right angles, we apply Euclid's Fifth Postulate and see that $\overleftrightarrow{AB}$ intersects l at a point E. Hence $\overleftrightarrow{AB}$ is not parallel to l if $\overleftrightarrow{AB} \neq l'$. Thus there is only one parallel to l at P. $\square$

Since the 18th century the conclusion of Theorem 7.1.6 has been used as the primary parallel axiom. Because this choice was first championed by Playfair, it is often referred to as Playfair's Parallel Postulate. We shall use the term "Euclidean Parallel Property." After we formulate the idea in a definition, we shall show that it is equivalent to Euclid's Fifth Postulate.

Definition. An incidence geometry satisfies the **Euclidean Parallel Property** (EPP) if for every line l and every point P, there is a unique line through P which is parallel to l.

Note that EPP is a property of an incidence geometry so that the Taxicab Plane, Euclidean Plane, and $\mathbb{R}^2$ with the max distance all satisfy EPP because they all have the same underlying incidence geometry, and it satisfies EPP. Of course, only the second is a neutral geometry. Note also that if $P \in l$ it is always true that there is only one line through P which is parallel to l whether the incidence geometry satisfies EPP or not. We now finish showing that EPP is equivalent to Euclid's Fifth Postulate.

Theorem 7.1.7. *If a neutral geometry satisfies* EPP *then it also satisfies* EFP.

PROOF. Let $\overleftrightarrow{BC}$ be a transversal of $\overleftrightarrow{AB}$ and $\overleftrightarrow{CD}$ with A and D on the same side of $\overleftrightarrow{BC}$. Suppose that $m(\angle ABC) + m(\angle BCD) < 180$. We want to show that $\overrightarrow{BA} \cap \overrightarrow{CD} \neq \varnothing$. Choose E on the same side of $\overleftrightarrow{BC}$ as A with $\angle EBC$ and $\angle BCD$ supplementary. Choose F with F—B—E. See Figure 7-9. Then $\angle FBC$ and $\angle EBC$ are supplementary so that $\angle FBC \simeq \angle DCB$.

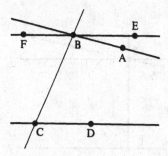

Figure 7-9

By Theorem 7.1.2, $\overleftrightarrow{BE} \| \overleftrightarrow{CD}$. Then by the hypothesis of EPP, $\overleftrightarrow{BA}$ is not parallel to $\overleftrightarrow{CD}$ because $\overleftrightarrow{BA} \neq \overleftrightarrow{BE}$ (Why?). Hence $\overleftrightarrow{BA} \cap \overleftrightarrow{CD} \neq \varnothing$. We now must show that actually $\overrightarrow{BA} \cap \overrightarrow{CD} \neq \varnothing$. Since

$$m(\angle CBA) + m(\angle BCD) < 180 = m(\angle CBE) + m(\angle BCD)$$

we have $\angle CBA < \angle CBE$. Thus $A \in \mathrm{int}(\angle CBE)$ because A and E are on the same side of $\overleftrightarrow{BC}$.

This is means that A is on the same side of $\overleftrightarrow{BE}$ as C. Thus all of $\mathrm{int}(\overrightarrow{BA})$ is on the same side of $\overleftrightarrow{BE}$ as $\overleftrightarrow{CD}$. ($\overleftrightarrow{CD}$ lies on one side of $\overleftrightarrow{BE}$ because $\overleftrightarrow{CD} \| \overleftrightarrow{BE}$.) Since $\overleftrightarrow{CD} \cap \overleftrightarrow{AB} \neq \varnothing$, the point of intersection must belong to $\overrightarrow{BA}$. Finally, since $\mathrm{int}(\overrightarrow{BA})$ and $\mathrm{int}(\overrightarrow{CD})$ lie on the same side of $\overleftrightarrow{BC}$ (Why?), the point of intersection must belong to $\overrightarrow{CD}$. Hence $\overrightarrow{BA} \cap \overrightarrow{CD} \neq \varnothing$. □

Because of Theorems 7.1.6 and 7.1.7, Euclid's Fifth Postulate and EPP are equivalent for neutral geometries.

Problem Set 7.1

Part A.

1. In a neutral geometry if $\angle ABC$ is acute then the foot of the perpendicular from A to $\overleftrightarrow{BC}$ is an element of $\text{int}(\overrightarrow{BC})$.

2. Given two lines and a transversal in a protractor geometry, prove that a pair of alternate interior angles are congruent if and only if a pair of corresponding angles are congruent.

*3. In a neutral geometry, if l is a transversal of l_1 and l_2 with a pair corresponding angles congruent, prove that $l_1 \| l_2$.

4. In a neutral geometry, if $\overleftrightarrow{BC}$ is a common perpendicular of $\overleftrightarrow{AB}$ and $\overleftrightarrow{CD}$, prove that if l is a transversal of $\overleftrightarrow{AB}$ and $\overleftrightarrow{CD}$ that contains the midpoint of $\overline{BC}$ then a pair of alternate interior angles for l are congruent.

5. Give an example of the following in the Poincaré Plane: Two lines l_1 and l_2 which have a common perpendicular and a transversal l for which a pair of alternate interior angles are not congruent. (Thus the converse of Theorem 7.1.1 is false.)

6. In the Poincaré Plane show that two distinct type I lines are parallel but do not have a common perpendicular.

7. Using vector notation for the Euclidean Plane prove that $L_{AB} \| L_{CD}$ if and only if there is a real number λ with $A - B = \lambda(C - D)$.

8. Let $\square ABCD$ be a quadrilateral in a neutral geometry. If $\overline{AB} \simeq \overline{CD}$ and $\overline{AD} \simeq \overline{BC}$ prove that $\overleftrightarrow{AB} \| \overleftrightarrow{CD}$ and $\overleftrightarrow{AD} \| \overleftrightarrow{BC}$.

9. Let $\square ABCD$ be the quadrilateral in $\mathscr{H}$ with $A = (0, 15)$, $B = (12, 9)$, $C = (12, 5)$, $D = (0, 13)$. Show that $\overleftrightarrow{AB} \| \overleftrightarrow{CD}$ and $\overleftrightarrow{AD} \| \overleftrightarrow{BC}$. Show that $\overline{AB}$ is not congruent to $\overline{CD}$. Hence the converse of Problem A8 is false in a neutral geometry.

10. In $\mathscr{H}$ let $l = {}_2L_5$ and let $P = (1, 2)$. Find a line l' through P parallel to l.

*11. Let $\{\mathscr{S}, \mathscr{L}, d, m\}$ be a neutral geometry that satisfies EPP. Prove that if $l_1 \| l_2$ and l is a transversal of l_1 and l_2, then a pair of alternate interior angles are congruent.

Part B. "Prove" may mean "find a counterexample".

12. Prove that the Moulton Plane satisfies EPP.

13. Prove that the Missing Strip Plane satisfies EPP.

14. Given a quadrilateral $\square ABCD$ in a neutral geometry with $\overrightarrow{AD} \| \overrightarrow{BC}$ and $\angle B \simeq \angle D$, prove that $\overrightarrow{AB} \| \overrightarrow{CD}$.

Part C. Expository exercises.

15. Using Bell [1937], Coolidge [1940] and Struik [1967], compare and contrast the lives of Bolyai and Lobachevski. What effect did their discovery of non-Euclidean geometry have on their lives?

16. Look up the list in Martin [1975] of twenty-six equivalent forms of Euclid's Fifth Postulate and describe them in words. Which are the most "geometric"? Which are the most dissimilar in content? Which do you find most "obvious"? Note that all of them are true in the Euclidean Plane. Find examples to show that in the Poincaré Plane these properties do not hold.

17. Write an essay which gives Gauss's view of the parallel controversy. See Hall [1970]. Do you admire, condone, or condemn his stand?

7.2 Saccheri Quadrilaterals

In the previous section we mentioned a number of attempts to "prove" that Euclid's Fifth Postulate followed from the other postulates of a neutral geometry. One of these deserves special mention because it contributed a direction for research in plane geometry.

In 1733 there appeared the book *Euclid Vindicated of All Flaw* by the Jesuit priest Gerolamo Saccheri. In it the author purported to prove Euclid's Fifth Postulate as a theorem. We now recognize basic flaws in his argument at certain crucial steps. However, the book was and is important in the development of the theory of parallels because it was the first to investigate the consequences of assuming the negation of Euclid's Fifth Postulate. A translation of the book is given by Halstead [1986].

Despite his failure to actually prove Euclid's Postulate as a theorem, Saccheri did contribute a substantial body of correct results. Did he know about the flaws in his proof? Certainly the erroneous proofs were unlike any of the rest of his carefully reasoned development. It has been suggested that Saccheri knew what he did was fallacious and that the "proof" was included so that the Church would approve the publication of his work. Whether he intended it or not, Saccheri did invent non-Euclidean geometry, although he gave no models. His contributions are remembered today in the following definition and a theorem which bears his name.

Definition. A quadrilateral $\square ABCD$ in a protractor geometry is a **Saccheri quadrilateral** if $\angle A$ and $\angle D$ are right angles and $\overline{AB} \simeq \overline{CD}$. In this case we write $\boxed{S}ABCD$. The **lower base** of $\boxed{S}ABCD$ is $\overline{AD}$, the **upper base** is $\overline{BC}$, the **legs** are $\overline{AB}$ and $\overline{CD}$, the **lower base angles** are $\angle A$ and $\angle D$, and the **upper base angles** are $\angle B$ and $\angle C$. (See Figure 7-10).

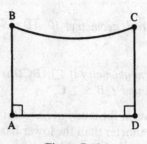

Figure 7-10

The basic approach of Saccheri (and those who followed him) was to try to prove something which turned out not to be true: that every Saccheri quadrilateral was actually a rectangle. If that were true it would not be hard to prove that EPP holds. Saccheri's main contribution comes from a careful investigation of three cases: (i) $\angle B$ is obtuse (which he showed was impossible), (ii) $\angle B$ is a right angle (which is equivalent to EPP) and (iii) $\angle B$ is acute (which he claimed to have proven is impossible, but in fact is possible).

Note that the order the vertices are listed for a Saccheri quadrilateral is important. If $\square ABCD$ is a Saccheri quadrilateral, so is $\square DCBA$, but $\square BCDA$ may not be. (In fact, $\square BCDA$ will also be a Saccheri quadrilateral if and only if it is a rectangle.) It must always be remembered that the first and last letters listed in the name of a Saccheri quadrilateral refer to the lower base angles which are right angles.

Before we prove Saccheri's Theorem (i.e., the sum of the measures of the angles of a triangle in a neutral geometry is less than or equal to 180) we will require several preliminary results.

Theorem 7.2.1. *In a neutral geometry a Saccheri quadrilateral* $\boxed{S}ABCD$ *is a convex quadrilateral.*

PROOF. Since $\overleftrightarrow{AB}$ and $\overleftrightarrow{CD}$ have a common perpendicular (namely $\overleftrightarrow{AD}$), Theorem 7.1.2 shows that $\overleftrightarrow{AB} \parallel \overleftrightarrow{CD}$. By theorem 4.5.5, $\boxed{S}ABCD$ is a convex quadrilateral.   □

Note that we did *not* use the fact that $\overline{AB} \simeq \overline{CD}$ in the above proof. The important point was that $\angle A$ and $\angle D$ were right angles.

Definition. Two convex quadrilaterals in a protractor geometry are **congruent** if the corresponding sides and angles are congruent. In this case we write $\square ABCD \simeq \square EFGH$.

We leave the proof of the next theorem and its corollary to Problems A3 and A4.

Theorem 7.2.2. *In a neutral geometry, if* $\overline{AD} \simeq \overline{PS}$ *and* $\overline{AB} \simeq \overline{PQ}$, *then* $\boxed{S}ABCD \simeq \boxed{S}PQRS$.

Corollary 7.2.3. *In a neutral geometry if* $\square ABCD$ *is a Saccheri quadrilateral then* $\boxed{S}ABCD \simeq \boxed{S}DCBA$ *and* $\angle B \simeq \angle C$.

One of the crucial points in the study of Saccheri quadrilaterals is the fact that the upper base is not shorter than the lower base. To prove this we need a generalization of the Triangle Inequality.

Theorem 7.2.4 (Polygon Inequality). *Suppose* $n \geq 3$. *If* $P_1, P_2, \ldots, P_n$ *are points in a neutral geometry then*

$$d(P_1, P_n) \leq d(P_1, P_2) + d(P_2, P_3) + \cdots + d(P_{n-1}, P_n).$$

PROOF. We use the Principle of Induction. If $n = 3$ then the result is just the Triangle Inequality (as given in Problem A6 of Section 6.3). Suppose that the result is true for $n = k$. Then

$$d(P_1, P_k) \leq d(P_1, P_2) + d(P_2, P_3) + \cdots + d(P_{k-1}, P_k).$$

By the Triangle Inequality again

$$d(P_1, P_{k+1}) \leq d(P_1, P_k) + d(P_k, P_{k+1}).$$

Combining these two inequalities we have

$$d(P_1, P_{k+1}) \leq d(P_1, P_2) + d(P_2, P_3) + \cdots + d(P_{k-1}, P_k) + d(P_k, P_{k+1})$$

so that the result is true for $n = k + 1$ whenever it is true for $n = k$. By the Principle of Induction the result is valid for all $n \geq 3$. ☐

Theorem 7.2.5. *In a neutral geometry, given* $\boxed{S}ABCD$, *then* $\overline{BC} \geq \overline{AD}$.

PROOF. We shall construct a chain of congruent Saccheri quadrilaterals as in Figure 7-11. Let $A_1 = A$, $A_2 = D$, $B_1 = B$, and $B_2 = C$. For each $k \geq 3$ let A_k be the unique point on $\overleftrightarrow{AD}$ such that $A_{k-2}\text{---}A_{k-1}\text{---}A_k$ and $\overline{A_{k-1}A_k} \simeq \overline{AD}$. Note that $d(A_1, A_{n+1}) = n \cdot d(A, D)$. For each $k \geq 3$ let B_k be the unique point on the same side of $\overleftrightarrow{AD}$ as B with $\overline{B_kA_k} \perp \overleftrightarrow{AD}$ and $\overline{B_kA_k} \simeq \overline{BA}$.

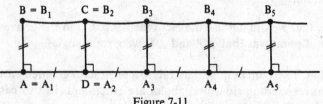

Figure 7-11

By Theorem 7.2.2, $\boxed{S}A_iB_iB_{i+1}A_{i+1} \simeq \boxed{S}ABCD$ for all $i \geq 1$. In particular $\overline{B_1B_2} \simeq \overline{B_2B_3} \simeq \cdots$. By the Polygon Inequality

$$d(A_1, A_{n+1}) \leq d(A_1, B_1) + d(B_1, B_2) + \cdots + d(B_n, B_{n+1}) + d(B_{n+1}, A_{n+1}).$$

Hence, since $d(A, B) = d(A_1, B_1)' = d(B_{n+1}, A_{n+1})$ and $d(B_i, B_{i+1}) = d(B, C)$,

$$n \cdot d(A, D) \le 2d(A, B) + n \cdot d(B, C) \quad \text{for } n \ge 1.$$

Then

$$d(A, D) - d(B, C) \le \frac{2}{n} d(A, B) \quad \text{for } n \ge 1. \tag{2-1}$$

Since Inequality (2-1) holds for all $n \ge 1$ and the right hand side can be made arbitrarily small by choosing large values for n, we must have $d(A, D) - d(B, C) \le 0$. Therefore

$$\overline{AD} \le \overline{BC}. \qquad \square$$

The previous result and the Open Mouth Theorem can be combined to prove the next result.

Theorem 7.2.6. *In a neutral geometry, given* $\boxed{\text{S}}ABCD$, $\angle ABD \le \angle BDC$.

Theorem 7.2.7. *In a neutral geometry the sum of the measures of the acute angles of a right triangle is less than or equal to* 90.

PROOF. Let $\triangle ABD$ have a right angle at A and let C be the unique point on the same side of $\overleftrightarrow{AD}$ as B with $\overline{CD} \perp \overleftrightarrow{AD}$ and $\overline{AB} \simeq \overline{DC}$. See Figure 7-12. Then we have $\boxed{\text{S}}ABCD$ and by Theorem 7.2.6

$$m(\angle ABD) + m(\angle ADB) \le m(\angle BDC) + m(\angle ADB).$$

Since $\boxed{\text{S}}ABCD$ is a convex quadrilateral (Theorem 7.2.1), $B \in \text{int}(\angle ADC)$ (Theorem 4.5.3) and so $m(\angle BDC) + m(\angle ADB) = m(\angle ADC) = 90$. Thus

$$m(\angle ABD) + m(\angle ADB) \le 90. \qquad \square$$

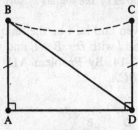

Figure 7-12

Theorem 7.2.8 (Saccheri's Theorem). *In a neutral geometry, the sum of the measures of the angles of a triangle is less than or equal to* 180.

PROOF. Let $\triangle ABC$ be any triangle and assume $\overline{AC}$ is a longest side. Then by Theorem 6.4.3, the foot of the perpendicular from B to $\overleftrightarrow{AC}$ is a point D with $A{-}D{-}C$. See Figure 7-13. $D \in \text{int}(\angle ABC)$ so that

$$m(\angle CAB) + m(\angle ABC) + m(\angle BCA) = m(\angle DAB) + m(\angle ABD)$$
$$+ m(\angle DBC) + m(\angle BCD)$$
$$\le 90 + 90 = 180. \qquad \square$$

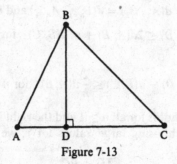

Figure 7-13

It must be remembered that Theorem 7.2.8 is the best possible result. We
have already seen an example of a triangle in $\mathcal{H}$ (Problem A1 of Section 5.1)
in which the sum of the measures of the angles is actually strictly less than
180. In your high school geometry course you learned that the sum of the
measures of the angles of a triangle was exactly 180. That result was correct
because you were dealing exclusively with a geometry which satisfied EPP.
In the Moulton Plane we saw an example (Problem A10 of Section 5.2) where
the "angle sum" was greater than 180. This does not contradict Theorem 7.2.8
since the Moulton Plane is not a neutral geometry.

The next result shows that the assumption of the Euclidean Parallel
Property forces the "angle sum" to be 180. We shall see in Chapter 9 that in a
neutral geometry EPP is actually equivalent to the assumption that the
"angle sum" is 180.

Theorem 7.2.9 (Euclidean Angle Sum). *In a neutral geometry which satisfies*
EPP, *the sum of the measures of the angles of any triangle is exactly* 180.

PROOF. Let $\triangle ABC$ be given and let l be the unique line through B parallel
to $\overleftrightarrow{AC}$. Choose D and E on l with D—B—E and with D and A on the same
side of $\overleftrightarrow{BC}$. See Figure 7-14. By Problem A11 of Section 7.1, $\angle DBA \simeq$
$\angle BAC$ and $\angle EBC \simeq \angle BCA$.

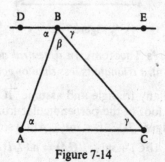

Figure 7-14

We know that $A \in \text{int}(\angle DBC)$ so that $m(\angle DBA) + m(\angle ABC) = m(\angle DBC)$. Thus

$$m(\angle CAB) + m(\angle ABC) + m(\angle BCA) = m(\angle DBA) + m(\angle ABC) + m(\angle EBC)$$
$$= m(\angle DBC) + m(\angle EBC)$$
$$= 180. \qquad \square$$

Definition. A quadrilateral $\square ABCD$ is a **parallelogram** if $\overleftrightarrow{AB} \parallel \overleftrightarrow{CD}$ and $\overleftrightarrow{AD} \parallel \overleftrightarrow{BC}$. A quadrilateral $\square ABCD$ is a **rectangle** if all four angles are right angles. A rectangle $\square ABCD$ is a **square** if all sides are congruent.

Theorem 7.2.10. *In a neutral geometry a Saccheri quadrilateral is a parallelogram.*

PROOF. In $\text{\small S}ABCD$, $\overleftrightarrow{AB} \parallel \overleftrightarrow{CD}$ since the two lines have a common perpendicular, namely $\overleftrightarrow{AD}$. By Problem A6 the line joining the midpoints of $\overline{AD}$ and $\overline{BC}$ is perpendicular to both. Hence $\overleftrightarrow{AD} \parallel \overleftrightarrow{BC}$. $\qquad \square$

As mentioned at the beginning of this chapter, there have been many attempts to prove that EPP is a theorem in neutral geometry. Some of the false proofs offered came from a basic misunderstanding of Saccheri quadrilaterals and in particular Theorem 7.2.10. It was erroneously assumed that since $\text{\small S}ABCD$ was a parallelogram with two right angles, "it must be a rectangle" (see Problem A2 for a counterexample), or "it must have opposite sides congruent" (see Problem A1 for a counterexample).

Another misunderstanding in attempts to prove that EPP followed from the axioms of a neutral geometry came from a misuse of the concept of equidistant lines. Recall from Section 6.4 that the distance from a point P to a line l, $d(P, l)$, is the perpendicular distance.

Definition. A set of points $\mathscr{A}$ in a neutral geometry is **equidistant** from the line l if $d(A, l)$ is the same for all $A \in \mathscr{A}$ (i.e., $d(A, l) = d(A', l)$ for all $A, A' \in \mathscr{A}$).

Certainly if two lines l and l' are equidistant then they are parallel. Quite a few incorrect proofs of EPP came about by assuming that parallel lines are equidistant. In Problem A23 there is a specific example in $\mathscr{H}$ to show that parallel lines need not be equidistant. In fact, in Chapters 8 and 9 we will show that the statement "parallel lines if and only if equidistant" is equivalent, for a neutral geometry, to the statement "satisfies EPP."

It would seem to be hard to show that one line is equidistant from another as there are infinitely many points to check. The last main result of this section, Giordano's Theorem (Theorem 7.2.13), shows that it is sufficient to check only three points.

Theorem 7.2.11. *Let $\square ABCD$ be a quadrilateral in a neutral geometry with right angles at A and D. If $\overline{AB} > \overline{DC}$ then $\angle ABC < \angle DCB$.*

PROOF. Choose $E \in \overline{DC}$ so that $D-C-E$ and $\overline{DE} \simeq \overline{AB}$. Then $\square ABED$ is a Saccheri quadrilateral so that $\angle ABE \simeq \angle DEB$ by Corollary 7.2.3. By the Exterior Angle Theorem, $\angle DEB < \angle DCB$. On the other hand $C \in$ int($\angle ABE$) (Why?) so that $\angle ABC < \angle ABE$. Thus $\angle ABC < \angle ABE \simeq \angle DEB < \angle DCB$. $\qquad\square$

Corollary 7.2.12. *In a neutral geometry, if* $\square ABCD$ *has right angles at* A *and* D *then*

(i) $\overline{AB} > \overline{CD}$ *if and only if* $\angle ABC < \angle DCB$
(ii) $\overline{AB} \simeq \overline{CD}$ *if and only if* $\angle ABC \simeq \angle DCB$
(iii) $\overline{AB} < \overline{CD}$ *if and only if* $\angle ABC > \angle DCB$.

Theorem 7.2.13 (Giordano's Theorem). *In a neutral geometry, if there are three distinct points on a line* l *which are the same distance from a line* l', *then* l *is equidistant from* l'.

PROOF. If $l = l'$ then l is equidistant from l'. Hence we may assume that $l \neq l'$. Let A, B, $C \in l$ with $d(A, l') = d(B, l') = d(C, l')$. Two of A, B, C must be on the same side of l'. By Problem A22, l is parallel to l' and hence all of l lies on one side of l'. In particular A, B, C all lie on the same side of l'. Since one of the three points must be between the other two, we may assume that $A-B-C$. Let D, E, F be the feet of the perpendiculars from A, B, C to l'. See Figure 7-15.

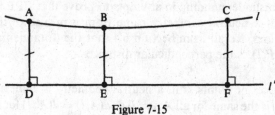

Figure 7-15

We have three Saccheri quadrilaterals: $\boxed{\text{S}}DABE$, $\boxed{\text{S}}DACF$, and $\boxed{\text{S}}EBCF$. Thus

$$\angle ABE \simeq \angle BAD \simeq \angle BCF \simeq \angle CBE \qquad (2\text{-}2)$$

by Corollary 7.2.3. Hence $\angle ABE$ is a right angle since $\angle ABE$ and $\angle CBE$ form a linear pair. Thus all the angles in Congruence (2-2) are right. Hence $\boxed{\text{S}}DACF$ is a rectangle. We must show that if $P \in l$ and if S is the foot of the perpendicular from P to l then $\overline{PS} \simeq \overline{AD}$.

Case 1. Suppose that P is between two of the points A, B, C. Then $A-P-C$. If PS is not perpendicular to l then one of $\angle APS$ and $\angle CPS$ is acute. Assume that $\angle APS$ is acute so that $\angle CPS$ is obtuse. See Figure 7-16. By Corollary 7.2.12, $AD < PS < CF$. Since this contradicts $\overline{AD} \simeq \overline{CF}$, we must have $\overline{PS} \perp l$. Hence $\angle APS \simeq \angle PAD$ and $\overline{PS} \simeq \overline{AD}$ by Corollary 7.2.12.

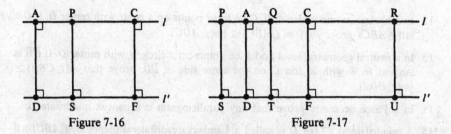

Figure 7-16   Figure 7-17

Case 2. Now suppose that $P \in l$ but $P \notin \overline{AC}$. Let Q be the unique point of l with $P\!-\!A\!-\!Q$ and $\overline{AQ} \simeq \overline{PA}$. Let T be the foot of the perpendicular from Q to l'. See Figure 7-17. $\triangle PAD \simeq \triangle QAD$ by SAS. $\triangle PDS \simeq \angle QDT$ by HA. Hence $\overline{PS} \simeq \overline{QT}$. Similarly, let R be the unique point on l with $P\!-\!C\!-\!R$ and $\overline{PC} \simeq \overline{CR}$ and let U be the foot of the perpendicular from R to l'. Then $\overline{PS} \simeq \overline{RU}$. Hence P, Q, R are three points of l equidistant from l'. Since $P\!-\!A\!-\!Q$, then by Case 1, $\overline{AD} \simeq \overline{PS}$ because A is between two of the three points P, Q, R which are equidistant from l'. Hence for all P, $\overline{PS} \simeq \overline{AD}$ and l is equidistant from l'.   □

Problem Set 7.2

Part A.

1. In $\mathscr{H}$ let $A = (0,2)$, $B = (1,\sqrt{3})$, $C = (\frac{1}{2},\sqrt{3}/2)$ and $D = (0,1)$. Prove that $\square ABCD$ is a Saccheri quadrilateral. Show that $\overline{BC} > \overline{AD}$.

2. For the Saccheri quadrilateral of Problem A1 show that $m_H(\angle B) < 90$.

3. Prove Theorem 7.2.2.

4. Prove Corollary 7.2.3.

5. Prove that the diagonals of a Saccheri quadrilateral are congruent in a neutral geometry.

*6. Prove that the line joining the midpoints of the bases of a Saccheri quadrilateral is perpendicular to both bases in a neutral geometry.

7. Prove Theorem 7.2.6.

8. Given $\triangle ABC$ in a neutral geometry, prove that $m(\angle A) + m(\angle B) < 180$ in two different ways.

9. Given a convex quadrilateral $\square ABCD$ in a neutral geometry, prove that $m(\angle A) + m(\angle B) + m(\angle C) + m(\angle D) \leq 360$.

10. Prove that the upper base angles of a Saccheri quadrilateral in a neutral geometry are not obtuse. (You are proving that Saccheri's "Hypothesis of the obtuse angle" is false.)

11. Prove Omar Khayam's Theorem: In a neutral geometry, if $\text{S}ABCD$ then $\overline{BC} > \overline{AD}$ if and only if $m(\angle B) < 90$. (Hint: Prove $\overline{BC} \simeq \overline{AD}$ if and only if $\angle B$ is right.) (Yes, this is the same Omar Khayam who wrote "A loaf of bread, a jug of wine, and thou . . .".)

12. In a neutral geometry, let A, B, C be three points on a circle with center D. If $D \in$ int($\triangle ABC$) prove that $m(\angle ABC) \leq \frac{1}{2}m(\angle ADC)$.

13. In a neutral geometry, let A and B be points on a circle $\mathscr{C}$ with center D. If $\overleftrightarrow{CB}$ is tangent to $\mathscr{C}$ with A and C on the same side of $\overleftrightarrow{BD}$, prove that $m(\angle CBA) \geq \frac{1}{2}m(\angle BDA)$.

14. In a Pasch geometry prove that any parallelogram is a convex quadrilateral.

*15. A quadrilateral $\square ABCD$ is called a **Lambert quadrilateral** (denoted $\boxed{L}ABCD$) if $\angle A$, $\angle B$, and $\angle C$ are right angles. Prove that $\boxed{L}ABCD$ is a parallelogram and is a convex quadrilateral.

*16. If $\boxed{L}ABCD$, prove that $m(\angle D) \leq 90$.

17. In a neutral geometry, if $\boxed{L}ABCD$ and $m(\angle D) < 90$ prove that $\overline{DB} > \overline{AC}$.

18. In a neutral geometry, if $\boxed{L}ABCD$ and $m(\angle D) = 90$ prove that $\overline{DB} \simeq \overline{AC}$.

*19. Let $\overleftrightarrow{AB}$ be perpendicular to both $\overleftrightarrow{BC}$ and $\overleftrightarrow{AD}$ in a neutral geometry. If C and D are on the same side of $\overleftrightarrow{AB}$, prove that $AB \leq CD$.

20. Prove Corollary 7.2.12.

21. In a neutral geometry, if l is equidistant from l', prove that l' is equidistant from l.

*22. In a neutral geometry if A and B are equidistant from l and lie on the same side of l, prove that $\overleftrightarrow{AB} \| l$.

23. Let A, B, C, D be as in Problem A1. Show that $\overleftrightarrow{AD} \| \overleftrightarrow{BC}$ but that $\overleftrightarrow{AD}$ is not equidistant from $\overleftrightarrow{BC}$.

24. In the Euclidean Plane show that if $l \| l'$ then l is equidistant from l'.

Part B. "Prove" may mean "find a counterexample".

25. Prove that the sum of the measures of the angles of any quadrilateral is ≤ 360 in a neutral geometry.

26. Show that it makes sense to talk about the distance from a point to a line in the Taxicab Plane. (But note Theorem 6.4.2 is false in $\mathscr{T}$.) Prove that in $\mathscr{T}$, $l \| l'$ implies that l is equidistant from l'.

27. Repeat Problem B26 for $\mathbb{R}^2$ with the max distance d_S.

28. A quadrilateral $\square ABCD$ is **equiangular** if $\angle A \simeq \angle B \simeq \angle C \simeq \angle D$. Prove that an equiangular quadrilateral is a convex quadrilateral in a neutral geometry.

29. In a neutral geometry prove that the angles of an equiangular quadrilateral are not obtuse.

30. In a neutral geometry prove that an equiangular quadrilateral is a parallelogram.

31. In a neutral geometry prove that the opposite sides of an equiangular quadrilateral are congruent.

32. In a neutral geometry, prove that the line joining the midpoints of opposite sides of an equiangular quadrilateral is perpendicular to both sides.

33. Prove that in a neutral geometry the diagonals of an equiangular quadrilateral bisect each other.

34. In a neutral geometry prove that the opposite angles of a parallelogram are congruent.

35. Prove that in a neutral geometry the diagonals of a Saccheri quadrilateral bisect each other.

Part C. Expository exercises.

36. Write an essay which describes the contribution of Saccheri to the theory of parallels. Discuss the suggestion that religious pressure on Saccheri may have affected his scholarly integrity.

37. Write a long history of the parallel controversy.

7.3 The Critical Function

In Example 7.1.3 we saw a pair of parallel lines that did not have a common perpendicular. We have also seen many examples of parallel lines which do have a common perpendicular (e.g., by Problem A6 of Section 7.2 the parallel lines that are determined by the bases of a Saccheri quadrilateral). Thus it seems that there are two types of parallel lines—those that possess a common perpendicular and those that do not.

To help understand the differences between these two types of parallel lines, we develop the idea of the critical function. It will help us determine when one line is "just barely parallel" to another line. We shall see in the next chapter that this property of being "just barely parallel" is equivalent to "there is no common perpendicular." The critical function is also the key to the surprising and very basic All or None Theorem (Theorem 7.3.10). It will tell us that in a neutral geometry if we have a unique parallel to l through $P \notin l$ for one choice of P and l, then we have a unique parallel for *all* choices of P and l.

Theorem 7.3.1. *Let l be a line in a neutral geometry and let $P \notin l$. Let D be the foot of the perpendicular from P to l. Then $\overrightarrow{PC} \cap l = \varnothing$ whenever $m(\angle DPC) \geq 90$.*

PROOF. If $m(\angle DPC) = 90$ then $\overrightarrow{PC} \| l$ by Theorem 7.1.2 and so the theorem is true in this case. If $m(\angle DPC) > 90$ let A be a point on the same side of $\overrightarrow{PD}$ as C such that $m(\angle DPA) = 90$. See Figure 7-18. Then $\overrightarrow{PA} \| l$ and $\mathrm{int}(\overrightarrow{PC})$ lies on the opposite side of $\overrightarrow{PA}$ as D. Since all of l lies on one side of $\overrightarrow{PA}$, $\overrightarrow{PC} \cap l = \varnothing$. $\qquad \square$

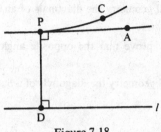

Figure 7-18

The preceding theorem tells us that if $m(\angle DPC)$ is large enough (for example, at least 90) then $\overrightarrow{PC}$ does not intersect l. On the other hand, certainly for some choices of C (say on l) $\overrightarrow{PC}$ does intersect l. Hence if $m(\angle DPC)$ is small enough then $\overrightarrow{PC}$ does intersect l. This dichotomy will lead us to the definition of the critical number for P and l. Because the definition uses the notion of a least upper bound, we first review that idea.

Definition. If $\mathscr{B}$ is a set of real numbers, then $r \in \mathbb{R}$ is a **least upper bound** of $\mathscr{B}$ (written $r = \text{lub } \mathscr{B}$) if

 (i) $b \leq r$ for all $b \in \mathscr{B}$; and
(ii) if $s < r$ then there is an element $b_s \in \mathscr{B}$ with $s < b_s$.

Thus lub $\mathscr{B}$ is the smallest number which is greater than or equal to every number in $\mathscr{B}$. In advanced calculus it is shown that if $\mathscr{B}$ is non-empty then $\mathscr{B}$ has a unique least upper bound if $\mathscr{B}$ is bounded (i.e., if there is some number N with $b \leq N$ for all $b \in \mathscr{B}$).

Example 7.3.2. Let $\mathscr{B} = \{-(1/n) \,|\, n \text{ is a positive integer}\}$. Show that $0 = \text{lub } \mathscr{B}$.

SOLUTION. Since $-1/n \leq 0$, part (i) of the definition is satisfied. Suppose that $s < 0$. Let k be an integer greater than $-1/s$. Since $s < 0$, k is positive and $-1/k \in \mathscr{B}$. Because $-1/s < k$ we see that $-s > 1/k$, or $s < -1/k$. Hence the second condition is satisfied. Thus $0 = \text{lub } \mathscr{B}$.   □

Definition. Let l be a line in a neutral geometry and let $P \notin l$. If D is the foot of the perpendicular from P to l let

$$K(P,l) = \{r \in \mathbb{R} \,|\, \text{there is a ray } \overrightarrow{PC} \text{ with } \overrightarrow{PC} \cap l \neq \varnothing \tag{3-1}$$
$$\text{and } r = m(\angle DPC)\}.$$

The **critical number** for P and l is

$$r(P,l) = \text{lub } K(P,l).$$

$K(P,l)$ contains the measures of all angles with vertex at P and such that one side of the angle is perpendicular to l and the other side intersects l.

$r(P, l)$ is the "largest" of these numbers and *a priori* may or may not be in this set. We will see in our first theorem that if $m(\angle DPC) = r(P, l)$ then $\overrightarrow{PC}$ is the "first" ray that does not intersect l so that $r(P, l) \notin K(P, l)$.

In Problem A3 you will show that in $\mathcal{E}$, $K(P, l) = \{r \in \mathbb{R} \mid 0 < r < 90\}$ and so $r(P, l) = 90$ for all lines and all points $P \notin l$. We shall postpone the calculation of $r(P, l)$ for a nontrivial example in $\mathcal{H}$ until after Theorem 7.3.3.

Before we can use the critical number, we must show that it exists, i.e., that lub $K(P, l)$ exists. But this is immediate because Theorem 7.3.1 shows that each of the numbers in $K(P, l)$ is less than 90. $K(P, l)$ is not empty because if $C \in l$, $C \neq D$, then $m(\angle DPC) \in K(P, l)$. Thus $K(P, l)$ is a nonempty, bounded set and so has a unique least upper bound (which is at most 90).

Theorem 7.3.3. *In a neutral geometry let $P \notin l$ and let D be the foot of the perpendicular from P to l. If $m(\angle DPC) \geq r(P, l)$ then $\overrightarrow{PC} \cap l = \emptyset$. If $m(\angle DPC) < r(P, l)$ then $\overrightarrow{PC} \cap l \neq \emptyset$.*

PROOF. First suppose that $m(\angle DPC) = r(P, l)$. We will show that $\overrightarrow{PC} \cap l = \emptyset$. Assume to the contrary that $\overrightarrow{PC}$ intersects l at a point R and let S be any point with D—R—S, as in Figure 7-19. Then $R \in \text{int}(\angle DPS)$ so that

$$m(\angle DPS) > m(\angle DPR) = m(\angle DPC) = r(P, l).$$

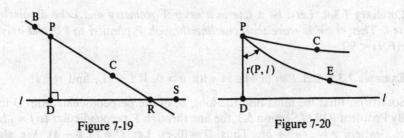

Figure 7-19   Figure 7-20

But $\overrightarrow{PS} \cap l = \{S\}$ so that $m(\angle DPS) \in K(P, l)$, which contradicts the fact that $r(P, l)$ is the least upper bound of $K(P, l)$. Hence if $m(\angle DPC) = r(P, l)$ we must have $\overrightarrow{PC} \cap l = \emptyset$. Note also if B—P—C then $m(\angle DPB) \geq 90$. By Theorem 7.3.1, $\overrightarrow{PB} \cap l = \emptyset$. Hence if $m(\angle DPC) = r(P, l)$ then $\overleftrightarrow{PC} \| l$.

Next suppose that $m(\angle DPC) > r(P, l)$. Let E be a point on the same side of $\overleftrightarrow{PD}$ as C with $m(\angle DPE) = r(P, l)$. See Figure 7-20. As noted at the end of the previous paragraph, $\overleftrightarrow{PE} \| l$. Int$(\overrightarrow{PC})$ and l lie on opposite sides of $\overleftrightarrow{PB}$. (Why?) Hence $\overrightarrow{PC} \cap l = \emptyset$ if $m(\angle DPC) > r(P, l)$.

Finally suppose that $m(\angle DPC) < r(P, l)$. We will show that $\overrightarrow{PC} \cap l \neq \emptyset$. By the definition of a least upper bound there exists a number $s = m(\angle DPF) \in K(P, l)$ with $m(\angle DPC) < s$. Since $s \in K(P, l)$, $\overrightarrow{PF}$ intersects l at a point A. See Figure 7-21.

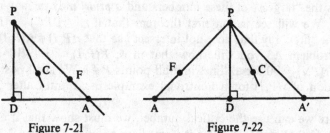

Figure 7-21   Figure 7-22

If A is on the same side of $\overleftrightarrow{PD}$ as C then $C \in \text{int}(\angle DPF)$ (Why?) so that by the Crossbar Theorem $\overrightarrow{PC}$ intersects $\overline{DA}$ and hence $\overrightarrow{PC} \cap l \neq \varnothing$. If A is on the opposite side of $\overleftrightarrow{PD}$ as C let A' be the unique point with A—D—A' and $\overline{AD} \simeq \overline{DA'}$. See Figure 7-22. Then by SAS, $\angle DPA' \simeq \angle DPA$. $C \in \text{int}(\angle DPA')$ and as before $\overrightarrow{PC}$ intersects $\overline{DA'}$ and hence l. Thus if $m(\angle DPC) < r(P,l)$ then $\overrightarrow{PC} \cap l \neq \varnothing$. $\square$

Note that Theorem 7.3.3 says that if $C \notin \overleftrightarrow{PD}$ then $\overrightarrow{PC} \cap l \neq \varnothing$ if and only if $m(\angle DPC) < r(P,l)$. In particular if $\overrightarrow{PC} \cap l = \varnothing$ then $r(P,l) \leq m(\angle DPC)$.

The proof of the next corollary is left to Problem A2. This result shows the connection between the critical numbers and EPP.

Corollary 7.3.4. *Let l be a line in a neutral geometry and P be a point not on l. Then there is more than one line through P parallel to l if and only if $r(P, l) < 90$.*

Example 7.3.5. Let $P = (a, b) \in \mathbb{H}$ with $a > 0$. If $l = {}_0L$, find $r(P, l)$.

SOLUTION. First we must find the foot, D, of the perpendicular from D to l. By Problem B19 of Section 5.3, the line through P perpendicular to $l = {}_0L$ is ${}_0L_r$, where $r = \sqrt{a^2 + b^2}$. Thus $D = (0, r)$. Let $C = (a, b + 1)$. We shall first shown that $r(P, l) = m_H(\angle DPC)$ and then compute $m_H(\angle DPC)$. See Figure 7-23.

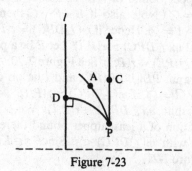

Figure 7-23

Clearly $\overline{PC} \cap l = \varnothing$ so that $r(P, l) \le m_H(\angle DPC)$. However, if $A \in$ int$(\angle DPC)$ then $\overrightarrow{PA}$ is a type II line and must intersect l. Hence $r(P, l) \ge m_H(\angle DPC)$ so that $r(P, l) = m_H(\angle DPC)$.

We now compute $m_H(\angle DPC)$.

$$T_{PD} = (-b, a) \quad \text{and} \quad T_{PC} = (0, 1)$$

so that

$$m_H(\angle DPC) = \cos^{-1}\left(\frac{a}{\sqrt{a^2 + b^2}}\right) = \cos^{-1}\left(\frac{a}{r}\right) = \tan^{-1}\left(\frac{b}{a}\right).$$

This gives a value for $r(P, l)$. It will be useful to see how this can be expressed in terms of the distance from P to l. Recall from Equation (2-10) of Section 2.2 that we can parametrize $_0L_r$ as

$$x = r \tanh(s) \qquad y = r \operatorname{sech}(s)$$

where s is the distance from $D = (0, r)$ to $B = (x, y)$. Thus if $t = d_H(P, D)$, we have

$$\frac{b}{a} = \frac{r \operatorname{sech}(t)}{r \tanh(t)} = \frac{1}{\sinh(t)}$$

so that

$$r(P, l) = \tan^{-1}\left(\frac{1}{\sinh(t)}\right). \tag{3-2}$$

$\square$

The reason we wanted to express $r(P, l)$ in terms of $d_H(P, D)$ in the previous example is given by the following theorem which says that $r(P, l)$ depends just on $d(P, l)$, the distance from P to l.

Theorem 7.3.6. *Let P and P' be points in a neutral geometry and let l and l' be lines with $P \notin l$ and $P' \notin l'$. If $d(P, l) = d(P', l')$ then $r(P, l) = r(P', l')$.*

PROOF. We shall show that $K(P, l) = K(P', l')$. This implies that $r(P, l) =$ lub $K(P, l) =$ lub $K(P', l') = r(P', l')$. Let D be the foot of the perpendicular from P to l and let D' be the foot of the perpendicular from P' to l'. By hypothesis $\overline{DP} \simeq \overline{D'P'}$.

If $s \in K(P, l)$ then there exists a point $C \in l$ with $m(\angle DPC) = s$. Choose $C' \in l'$ so that $\overline{DC} \simeq \overline{D'C'}$. See Figure 7-24. Then $\triangle PDC \simeq \triangle P'D'C'$ by SAS so that $m(\angle DPC) = m(\angle D'P'C')$. Hence $s \in K(P', l')$ so that $K(P, l) \subset K(P', l')$. Similarly, $K(P', l') \subset K(P, l)$ so that $K(P, l) = K(P', l')$. Thus $r(P, l) = r(P', l')$. $\square$

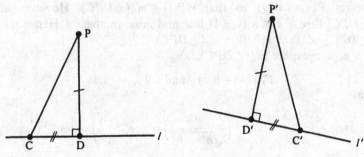

Figure 7-24

As a consequence of Theorem 7.3.6 we may make the following definition.

Definition. The **critical function** of a neutral geometry is the function $\Pi: \{t | t > 0\} \to \{r | 0 < r \le 90\}$ given by

$$\Pi(t) = r(P, l)$$

where l is any line and P is any point whose distance from l is t.

In Example 7.3.5 we saw that the critical function for $\mathscr{H}$ was $\Pi(t) = \tan^{-1}(1/\sinh(t))$. The formula was derived for a particular line and point, but according to Theorem 7.3.6 the computed result depends only on the distance. In Problem A4 you will show that in $\mathscr{H}$, $\Pi(t)$ is a strictly decreasing function. This is a special case of the next theorem.

Theorem 7.3.7. *In a neutral geometry, the critical function is nonincreasing, i.e.,*

$$\text{if } t' > t \quad \text{then } \Pi(t') \le \Pi(t).$$

PROOF. Let l be a line, $D \in l$, and let P, P' be points so that $P'-P-D$, $\overline{P'D} \perp l$, $P'D = t'$, and $PD = t$. Choose C, C' on the same side of $\overline{PD}$ so that $m(\angle DPC) = m(\angle DP'C') = \Pi(t)$. See Figure 7-25.

By the proof of Theorem 7.3.3, $\overrightarrow{PC} \| l$. By Problem A3 of Section 7.1, $\overrightarrow{P'C'} \| \overrightarrow{PC}$. Since P' and D lie on opposite sides of $\overleftrightarrow{PC}$, $\overrightarrow{P'C'} \cap l = \varnothing$. Thus $\overrightarrow{P'C'} \| l$ so that $\Pi(t') = r(P', l) \le m(\angle DP'C') = \Pi(t)$. $\qquad \square$

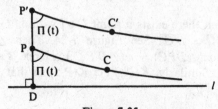

Figure 7-25

Among other things, Theorem 7.3.7 implies that if $\Pi(t_0) < 90$ for some value of t_0, then $\Pi(t) < 90$ for all $t \geq t_0$. However, it might be possible that $\Pi(t) = 90$ for some small values of t. Our next big task is to show that if $\Pi(a) < 90$ for some value a, then $\Pi(t) < 90$ for all $t > 0$. This requires a preliminary result.

Theorem 7.3.8. *In a neutral geometry, if* $\Pi(a) < 90$ *then* $\Pi(a/2) < 90$.

PROOF. Let l be a line, $D \in l$, and choose P, P' so that $P - P' - D$, $\overline{PD} \perp l$, $PP' = P'D = a/2$. Choose C with $m(\angle DPC) = \Pi(a) < 90$. Finally let l' be the unique line perpendicular to $\overline{PD}$ at P'. There are two possibilities: either $\overrightarrow{PC} \cap l' \neq \varnothing$ or $\overrightarrow{PC} \cap l' = \varnothing$. See Figures 7-26 and 7-27.

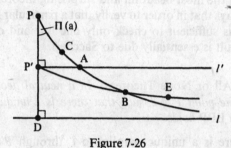

Figure 7-26

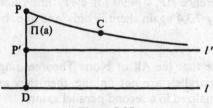

Figure 7-27

First suppose that $\overrightarrow{PC} \cap l' = \{A\}$. Choose B with $P - A - B$. Since $\angle DP'A$ is a right angle and $B \in \text{int}(\angle DP'A)$ (Why?), $m(\angle DP'B) < 90$. We will show that $\Pi(a/2) < m(\angle DP'B)$. We do this by showing that $\overrightarrow{P'B} \cap l = \varnothing$. Since $m(\angle DPA) = \Pi(a)$, $\overrightarrow{PA} \cap l = \varnothing$. Hence P and B are on the same side of l. Since $P - P' - D$, P and P' are on the same side of l. Hence P' and B are on the same side of l and $\overrightarrow{P'B} \cap l = \varnothing$.

If $P' - B - E$ then P' and E are on opposite sides of $\overrightarrow{PC}$. Thus D and E are on opposite sides of $\overrightarrow{PC}$ (Why?). Hence $\overline{BE} \cap l = \varnothing$ so that $\overrightarrow{P'B} \cap l = \varnothing$. This means that $\Pi(a/2) = r(P', l) \leq m(\angle DP'B) < 90$.

On the other hand, if $\overrightarrow{PC} \cap l' = \varnothing$, then $\Pi(a/2) = r(P, l') \leq m(\angle P'PC) = \Pi(a) < 90$. In either case, $\Pi(a/2) < 90$. $\qquad\square$

Theorem 7.3.9. *In a neutral geometry, if $\Pi(a) < 90$ for some real number a, then $\Pi(t) < 90$ for all $t > 0$.*

PROOF. For each positive integer n, let $a_n = a/2^n$. Then by Theorem 7.3.8, $\Pi(a_1) < 90$ since $\Pi(a) < 90$. By induction $\Pi(a_n) < 90$ for each n. Now suppose that t has been given. Choose n large enough so that

$$a_n = \frac{a}{2^n} < t.$$

Then by Theorem 7.3.8, $\Pi(t) \leq \Pi(a_n) < 90$.   □

The Euclidean Parallel Property (EPP) assumes that for *each* line l and each point $P \notin l$ there is only one line through P parallel to l. The next theorem is one of the most beautiful and surprising theorems in elementary mathematics. It says that in order to verify that a particular neutral geometry satisfies EPP, it is sufficient to check only one line and one point not on that line. The result is essentially due to Saccheri.

Theorem 7.3.10 (All or None Theorem). *In a neutral geometry, if there is one line l' and one $\cdot$point $P' \notin l'$ such that there is a unique line through P' parallel to l', then EPP holds.*

PROOF. Since there is a unique parallel to l' through P', $r(P', l') = 90$ by Corollary 7.3.4. Thus $\Pi(a) = 90$ for $a = d(P', l')$. By Theorem 7.3.9, $\Pi(t) = 90$ for all $t > 0$. Hence $r(P, l) = 90$ for every line l and every point $P \notin l$. Thus by Corollary 7.3.4 again there is only one line through P parallel to l.   □

We should note that the All or None Theorem implies that if there is one point where parallels are not unique, then they are not unique anywhere. This is formalized in a second parallel axiom.

Definition. A neutral geometry satisfies the **Hyperbolic Parallel Property** (HPP) if for each line l and each point $P \notin l$ there is more than one line through P parallel to l.

The All or None Theorem says that *exactly* one of EPP and HPP holds in each neutral geometry. Furthermore, to see whether HPP or EPP holds it suffices to check exactly one line and one point not on the line.

Definition. A **Euclidean geometry** is a neutral geometry that satisfies EPP. A **hyperbolic geometry** is a neutral geometry that satisfies HPP.

We shall investigate some of the properties of hyperbolic and Euclidean geometries in the next two chapters. We will see that $\mathscr{E}$ is a Euclidean geometry and $\mathscr{H}$ is a hyperbolic geometry in Problems A5 and A6 below.

PROBLEM SET 7.3

Part A.

1. Find the least upper bound for each of the sets:
   i. $\mathcal{B}_1 = \{\sin(x) \,|\, x \in \mathbb{R}\}$
   ii. $\mathcal{B}_2 = \{(-1)^n \,|\, n$ is an integer$\}$
   iii. $\mathcal{B}_3 = \{r \,|\, r$ is a rational number and $r^2 < 2\}$

2. Prove Corollary 7.3.4.

3. Prove that in the Euclidean Plane $r(P, l) = 90$ for every line l and every point $P \notin l$. Hence $\Pi(t) = 90$ for all t.

4. Prove that in the Poincaré Plane $\Pi(t) = \tan^{-1}(1/\sinh(t))$ is strictly decreasing: if $t < t'$ then $\Pi(t) > \Pi(t')$. (Use calculus.)

5. Prove that $\{\mathbb{R}^2, \mathcal{L}_E, d_E, m_E\}$ is a Euclidean geometry.

6. Prove that $\{\mathbb{H}, \mathcal{L}_H, d_H, m_H\}$ is a hyperbolic geometry.

7. In $\mathcal{H}$ let $l = {}_0L$. Let $\mathcal{A}$ be the intersection of $\mathbb{H}$ with the Euclidean line through $O = (0,0)$ and $P = (a, b)$ where $a > 0$, $b > 0$. See Figure 7-28. Prove that $\mathcal{A}$ is equidistant from l in $\mathcal{H}$. (Note $\mathcal{A}$ is *not* a line in $\mathcal{H}$.)

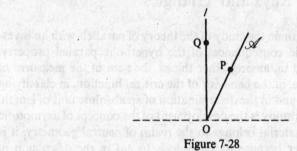

Figure 7-28

8. In Problem A7 let $Q = (0, 1)$. Prove that $m_E(\angle POQ) = 90 - \Pi(t_0)$ where t_0 is the hyperbolic distance from $\mathcal{A}$ to l.

9. In Problems A7 and A8 prove that $\mathcal{A} = \{R = (r, s) \in \mathbb{H} \,|\, r > 0$ and $d(R, l) = t_0\}$.

10. If l is a line and $P \notin l$ is a point in a neutral geometry which satisfies HPP, prove that there are infinitely many lines through P parallel to l.

11. Prove that in a Euclidean geometry every Saccheri quadrilateral is a rectangle.

Part C. Expository exercises.

12. Why is the All or None Theorem so surprising? What other names might be given to this important result, and why?

CHAPTER 8
Hyperbolic Geometry

8.1 Asymptotic Rays and Triangles

In this chapter we continue the study of the theory of parallels with an investigation of some basic consequences of the hyperbolic parallel property. We shall be interested in, among other things, the sum of the measures of the angles of a triangle, in the behavior of the critical function, in classifying types of parallel lines, and in the determination of an absolute unit of length.

The key step in this study is the development of the concept of asymptotic rays. Although this material belongs to the realm of neutral geometry, it is studied in this chapter because its purpose is to aid in the discussion of hyperbolic geometry. Furthermore, in a Euclidean geometry, the concept of asymptotic rays is superfluous—it adds nothing to the concept of parallelism. (See Problem A3.)

Definition. Let A, B, C, D be four points in a neutral geometry such that no three are collinear, with C and D on the same side of $\overleftrightarrow{AB}$, and $\overrightarrow{AD} \| \overrightarrow{BC}$. Then the set

$$\triangle DABC = \overrightarrow{AD} \cup \overline{AB} \cup \overrightarrow{BC}$$

is an **open triangle** (or a **biangle**).

We have sketched some open triangles in Figure 8-1. The open triangle in part (a) is in $\mathbb{R}^2$ while those in (b) and (c) are in $\mathbb{H}$. Part (d) illustrates the standard "pictorial" representation of an open triangle which we shall use.

Definition. Let $\triangle DABC$ be an open triangle. $\overrightarrow{BC}$ is **strictly asymptotic** to $\overrightarrow{AD}$ if for every $E \in \text{int}(\angle ABC)$, $\overrightarrow{BE}$ intersects $\overrightarrow{AD}$.

196

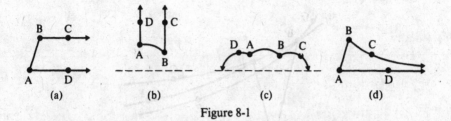

Figure 8-1

In Figures 8-1(a) and 8-1(b), $\overrightarrow{BC}$ is strictly asymptotic to $\overrightarrow{AD}$. In Figure 8-1(c), $\overrightarrow{BC}$ is not strictly asymptotic to $\overrightarrow{AD}$. Because of the way the definition is worded, a ray is never strictly asymptotic to itself. The idea is that $\overrightarrow{BC}$ is strictly asymptotic to $\overrightarrow{AD}$ if there are no rays interior to $\angle ABC$ which are parallel to $\overrightarrow{AD}$. Two rays will be asymptotic if they are either strictly asymptotic or one is a subset of the other. More formally,

Definition. Two rays $\overrightarrow{PQ}$ and $\overrightarrow{RS}$ are **equivalent** (written $\overrightarrow{PQ} \sim \overrightarrow{RS}$) if either $\overrightarrow{PQ} \subset \overrightarrow{RS}$ or $\overrightarrow{RS} \subset \overrightarrow{PQ}$.

The ray $\overrightarrow{BC}$ is **asymptotic** to the ray $\overrightarrow{AD}$ (written $\overrightarrow{BC} \mid \overrightarrow{AD}$) if either $\overrightarrow{BC}$ is strictly asymptotic to $\overrightarrow{AD}$ or $\overrightarrow{BC} \sim \overrightarrow{AD}$.

Our first goal is to show that "asymptotic to" is an equivalence relation. Except for the reflexive condition, this result is quite technical. On first reading you may wish to skip the proofs of Theorems 8.1.1, 8.1.5, 8.1.6 and 8.1.7.

The first step is to show that the notion of asymptotic rays depends only on the directions of the rays and not their endpoints. This idea is made precise by the definition of equivalent rays and the next two results.

Theorem 8.1.1. *In a neutral geometry if* $\overrightarrow{BC} \sim \overrightarrow{B'C'}$, *and* $\overrightarrow{BC} \mid \overrightarrow{AD}$, *then* $\overrightarrow{B'C'} \mid \overrightarrow{AD}$.

PROOF. By Problem A1, $\sim$ is an equivalence relation. Hence if $\overrightarrow{BC} \sim \overrightarrow{AD}$, then $\overrightarrow{B'C'} \sim \overrightarrow{AD}$ also and $\overrightarrow{B'C'} \mid \overrightarrow{AD}$. Thus we may assume that $\overrightarrow{BC} \not\sim \overrightarrow{AD}$; that is, $\overrightarrow{BC}$ is strictly asymptotic to $\overrightarrow{AD}$.

Case 1. Assume that $\overrightarrow{BC} \subset \overrightarrow{B'C'}$. If $B = B'$ then $\overrightarrow{BC} = \overrightarrow{B'C'}$ and so we are done. We therefore may consider the case $B \neq B'$. In this case $B'—B—C$ so that $\overrightarrow{B'C'} = \overrightarrow{B'C}$. By hypothesis $\overrightarrow{BC}$ is strictly asymptotic to $\overrightarrow{AD}$.

Let $E \in \text{int}(\angle AB'C)$. We must show that $\overrightarrow{B'E} \cap \overrightarrow{AD} \neq \varnothing$. Suppose to the contrary that $\overrightarrow{B'E} \cap \overrightarrow{AD} = \varnothing$. We claim that then $\overrightarrow{B'E} \parallel \overrightarrow{AD}$. See Figure 8-2. First note that if $G—B'—E$ then $\overrightarrow{B'G} \cap \overrightarrow{AD} = \varnothing$ because $\overrightarrow{B'C} \parallel \overrightarrow{AD}$ and G lies on the opposite side of $\overrightarrow{B'C}$ as A and D. Next if $H—A—D$, then $\overrightarrow{B'E} \cap \overrightarrow{AH} = \varnothing$ since E and H are on opposite sides of $\overrightarrow{AB'}$. Thus $\overleftrightarrow{B'E} \cap \overleftrightarrow{AD} = (\overrightarrow{B'E} \cap \overrightarrow{AD}) \cup (\overrightarrow{B'E} \cap \overrightarrow{AH}) \cup (\overrightarrow{B'G} \cap \overrightarrow{AD}) = \varnothing$ and $\overleftrightarrow{B'E} \parallel \overleftrightarrow{AD}$.

By the Exterior Angle Theorem applied to $\triangle B'BA$, $\angle CBA > \angle CB'A > \angle CB'E$ so that there exists a point $F \in \text{int}(\angle CBA)$ with $\angle CBF \simeq \angle CB'E$. Then $\overrightarrow{BF} \parallel \overrightarrow{B'E}$ (Why?). Thus $\overrightarrow{BF}$ lies all on one side of $\overleftrightarrow{B'E}$, as does $\overrightarrow{AD}$. Since

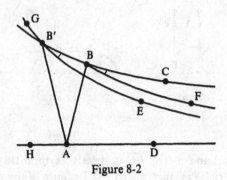

Figure 8-2

B and A are on opposite sides of $\overleftrightarrow{B'E}$ (Why?), $\overrightarrow{BF} \cap \overline{AD} = \varnothing$. However, this contradicts $\overrightarrow{BC} \mid \overline{AD}$, so that $\overrightarrow{B'E} \cap \overline{AD} \neq \varnothing$. Hence $\overrightarrow{B'C} \mid \overline{AD}$.

Case 2. Now assume $\overrightarrow{B'C'} \subset \overrightarrow{BC}$ and $B \neq B'$. Then $B—B'—C'$ and $\overrightarrow{BC} = \overrightarrow{BC'}$. Let $E \in \mathrm{int}(\angle AB'C')$. See Figure 8-3. We must show that $\overrightarrow{B'E} \cap \overline{AD} \neq \varnothing$. Assume to the contrary that $\overrightarrow{B'E} \cap \overline{AD} = \varnothing$. By an argument similar to the one used in Case 1, $\overrightarrow{B'E} \| \overline{AD}$. We first shall show that $E \in \mathrm{int}(\angle ABC')$.

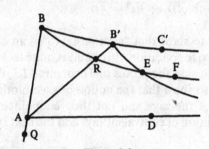

Figure 8-3

Since $E \in \mathrm{int}(\angle AB'C')$, E and A are on the same side of $\overleftrightarrow{B'C'} = \overleftrightarrow{BC}$. E and C' are on the same side of $\overleftrightarrow{A\,B'}$. Since $B—B'—C'$, E and B are on opposite sides of $\overleftrightarrow{A\,B'}$. Thus $\overrightarrow{B'E} \cap \overrightarrow{AB} = \varnothing$ by the Z Theorem. $\overrightarrow{B'E}$ lies on one side of $\overleftrightarrow{AD}$ since $\overrightarrow{B'E} \| \overline{AD}$. B and B' are on the same side of $\overleftrightarrow{AD}$. Thus $\overrightarrow{B'E}$ lies on the opposite side of $\overleftrightarrow{AD}$ as Q if $B—A—Q$. Hence $\overrightarrow{B'E} \cap \overrightarrow{AQ} = \varnothing$. This means that $\overrightarrow{B'E} \cap \overrightarrow{AB} = \varnothing$ so that B' and E lie on the same side of $\overleftrightarrow{AB}$. Thus E and C' lie on the same side of $\overleftrightarrow{AB}$ and so $E \in \mathrm{int}(\angle ABC')$.

Now choose F with $B—E—F$. If $\overline{BE} \cap \overline{AD} \neq \varnothing$, then $\overleftrightarrow{AD}$ intersects either $\overline{BB'}$ or $\overline{B'E}$ by Pasch's Theorem. But this is impossible since $\overleftrightarrow{AD} \| \overleftrightarrow{BB'}$ and $\overleftrightarrow{AD} \| \overrightarrow{B'E}$. Hence $\overline{BE} \cap \overline{AD} = \varnothing$.

Since $E \in \mathrm{int}(\angle ABB')$, $\overline{BE}$ intersects $\overline{AB'}$ at a point R. Since $E \in \mathrm{int}(\angle AB'C')$, E and B are on opposite sides of $\overleftrightarrow{AB'}$ so that $B—R—E—F$. A and R are on the same side of $\overleftrightarrow{B'E}$ while R and F are on opposite sides of $\overleftrightarrow{B'E}$. Hence F and A are on opposite sides of $\overleftrightarrow{B'E}$. Since $\overrightarrow{B'E} \| \overrightarrow{AD}$, we must have $\overrightarrow{EF} \cap \overline{AD} = \varnothing$.

Thus since $\overline{BE} \cap \overrightarrow{AD} = \varnothing$ and $\overline{EF} \cap \overrightarrow{AD} = \varnothing$, and $B-E-F$, we have $\overrightarrow{BE} \cap \overrightarrow{AD} = \varnothing$. However, this contradicts $\overrightarrow{BC'} | \overrightarrow{AD}$. What caused the contradiction? It was the assumption that $\overrightarrow{B'E} \cap \overrightarrow{AD} = \varnothing$. Hence it must be that $\overrightarrow{B'E} \cap \overrightarrow{AD} \neq \varnothing$ when $E \in \text{int}(\angle AB'C')$, so that $\overrightarrow{B'C'} | \overrightarrow{AD}$. $\qquad\square$

The proof of Theorem 8.1.1 became quite involved as we carefully verified different cases and kept track of which side of a given line a certain pair of points were on. The next result is much simpler to prove and is left as Problem A2.

Theorem 8.1.2. *In a neutral geometry, if* $\overrightarrow{AD} \sim \overrightarrow{A'D'}$ *and* $\overrightarrow{BC} | \overrightarrow{AD}$, *then* $\overrightarrow{BC} | \overrightarrow{A'D'}$.

Theorem 8.1.3. *In a neutral geometry, if* $\overrightarrow{AD} \sim \overrightarrow{A'D'}$, $\overrightarrow{BC} \sim \overrightarrow{B'C'}$, *and* $\overrightarrow{BC} | \overrightarrow{AD}$, *then* $\overrightarrow{B'C'} | \overrightarrow{A'D'}$.

PROOF. By Theorem 8.1.1, $\overrightarrow{B'C'} | \overrightarrow{AD}$. By Theorem 8.1.2, $\overrightarrow{B'C'} | \overrightarrow{A'D'}$. $\qquad\square$

As we have mentioned before, if we are given a line l and a point P there may not be a unique line through P which is parallel to l. However, the next result shows that there is a uniqueness result in the case of asymptotic rays.

Theorem 8.1.4. *In a neutral geometry, given a ray* $\overrightarrow{AD}$ *and a point* $B \notin \overrightarrow{AD}$, *there is a unique ray* $\overrightarrow{BC}$ *with* $\overrightarrow{BC} | \overrightarrow{AD}$.

PROOF. Let A' be the foot of the perpendicular from B to $\overleftrightarrow{AD}$ and choose $D' \in \overrightarrow{AD}$ so that $\overrightarrow{A'D'} \sim \overrightarrow{AD}$. Note that $\overrightarrow{BC}$ can be asymptotic to $\overrightarrow{AD} \sim \overrightarrow{A'D'}$ only if C and D' lie on the same side of $\overleftrightarrow{A'B}$. See Figure 8-4.

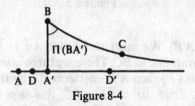

Figure 8-4

If C is on the same side of $\overleftrightarrow{A'B}$ as D', then by the definition of the critical function, $\overrightarrow{BC} | \overrightarrow{A'D'}$ if and only if $m(\angle A'BC) = \Pi(BA') = r(B, \overrightarrow{AD})$. Since there is a unique ray $\overrightarrow{BC}$ with C on the same side of $\overleftrightarrow{A'B}$ as D' and $m(\angle A'BC) = \Pi(BA')$, the result is immediate. $\qquad\square$

The proof of Theorem 8.1.4 is important on a philosophical level because it shows that there is a relationship between the critical function Π and the existence of asymptotic rays. This relationship will be exploited in the next two sections.

We now show that the relation "is asymptotic to" is symmetric. Note that the only case to prove is when $\overrightarrow{BC}$ is strictly asymptotic to $\overrightarrow{AD}$, because if $\overrightarrow{BC}|\overrightarrow{AD}$ with $\overrightarrow{BC} \sim \overrightarrow{AD}$ then $\overrightarrow{AD} \sim \overrightarrow{BC}$ and $\overrightarrow{AD}|\overrightarrow{BC}$.

Theorem 8.1.5. *In a neutral geometry, if* $\overrightarrow{BC}|\overrightarrow{AD}$ *then* $\overrightarrow{AD}|\overrightarrow{BC}$ *also.*

PROOF. We may assume that $\overrightarrow{BC}$ is strictly asymptotic to $\overrightarrow{AD}$. If $\Pi(t) = 90$ for some t then the Euclidean Parallel Property holds by the All or None Theorem. By Problem A3 the concept of asymptotic is the same as parallel so that the result follows immediately. Hence we assume that $\Pi(t) < 90$ for all $t > 0$.

Let A' be the foot of the perpendicular from B to $\overrightarrow{AD}$. Choose $D' \in \overrightarrow{AD}$ so that $\overrightarrow{A'D'} \sim \overrightarrow{AD}$. Let F be the foot of the perpendicular from A' to $\overrightarrow{BC}$. See Figure 8-5. Now $m(\angle A'BC) = \Pi(A'B) < 90$ so that by Problem A1 of Section 7.1, $F \in \text{int}(\overrightarrow{BC})$.

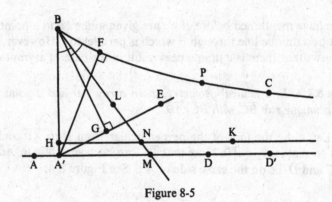

Figure 8-5

Let $E \in \text{int}(\angle D'A'B)$. We must show that $\overrightarrow{A'E} \cap \overrightarrow{BC} \neq \varnothing$ in order for $\overrightarrow{AD}$ to be strictly asymptotic to $\overrightarrow{BC}$. There are three cases depending on where E lies. If $E \in \text{int}(\angle BA'F)$ then $\overrightarrow{A'E}$ intersects $\overrightarrow{BF}$ by the Crossbar Theorem applied to $\triangle A'BF$. Since $\overrightarrow{BF} \subset \overrightarrow{BF} = \overrightarrow{BC}$, this says that $\overrightarrow{A'E} \cap \overrightarrow{BC} \neq \varnothing$.

The second case is $E \in \text{int}(\overrightarrow{A'F})$. However, in this situation $\overrightarrow{A'E} \cap \overrightarrow{BC} = \{F\}$. The last case occurs when $E \in \text{int}(\angle D'A'F)$. This is the one illustrated in Figure 8-5.

Let G be the foot of the perpendicular from B to $\overrightarrow{A'E}$. Since $\angle BA'G$ is acute (Why?), $G \in \overrightarrow{A'E}$. If $G \notin \text{Int} \angle A'BC$ then either $G \in \overrightarrow{BC}$ or G is on the opposite side of $\overrightarrow{BC}$ as A'. Either way $\overrightarrow{A'E}$ intersects $\overrightarrow{BC}$ and $\overrightarrow{A'D'}|\overrightarrow{BC}$. Hence $\overrightarrow{AD}|\overrightarrow{BC}$.

On the other hand, if $G \in \text{Int} \angle A'BC$ then since $BG < BA'$ there is a unique point H with B—H—A' and $\overline{BH} \simeq \overline{BG}$. Choose K on the same side of $\overleftrightarrow{BA'}$ as D' so that $\overrightarrow{HK} \perp \overleftrightarrow{BA'}$. Choose L on the same side of $\overleftrightarrow{BA'}$ as C with $\angle HBL \simeq \angle GBC < \angle A'BC$. Because $\overrightarrow{BC}|\overrightarrow{A'D'}$, it must be that $\overrightarrow{BL}$ intersects $\overrightarrow{A'D'}$ at some point M.

Since $\overrightarrow{HK} \parallel \overrightarrow{A'D'}$, Pasch's Theorem applied to $\triangle BA'M$ implies that $\overrightarrow{HK} \cap \overline{BM} = \{N\}$ for some N. Let $P \in \overrightarrow{BC}$ so that $\overline{BP} \simeq \overline{BN}$. Then $\triangle NBH \simeq \triangle PBG$ by SAS. But this means that $\angle BGP$ is a right angle since $\angle BHN$ is a right angle. Hence $P \in \overrightarrow{A'E}$. Since P and E are on the same side of $\overleftrightarrow{A'B}$ (namely the side that contains D'), $P \in \overrightarrow{A'E}$. Thus $\overrightarrow{A'E} \cap \overrightarrow{BC} \neq \varnothing$ and $\overrightarrow{A'D'} | \overrightarrow{BC}$. Since $\overrightarrow{AD} \sim \overrightarrow{A'D'}$, we have $\overrightarrow{AD} | \overrightarrow{BC}$.   □

In order to prove the transitivity of the asymptotic relation we need the next result. It tells us that if three rays are asymptotic, then there is a common transversal to the lines that contain them.

Theorem 8.1.6. *Let $\overleftrightarrow{AB}$, $\overleftrightarrow{CD}$ and $\overleftrightarrow{EF}$ be distinct lines in a neutral geometry. If $\overrightarrow{AB} | \overrightarrow{CD}$ and $\overrightarrow{CD} | \overrightarrow{EF}$ then there is a line l which intersects all three lines $\overleftrightarrow{AB}$, $\overleftrightarrow{CD}$ and $\overleftrightarrow{EF}$.*

PROOF. Since the lines are distinct, the rays cannot be equivalent. Thus $\overrightarrow{AB}$ is strictly asymptotic to $\overrightarrow{CD}$, and $\overrightarrow{CD}$ is strictly asymptotic to $\overrightarrow{EF}$.

If A and E are on opposite sides of $\overleftrightarrow{CD}$ then $\overline{AE}$ intersects $\overleftrightarrow{CD}$ (as well as $\overleftrightarrow{AB}$ and $\overleftrightarrow{EF}$). Hence in this case we may let $l = \overleftrightarrow{AE}$. See Figure 8-6.

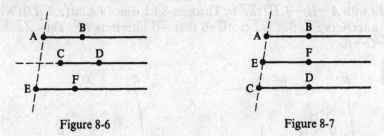

Figure 8-6   Figure 8-7

Now assume A and E are on the same side of $\overleftrightarrow{CD}$. If $A \in \overleftrightarrow{CE}$ we may let $l = \overleftrightarrow{CE}$ and be done. See Figure 8-7. Hence we assume $A \notin \overleftrightarrow{CE}$. Now $D \notin \overleftrightarrow{CE}$ or else $\overrightarrow{EF} \cap \overrightarrow{CD} \neq \varnothing$, which contradicts $\overrightarrow{CD} | \overrightarrow{EF}$. Thus A and D are either on the same side of $\overleftrightarrow{CE}$ or on opposite sides of $\overleftrightarrow{CE}$. If they are on the same side then $A \in \text{int}(\angle DCE)$. Since $\overrightarrow{CD} | \overrightarrow{EF}$, $\overrightarrow{CA} \cap \overrightarrow{EF} \neq \varnothing$, and we may let $l = \overleftrightarrow{CA}$. See Figure 8-8.

Thus we are left with the case A and E are on the same side of $\overleftrightarrow{CD}$ while A and D are on opposite sides of $\overleftrightarrow{CE}$. We will show that $l = \overleftrightarrow{CE}$ is a common transversal. See Figure 8-9.

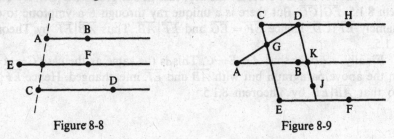

Figure 8-8   Figure 8-9

$\overleftrightarrow{AD}$ intersects $\overleftrightarrow{CE}$ at a point G. Choose H with C—D—H so that $\overline{DH} \sim \overline{CD}$ and thus $\overrightarrow{DH} | \overrightarrow{AB}$. $\angle HDG > \angle DCG$ by the Exterior Angle Theorem. Hence we may find $J \in \text{int}(\angle HDG)$ with $\angle HDJ \simeq \angle DCG$. Then $\overrightarrow{DJ} \| \overrightarrow{CE}$. $\overrightarrow{DJ}$ intersects $\overline{AB}$ at a point K since $\overrightarrow{DH} | \overrightarrow{AB}$.

We now apply Pasch's Theorem to $\triangle ADK$. Since $\overleftrightarrow{CE}$ intersects $\overline{AD}$, it must intersect $\overline{AK}$ (because $\overrightarrow{CE} \| \overrightarrow{DK}$). Thus $\overleftrightarrow{CE}$ intersects $\overline{AB}$ and we may let $l = \overleftrightarrow{CE}$.   □

Now we can prove that $|$ is transitive.

Theorem 8.1.7. *In a neutral geometry if* $\overrightarrow{AB} | \overrightarrow{CD}$ *and* $\overrightarrow{CD} | \overrightarrow{EF}$ *then* $\overrightarrow{AB} | \overrightarrow{EF}$.

PROOF. If any two of the three rays are equivalent, the result is immediate. Hence we may assume that the three lines $\overleftrightarrow{AB}$, $\overleftrightarrow{CD}$, and $\overleftrightarrow{EF}$ are distinct. By Theorem 8.1.6 there is a line l that intersects all three of these lines. We may replace the original rays with equivalent rays whose endpoints lie on l. That is, we may as well assume that A, C, and E lie on a single line l. Thus either A—C—E, C—A—E, or A—E—C.

Suppose that A—C—E and let $G \in \text{int}(\angle EAB)$. Since $\overrightarrow{AB} | \overrightarrow{CD}$, $\overrightarrow{AG}$ intersects $\overleftrightarrow{CD}$ at some point H. See Figure 8-10. Choose I with C—H—I and J with A—H—J. $\overrightarrow{HI} | \overrightarrow{EF}$ by Theorem 8.1.1. Since $J \in \text{int}(\angle EHI)$ (Why?) $\overrightarrow{HJ}$ intersects $\overrightarrow{EF}$. But $\overrightarrow{HJ} \subset \overrightarrow{AG}$ so that $\overrightarrow{AG}$ intersects $\overrightarrow{EF}$. Thus $\overrightarrow{AB} | \overrightarrow{EF}$ if A—C—E.

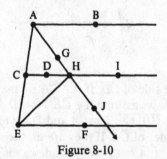

Figure 8-10

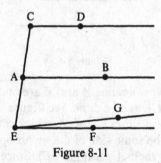

Figure 8-11

Now suppose C—A—E. Through the point E there is a unique ray $\overrightarrow{EG}$ such that $\overrightarrow{EG} | \overrightarrow{AB}$ by Theorem 8.1.4. Thus we have $\overrightarrow{CD} | \overrightarrow{AB}$ and $\overrightarrow{AB} | \overrightarrow{EG}$ and C—A—E. See Figure 8-11. By the first case in our proof, $\overrightarrow{CD} | \overrightarrow{EG}$. By Theorem 8.1.5, $\overrightarrow{EG} | \overrightarrow{CD}$. But there is a unique ray through E asymptotic to $\overrightarrow{CD}$, namely $\overrightarrow{EF} | \overrightarrow{CD}$. Hence $\overrightarrow{EF} = \overrightarrow{EG}$ and $\overrightarrow{EF} | \overrightarrow{AB}$. Thus $\overrightarrow{AB} | \overrightarrow{EF}$ by Theorem 8.1.5.

Finally, suppose that A—E—C. This is the same as the case C—A—E in the above paragraph but with $\overrightarrow{AB}$ and $\overrightarrow{EF}$ interchanged. Hence $\overrightarrow{EF} | \overrightarrow{AB}$ so that $\overrightarrow{AB} | \overrightarrow{EF}$ by Theorem 8.1.5.   □

Since by definition $\overrightarrow{AB} \,|\, \overrightarrow{AB}$, Theorems 8.1.5 and 8.1.7 prove that "asymptotic to" is an equivalence relation on the set of rays.

Definition. The open triangle $\triangle DABC$ is called an **asymptotic (or closed) triangle** if $\overrightarrow{AD} \,|\, \overrightarrow{BC}$.

Suppose that $\triangle DABC$ is an asymptotic triangle. If Ω denotes the equivalence class (under $|$) of $\overrightarrow{AD}$ (and of $\overrightarrow{BC}$), then some authors would write the triangle as $\triangle AB\Omega$. Ω is a "point at infinity" or an "ideal point." We shall not use this terminology except in Problem B13.

Theorem 8.1.8 (Congruence Theorem for Asymptotic Triangles). *In a neutral geometry, if $\triangle DABC$ and $\triangle SPQR$ are two asymptotic triangles with $\overline{AB} \simeq \overline{PQ}$ and $\angle ABC \simeq \angle PQR$, then $\angle BAD \simeq \angle QPS$.*

PROOF. If the angles are not congruent, then one is larger than the other. We may assume that $\angle BAD > \angle QPS$. Choose $E \in \mathrm{int}(\angle BAD)$ with $\angle BAE \simeq \angle QPS$. Since $\overrightarrow{AD} \,|\, \overrightarrow{BC}$, $\overrightarrow{AE}$ intersects $\overrightarrow{BC}$ at a point F. See Figure 8-12.

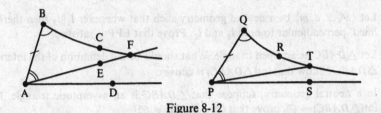

Figure 8-12

Let $T \in \overrightarrow{QR}$ with $\overline{QT} \simeq \overline{BF}$. Then $\triangle ABF \simeq \triangle PQT$ by SAS so that $\angle BAF \simeq \angle QPT$. But $\angle QPT < \angle QPS$ (Why?) which means that $\angle QPS \simeq \angle BAE \simeq \angle QPT < \angle QPS$, a contradiction. Thus we must have $\angle BAD \simeq \angle QPS$. $\qquad\square$

Definition. Two lines l and l' are **asymptotic**, or **asymptotically parallel** (written $l \,|\, l'$), if there are rays $\overrightarrow{AD} \subset l$ and $\overrightarrow{BC} \subset l'$ with $\overrightarrow{AD} \,|\, \overrightarrow{BC}$.

If $\overrightarrow{AB} \,|\, \overrightarrow{CD}$ then it certainly must be true that $\overleftrightarrow{AB} \,\|\, \overleftrightarrow{CD}$. Hence asymptotic lines are parallel. If a geometry satisfies EPP then the converse is also true (Problem A3). However, the situation is quite different in a geometry which satisfies HPP.

Theorem 8.1.9. *In a neutral geometry which satisfies HPP, if two distinct lines l and l' have a common perpendicular, then the lines are parallel but not asymptotic.*

PROOF. Suppose that $\overrightarrow{AB}$ is perpendicular to l at A and l' at B. By Theorem 7.1.2, $l \| l'$. Since we assume HPP, $\Pi(AB) < 90$. Thus l cannot contain a ray that is asymptotic to a ray in l'. Hence l is not asymptotic to l'.   □

In Section 8.3 we shall show that the converse of Theorem 8.1.9 is true: In a hyperbolic geometry if $l \| l'$ but $l \not\chi l'$ then l and l' have a common perpendicular. We shall also see that if $l | l'$ then there are points on l and l' arbitrarily close together so that l approaches l' if l is asymptotic to l. On the other hand if l is parallel to l' but not asymptotic then the lines l and l' actually pull apart. Note how this contrasts with the situation in $\mathscr{E}$ where two lines are parallel if and only if they are equidistant.

PROBLEM SET 8.1

Part A.

1. Prove that $\sim$ is an equivalence relation on the set of rays in a metric geometry.

2. Prove Theorem 8.1.2.

3. Prove that in a neutral geometry which satisfies EPP, $l \| l'$ if and only if $l \| l'$.

4. Let $\{\mathscr{S}, \mathscr{L}, d, m\}$ be a neutral geometry such that whenever $l_1 \| l_2$ then there is a line l' perpendicular to both l_1 and l_2. Prove that EPP is satisfied.

5. Let $\triangle DABC$ be an open triangle. What should be the definition of the interior of $\triangle DABC$? Show that int($\triangle DABC$) is convex.

6. In a neutral geometry, suppose that $\triangle DABC$ is an asymptotic triangle. If $l \cap$ int($\triangle DABC$) $\neq \varnothing$, prove that $l \cap \triangle DABC \neq \varnothing$.

7. In a neutral geometry, if $\overrightarrow{AB} \| \overrightarrow{CD}$, $\overrightarrow{CD} \| \overrightarrow{EF}$ and $A—C—E$ prove that $\overrightarrow{AB} \| \overrightarrow{EF}$.

8. Let $A = (0, 1)$ and $D = (0, 2)$. Sketch two different asymptotic triangles $\triangle DABC$ in $\mathscr{H}$ for some choices of B and C. How many are there? If $E = (1, 1)$ find the unique ray $\overrightarrow{EF}$ with $\overrightarrow{EF} | \overrightarrow{AD}$. (See Theorem 8.1.4.)

9. Let $A = (0, 1)$, $D = (1/\sqrt{2}, 1/\sqrt{2})$ and $E = (0, \frac{1}{2})$ and repeat Problem A8.

10. In the Poincaré Plane let $A = (1, 1)$ and $B = (1, 5)$.
    (a) Sketch five rays asymptotic to $\overrightarrow{AB}$;
    (b) Sketch five rays asymptotic to $\overrightarrow{BA}$.

Part B. "Prove" may mean "find a counterexample".

11. In a neutral geometry prove that "asymptotic to" is an equivalence relation on the set of lines.

12. In a neutral geometry suppose that $\triangle DABC$ is an open triangle. If $l \cap$ int($\triangle DABC$) $\neq \varnothing$ prove that $l \cap \triangle DABC \neq \varnothing$. See Problem A6.

13. Show that there is a bijection between the set of ideal points in $\mathscr{H}$ (that is, the set of equivalence classes of asymptotic rays) and the set $\mathbb{R} \cup \{*\}$, where $*$ denotes an extra point not in $\mathbb{R}$. (Hint: $*$ will correspond to the class of an upward pointing type I ray.)

8.2 Angle Sum and the Defect of a Triangle

Throughout the history of geometry, the Euclidean parallel postulate sparked an enormous amount of interest. It gradually became apparent that this postulate was intimately tied to a concept called the angle defect which we define and investigate in this section.

Definition. Let $\triangle ABC$ be a triangle in a protractor geometry. The **defect** of $\triangle ABC$ is

$$\delta(\triangle ABC) = 180 - (m(\angle A) + m(\angle B) + m(\angle C)).$$

We already know that for a Euclidean geometry $\delta(\triangle ABC) = 0$ for all triangles (Theorem 7.2.9). We have seen examples in which $\delta(\triangle ABC) < 0$ (the Moulton Plane) and in which $\delta(\triangle ABC) > 0$ (the Poincaré Plane). We also know that for a neutral geometry $\delta(\triangle ABC) \geq 0$ (Theorem 7.2.8). Gauss recognized that in order to prove EPP is satisfied it was sufficient to prove that $\delta(\triangle ABC) = 0$ for one triangle. In fact, he actually tried to compute the defect of a large triangle on earth but could not be sure of the exact value due to experimental error.

In this section we will investigate the properties of the defect of a triangle under the assumptions of HPP. We will show that for any $\triangle ABC$ in a hyperbolic geometry, $\delta(\triangle ABC) > 0$. In fact we will show that if t is any number between 0 and 180 then we can find a triangle whose defect is exactly t! To do this requires a detailed study of the critical function Π. The first step in this program is to generalize the Exterior Angle Theorem to asymptotic triangles.

Definition. Let $\triangle DABC$ be an open triangle and let P and Q be points in the neutral geometry with $P-A-D$ and $Q-A-B$. Then both $\angle PAB$ and $\angle QAD$ are **exterior angles** of $\triangle DABC$ whose remote interior angle is $\angle ABC$. (See Figure 8-13. Of course, $\angle PAB \simeq \angle QAD$.)

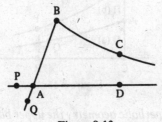

Figure 8-13

One word about our terminology. We will now speak of "a hyperbolic geometry." Remember that this is any neutral geometry which satisfies HPP. It need not refer specifically to the model which we call the Poincaré Plane.

Theorem 8.2.1. *In a hyperbolic geometry, an exterior angle of an asymptotic triangle is greater than its remote interior angle.*

PROOF. Let $\triangle DABC$ be an asymptotic triangle and choose P so that P—A—D. We must show that $\angle PAB > \angle ABC$. Choose E on the same side of $\overleftrightarrow{AB}$ as C with $\angle ABE \simeq \angle PAB$. See Figure 8-14. By Theorem 7.1.1 there is a line perpendicular to both $\overleftrightarrow{AD}$ and $\overleftrightarrow{BE}$. By Theorem 8.1.9, $\overrightarrow{BE} \| \overrightarrow{AD}$ but $\overrightarrow{BE} \nmid \overrightarrow{AD}$. Hence $\overrightarrow{BE} \cap \overrightarrow{AD} = \varnothing$ so that $E \notin \mathrm{int}(\angle ABC)$. Because $\overrightarrow{BC}|\overrightarrow{AD}$, $E \notin \mathrm{int}(\overrightarrow{BC})$. Since both C and E are on the same side of $\overleftrightarrow{AB}$, $C \in \mathrm{int}(\angle ABE)$, so that $\angle ABC < \angle ABE \simeq \angle PAB$. ☐

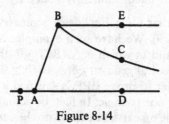

Figure 8-14

Theorem 8.2.2. *In a hyperbolic geometry the critical function Π is strictly decreasing.*

PROOF. We must show that $\Pi(a) > \Pi(b)$ if $0 < a < b$. Let B—A—C with $AC = a$ and $BC = b$. Choose P, Q, R all on the same side of $\overleftrightarrow{BC}$ with $m(\angle CBP) = \Pi(b)$, $m(\angle CAQ) = \Pi(a)$, and $m(\angle ACR) = 90$. See Figure 8-15. Then $\overrightarrow{BP}|\overrightarrow{CR}$ and $\overrightarrow{AQ}|\overrightarrow{CR}$. In the last section we showed that $|$ is an equivalence relation, so that $\overrightarrow{BP}|\overrightarrow{AQ}$. $\triangle QABP$ is therefore an asymptotic triangle. By Theorem 8.2.1 $\angle CAQ > \angle CBP$ so that $\Pi(a) = m(\angle CAQ) > m(\angle CBP) = \Pi(b)$. ☐

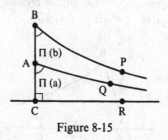

Figure 8-15

Theorem 8.2.3. *In a hyperbolic geometry the upper base angles of any Saccheri quadrilateral are acute.*

PROOF. Let $\boxed{S}ABCD$ be given and choose E and F with A—D—E and B—C—F. Choose P on the same side of $\overleftrightarrow{AB}$ as E, and Q on the same side of $\overleftrightarrow{CD}$ as E with $\overrightarrow{BP}|\overrightarrow{AE}$ and $\overrightarrow{CQ}|\overrightarrow{AE}$ as in Figure 8-16. Note $\overrightarrow{BP}|\overrightarrow{CQ}$ so that $\triangle PBCQ$ is an asymptotic triangle. Furthermore, $Q \in \mathrm{int}(\angle DCF)$.

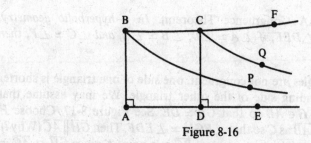

Figure 8-16

Since $\overline{AB} \simeq \overline{DC}$, we have

$$m(\angle ABP) = \Pi(AB) = \Pi(DC) = m(\angle DCQ).$$

By Theorem 8.2.1, $\angle QCF > \angle PBC$ so that

$$\begin{aligned}
m(\angle ABC) = m(\angle BCD) &= 180 - m(\angle DCF) \\
&= 180 - (m(\angle DCQ) + m(\angle QCF)) \\
&= 180 - (\Pi(CD) + m(\angle QCF)) \\
&< 180 - (\Pi(AB) + m(\angle PBC)) \\
&= 180 - (m(\angle ABP) + m(\angle PBC)) \\
&= 180 - m(\angle ABC).
\end{aligned}$$

Hence $2m(\angle ABC) < 180$ or $m(\angle ABC) < 90$. $\qquad\square$

We leave the proof of the next theorem as Problem A1. You might want to recall the proof of Saccheri's Theorem (Theorem 7.2.8) before attacking it.

Theorem 8.2.4. *In a hyperbolic geometry, the sum of the measures of the angles of any triangle is strictly less than* 180.

Note that the defect of any triangle in a hyperbolic geometry is strictly positive. This contrasts with our earlier result that in a neutral geometry $\delta(\triangle ABC) \geq 0$ for all triangles (Theorem 7.2.8). Before proceeding further we state the Defect Addition Theorem whose proof is left as Problem A2.

Theorem 8.2.5 (Defect Addition). *In a protractor geometry, if* $\triangle ABC$ *and* A—D—C *then*

$$\delta(\triangle ABC) = \delta(\triangle ABD) + \delta(\triangle DBC).$$

The next result is surprising because it runs contrary to our intuition. We already know from the SSS Congruence Theorem that the lengths of the three sides of a triangle determine the triangle up to a congruence. In a hyperbolic geometry, the measures of the three angles completely determine the triangle up to a congruence!

Theorem 8.2.6 (AAA Congruence Theorem). *In a hyperbolic geometry, given $\triangle ABC$ and $\triangle DEF$, if $\angle A \simeq \angle D$, $\angle B \simeq \angle E$, and $\angle C \simeq \angle F$, then $\triangle ABC \simeq \triangle DEF$.*

PROOF. If the triangles are not congruent, one side of one triangle is shorter than the corresponding side of the other triangle. We may assume that $\overline{DE} < \overline{AB}$. Choose $G \in \overline{AB}$ so that $\overline{GB} \simeq \overline{DE}$. See Figure 8-17. Choose H on the same side of $\overleftrightarrow{AB}$ as C so that $\angle BGH \simeq \angle EDF$. Then $\overleftrightarrow{GH} \| \overleftrightarrow{AC}$ (Why?). By Pasch's Theorem, $\overleftrightarrow{GH}$ must intersect $\overline{BC}$ in a point K since $\overleftrightarrow{GH} \cap \overleftrightarrow{AC} = \varnothing$. Then $\triangle GBK \simeq \triangle DEF$ by ASA. Hence $\delta(\triangle GBK) = \delta(\triangle DEF)$.

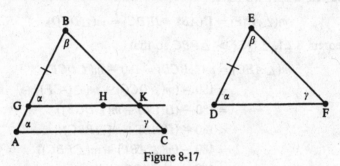

Figure 8-17

On the other hand, by Theorem 8.2.5 we have

$$\delta(\triangle ABC) = \delta(\triangle ABK) + \delta(\triangle AKC)$$
$$= \delta(\triangle AGK) + \delta(\triangle GBK) + \delta(\triangle AKC)$$
$$= \delta(\triangle AGK) + \delta(\triangle DEF) + \delta(\triangle AKC)$$
$$> \delta(\triangle DEF)$$

where the last inequality follows from Theorem 8.2.4. But by hypothesis we have $\delta(\triangle ABC) = \delta(\triangle DEF)$. Hence if $\triangle ABC \not\simeq \triangle DEF$ we have a contradiction. Thus $\triangle ABC \simeq \triangle DEF$. $\qquad\square$

Our next goal is to show that, for a hyperbolic geometry, the critical function takes on all values between 0 and 90. The first step is to show that $\lim_{x \to \infty} \Pi(x) = 0$.

Theorem 8.2.7. *In a hyperbolic geometry, $\lim_{x \to \infty} \Pi(x) = 0$.*

PROOF. Since $\Pi(x)$ is a decreasing, positive function the only way in which the conclusion could be false is if there is a positive number r with $\Pi(x) > r$ for all x. We will show that this assumption leads to the existence of a triangle of defect larger than 180, which is impossible. This triangle is found as a large triangle whose interior contains a large number of congruent triangles (which will each have the same defect).

Let l be a line. For each integer $n \geq 0$ choose a point A_n on l so that $A_n—A_{n+1}—A_{n+2}$ and $d(A_n, A_{n+1}) = 1$. (This could be done by choosing a ruler f for l and letting A_n be the point whose coordinate is n.) Let $l' = \overrightarrow{A_0B}$ be the unique perpendicular to l at A_0. For each $n > 0$ let B_n be a point on the same side of l as B with $m(\angle A_0A_nB_n) = r$. See Figure 8-18.

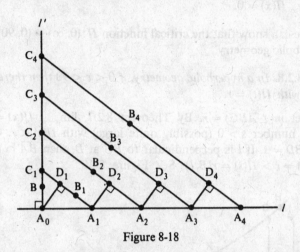

Figure 8-18

If $\Pi(n) > r$ for all n, $\overrightarrow{A_nB_n}$ intersects $\overrightarrow{A_0B}$ at a point C_n by the definition of $r(A_n, l') = \Pi(n)$. For each $n > 0$ let D_n be the foot of the perpendicular from A_{n-1} to $\overrightarrow{A_nC_n}$. Note $A_0—C_1—C_2—C_3 \cdots$.

By HA, $\triangle A_0A_1D_1 \simeq \triangle A_1A_2D_2 \simeq \triangle A_2A_3D_3$, and so on. Thus for each $n \geq 0$, each of the right triangles $\triangle A_nA_{n+1}D_{n+1}$ has the same defect Let this defect be $\delta(\triangle A_0A_1D_1) = a > 0$. We now compare the defects of $\triangle A_0A_nC_n$ and $\triangle A_0A_{n+1}C_{n+1}$. By the Defect Addition Theorem,

$$\delta(\triangle A_0A_{n+1}C_{n+1}) = \delta(\triangle A_0A_nC_{n+1}) + \delta(\triangle A_nC_{n+1}A_{n+1})$$
$$= \delta(\triangle A_0A_nC_n) + \delta(\triangle A_nC_nC_{n+1})$$
$$\quad + \delta(\triangle A_nC_{n+1}D_{n+1}) + \delta(\triangle A_nA_{n+1}D_{n+1})$$
$$> \delta(\triangle A_0A_nC_n) + \delta(\triangle A_nA_{n+1}D_{n+1})$$
$$= \delta(\triangle A_0A_nC_n) + a.$$

Let $d_n = \delta(\triangle A_0A_{n+1}C_{n+1})$ for $n \geq 0$. Then the previous inequality is

$$d_n > d_{n-1} + a \quad \text{for all } n \geq 1.$$

Using this repeatedly we get

$$d_1 > d_0 + a$$
$$d_2 > d_1 + a > d_0 + 2a$$
$$d_3 > d_2 + a > d_0 + 3a$$

and so on. In general,

$$d_n > d_0 + na.$$

If n is large enough then $na > 180$ so that $\delta(\triangle A_0 A_{n+1} C_{n+1}) = d_n > 180$, which is impossible. Hence the assumption $\Pi(x) > r$ for all x must be false and $\lim_{x \to \infty} \Pi(x) = 0$.   □

Now we can show that the critical function $\Pi : (0, \infty) \to (0, 90)$ is bijective in a hyperbolic geometry.

Theorem 8.2.8. *In a hyperbolic geometry, if $0 < r < 90$ then there is a unique number t with $\Pi(t) = r$.*

PROOF. Let $m(\angle ABC) = r$. By Theorem 8.2.7, $\lim_{x \to \infty} \Pi(x) = 0$ so that there is a number $s > 0$ (possibly quite large) with $\Pi(s) < r$. Choose $D \in \overline{BC}$ with $BD = s$. If l is perpendicular to $\overleftrightarrow{BC}$ at D, then $\overrightarrow{BA} \cap l = \varnothing$ since $m(\angle ABC) = r > \Pi(s) = r(B, l)$. See Figure 8-19.

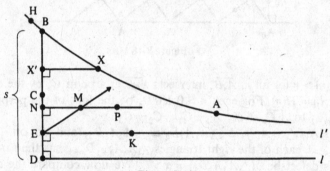

Figure 8-19

For each point $X \in \text{int}(\overrightarrow{BA})$ let X' be the foot of the perpendicular from X to $\overleftrightarrow{BC}$. $\overleftrightarrow{XX'} \| l$ (Why?) and must lie on the same side of l as B. Thus if $X \in \text{int}(\overrightarrow{BA})$, $X' \in \overline{BD}$. Let

$$\mathscr{F} = \{u = d(B, X') = BX' \,|\, X \in \text{int}(\overrightarrow{BA})\}.$$

$\mathscr{F}$ is non-empty and is bounded by $s = BD$. Thus $\mathscr{F}$ has a least upper bound t. We claim that for this t, $\Pi(t) = r$.

Choose $E \in \overline{BC}$ with $BE = t$. Let l' be the perpendicular to $\overleftrightarrow{BC}$ at E. We now show that $l' \cap \overrightarrow{BA} = \varnothing$. Suppose to the contrary that l' intersects $\overrightarrow{BA}$ at a point F and choose G with B—F—G. Since $G \in \text{int}(\overrightarrow{BA})$ we have a corresponding point G' on $\overline{BD}$. $\overleftrightarrow{GG'} \| l'$ and $\overleftrightarrow{GG'}$ lies on the opposite side of l' as B does. Hence $BG' > BE = t$, which contradicts the fact that t is the least upper bound of $\mathscr{F}$. Thus $l' \cap \overrightarrow{BA} = \varnothing$.

If A—B—H then H and E lie on opposite sides of the perpendicular to $\overleftrightarrow{BC}$ through B. Since this perpendicular line is parallel to l', $\overrightarrow{BH} \cap l' = \varnothing$ and

$\overleftrightarrow{AB} \| l'$. Hence $\triangle KEBA$ is an open triangle, where K is a point on l' on the same side of $\overleftrightarrow{BE}$ as A. If we show that $\triangle KEBA$ is an asymptotic triangle then $r = m(\angle ABE) = \Pi(BE) = \Pi(t)$. This is done in the next paragraphs.

Let $M \in \text{int}(\angle BEK)$. We need to prove that $\overrightarrow{EM} \cap \overrightarrow{BA} \neq \varnothing$. If $\overrightarrow{EM} \cap \overrightarrow{BA} = \varnothing$, let N be the foot of the perpendicular from M to $\overleftrightarrow{BC}$ and let P be a point with $N-M-P$. We now show that M and N are on the same side of $\overleftrightarrow{BA}$. $\overrightarrow{EM} \cap \overrightarrow{BH} = \varnothing$ (Why?) and $\overrightarrow{EM} \cap \overrightarrow{BA} = \varnothing$ so that $\overrightarrow{EM} \cap \overleftrightarrow{BA} = \varnothing$. Thus E and M are on the same side of $\overleftrightarrow{BA}$. Since E and N are on the same side of $\overleftrightarrow{BA}$ (because both are in $\text{int}(\overleftrightarrow{BD})$), M and N are on the same side of $\overleftrightarrow{BA}$. Therefore, $\overleftrightarrow{BA} \cap \overline{MN} = \varnothing$.

Since $\overrightarrow{EM} \cap \overrightarrow{BA}$ is assumed to be empty, $\text{int}(\overrightarrow{MP})$ and $\overrightarrow{BA}$ lie on opposite sides of $\overleftrightarrow{EM}$ and so $\text{int}(\overrightarrow{MP}) \cap \overrightarrow{BA} = \varnothing$. Thus $\overrightarrow{NM} \cap \overrightarrow{BA} = \varnothing$ which means that lub $\mathscr{F} \leq BN$. But since $M \in \text{int}(\angle BEK)$ we have $B-N-E$ so that $BN < BE$. This contradicts the fact that lub $\mathscr{F} = BE$. Hence it must be that $\overrightarrow{EM} \cap \overrightarrow{BA} \neq \varnothing$. Thus $\overrightarrow{EK} | \overrightarrow{BA}$ so that $\triangle KEBA$ really is an asymptotic triangle and $\Pi(t) = r$.

Since $\Pi(x)$ is strictly decreasing there can be only one value of x such that $\Pi(x) = r$. Hence there is a unique value t such that $\Pi(t) = r$. $\square$

Corollary 8.2.9. *In a hyperbolic geometry* $\lim_{x \to 0^+} \Pi(x) = 90$.

As our final result we would like to show that if $0 < r < 180$ then there is a triangle whose defect is exactly r. It should not surprise you that the proof will be based on a continuity argument.

Theorem 8.2.10. *In a hyperbolic geometry, if* $0 < r < 180$, *then there is a triangle whose defect is exactly* r.

PROOF. First we construct a triangle $\triangle ABC$ whose defect is greater than r. Let t be the (unique) number such that $\Pi(t) = \frac{1}{4}(180 - r)$. Choose points A, B, C and D with $A-D-C$, $\overline{BD} \perp \overline{AC}$, and $AD = BD = CD = t$. See Figure 8-20. Then $\angle DAB \simeq \angle DBA \simeq \angle DBC \simeq \angle DCB$. Since $\overrightarrow{AB} \cap \overrightarrow{DB} \neq \varnothing$, $m(\angle DAB) < \Pi(AD) = \Pi(t) = \frac{1}{4}(180 - r)$. Hence

$$\delta(\triangle ABC) = 180 - (m(\angle DAB) + m(\angle DBA) + m(\angle DBC) + m(\angle DCB))$$
$$= 180 - 4(m(\angle DAB))$$
$$> 180 - 4 \cdot \tfrac{1}{4}(180 - r) = r.$$

Let $s = m(\angle BAC)$. For each number x with $0 < x < s$ there is a point $P_x \in \text{int}(\angle BAC)$ with $m(\angle BAP_x) = x$. By the Crossbar Theorem $\overrightarrow{AP_x}$ intersects $\overline{BC}$ at a point Q_x. See Figure 8-21. We define a function $g : [0, s] \to \mathbb{R}$ by the rule

$$g(x) = \begin{cases} 0 & \text{if } x = 0 \\ \delta(\triangle ABQ_x) & \text{if } 0 < x < s \\ \delta(\triangle ABC) & \text{if } x = s. \end{cases}$$

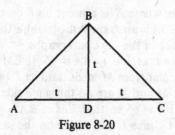

Figure 8-20

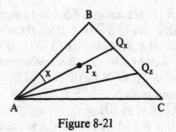

Figure 8-21

Suppose that we are able to show that g is a continuous function. Then since

$$g(0) = 0 < r < \delta(\triangle ABC) = g(s)$$

the Intermediate Value Theorem says that there is a number x with $0 < x < s$ and $g(x) = r$. Then $\delta(\triangle ABQ_x) = r$ and we would be done.

To show that g is continuous at z we must show that if $\varepsilon > 0$ then there is a number $\delta > 0$ such that whenever $x \in [0,s]$ and $|x - z| < \delta$ then $|g(x) - g(z)| < \varepsilon$. (This δ is not a defect!) For the sake of notation we write $B = Q_0$ and $C = Q_s$. Then if $x \neq z$, $|g(x) - g(z)| = |\delta(\triangle ABQ_x) - \delta(\triangle ABQ_z)| = \delta(\triangle Q_xAQ_z)$ and $|x - z| = m(\angle Q_xAQ_z)$. See Figure 8-21.

Let $\varepsilon > 0$ be given. Let f be a coordinate system for $\overleftrightarrow{BC}$ with origin at C and $f(B) < 0$. For each integer n let E_n be the point with coordinate $f(E_n) = n$. (Note $E_0 = C$.) We claim that for some positive n, $\delta(\triangle E_nAE_{n+1}) < \varepsilon$.

Assume to the contrary that for all $n \geq 0$, $\delta(\triangle E_nAE_{n+1}) \geq \varepsilon$. Then by the Defect Addition Theorem, $\delta(\triangle CAE_n)) \geq n\varepsilon$. For n large enough, $n\varepsilon > 180$. Thus for large values of n, $\delta(\triangle CAE_n) > 180$, which is impossible. Hence for some value of n, $\delta(\triangle E_nAE_{n+1}) < \varepsilon$. Fix this value of n.

Let $\delta = m(\angle E_nAE_{n+1})$. Suppose that $x \in [0,s]$ with $|x - z| < \delta$, so that $m(\angle Q_xAQ_z) < \delta = m(\angle E_nAE_{n+1})$. Then there exists a point $F \in \text{int}(\angle E_nAE_{n+1})$ with $\angle Q_xAQ_z \simeq \angle E_nAF$. Let $\overrightarrow{AF} \cap \overline{E_nE_{n+1}} = \{G\}$. See Figure 8-22. Now $AB < AD + DB$ by the Triangle Inequality. Since $\overline{DB} \simeq \overline{DC}$, $AB < AD + DC = AC$. See Figure 8-20. By Problem A8 of Section 6.3, if $P \in \overline{BC}$ then $AP \leq AC$. By the same problem (applied to $\triangle ABE_n$) $AC \leq AE_n$

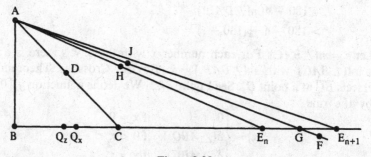

Figure 8-22

and (applied to $\triangle ABG$) $AC \le AG$. Thus for any $P \in \overline{BC}$, $AP \le AE_n$ and $AP \le AG$. In particular this is true for $P = Q_x$ and for $P = Q_z$.

Choose $H \in \overline{AE_n}$ and $J \in \overline{AG}$ with $\overline{AQ_z} \simeq \overline{AH}$ and $\overline{AQ_x} \simeq \overline{AJ}$. Then $\triangle Q_z A Q_x \simeq \triangle HAJ$. By Problem A11, $\delta(\triangle HAJ) < \delta(\triangle E_n A E_{n+1}) < \varepsilon$. Hence $\delta(\triangle Q_z A Q_x) < \delta(\triangle E_n A E_{n+1}) < \varepsilon$ so that $|g(x) - g(z)| < \varepsilon$. Thus we have shown that if $\varepsilon > 0$ is given, there is a number $\delta > 0$ such that $|g(x) - g(z)| < \varepsilon$ whenever $|x - z| < \delta$. Hence g is continuous at z, for each $z \in [0, s]$. $\square$

Problem Set 8.2

Part A.

1. Prove Theorem 8.2.4.

2. Prove the Defect Addition Theorem (Theorem 8.2.5).

3. Without using the results of this section prove that in $\mathcal{H}$ the critical function takes on all values between 0 and 90. (Hint: See Example 7.3.5.)

4. Prove that congruent triangles have the same defect.

5. Prove Corollary 8.2.9.

*6. Let $\angle ABC$ be given in a hyperbolic geometry. Prove that there is a unique line $l = \overleftrightarrow{DE}$ with $\overrightarrow{DE} \,|\, \overrightarrow{BC}$ and $\overrightarrow{ED} \,|\, \overrightarrow{BA}$. (Hint: Let $\overrightarrow{BF}$ be the bisector of $\angle ABC$. Choose G on $\overrightarrow{BF}$ so that $\Pi(BG) = m(\angle ABF)$. Let l be perpendicular to $\overrightarrow{BF}$ at G.) l is called the **line of enclosure** for $\angle ABC$. The set $\angle ABC \cup l$ is called a **doubly asymptotic triangle**.

7. Illustrate the line of enclosure for various angles in $\mathcal{H}$ using both type I and type II rays.

8. Consider $\angle ABC$ in $\mathcal{H}$ where $\overrightarrow{AB} = {}_0L$ and $\overrightarrow{BC} = {}_cL_r$, as in Figure 8-23. Prove that the bisector of $\angle ABC$ is part of ${}_dL_s$ where $d = c + r$.

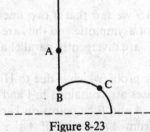

Figure 8-23

9. Prove that in a hyperbolic geometry there are no rectangles.

*10. Prove that in a hyperbolic geometry if $\square ABCD$ is a Lambert quadrilateral then $\angle D$ is acute.

11. In a protractor geometry, suppose that $\triangle ABC \subset (\triangle DEF \cup \text{int}(\triangle DEF))$. Prove that $\delta(\triangle ABC) \le \delta(\triangle DEF)$.

Part B. "Prove" may mean "find a counterexample".

12. Prove the AAA Congruence Theorem for a Euclidean geometry.

13. In a hyperbolic geometry, if $0 < t < 180$ prove there is a number $d > 0$, which depends on t, such that $\delta(\triangle PQR) < t$ for all triangles whose sides have length less than d.

14. Prove that in any hyperbolic geometry, the critical function $\Pi(x)$ is continuous.

15. In a hyperbolic geometry, prove that an exterior angle of an open triangle is greater than its remote interior angle.

8.3 The Distance Between Parallel Lines

We have seen that there are two types of parallel lines in a hyperbolic geometry: those that have a common perpendicular and those that don't. After proving that the property of two lines having a common perpendicular is equivalent to the lines not being asymptotic, we will investigate properties which deal with the distance between parallel lines. In the Euclidean plane two lines are parallel if and only if they are equidistant (Problem A4). This contrasts considerably to the hyperbolic situation. We will see that either two parallel lines are asymptotic or the perpendicular distance from a point on one to the other can be made arbitrarily large! This is the reason for the terminology in the next definition.

Definition. Two lines in a hyperbolic geometry are **divergently parallel** if they are parallel but not asymptotic.

Theorem 8.3.1. *In a hyperbolic geometry, two lines l and l' are divergently parallel if and only if they have a common perpendicular.*

PROOF. In Theorem 8.1.9 we saw that if two lines have a common perpendicular then they are not asymptotic and thus are divergently parallel. Thus we assume that l and l' are divergently parallel and show that they have a common perpendicular.

The basic idea of the proof (which is due to Hilbert) is to find a Saccheri quadrilateral whose bases are contained in l and l'. This is done in steps 1 and 2 of the proof. Let A and B be points on l and let A' and B' be the feet of the perpendiculars from A to B to l'. If $\overline{AA'} \simeq \overline{BB'}$ then $\square A'ABB'$ is the desired Saccheri quadrilateral so that we may proceed directly to step 3. Otherwise we may assume that $\overline{AA'} > \overline{BB'}$.

Step 1. Choose C with A—C—A' and $\overline{CA'} \simeq \overline{BB'}$, choose D with A—B—D, and choose D' with A'—B'—D'. Finally let $\overrightarrow{CE}$ be the unique ray with $\angle A'CE \simeq \angle B'BD$ and E on the same side of $\overleftrightarrow{AA'}$ as B. See Figure 8-24. In this first step we will show that $\overrightarrow{CE} \cap \overrightarrow{AD} \neq \varnothing$. This will involve finding a ray $\overrightarrow{A'P}$ which is asymptotic to $\overrightarrow{CE}$ and which does intersect $\overrightarrow{AD}$.

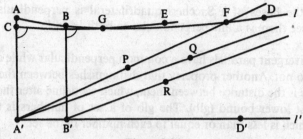

Figure 8-24

Let $\overrightarrow{A'P}$ be the unique ray through A' asymptotic to $\overleftrightarrow{CE}$, let $\overrightarrow{A'Q}$ be the ray through A' asymptotic to $\overleftrightarrow{AD}$, and let $\overrightarrow{B'R}$ be the ray through B' asymptotic to $\overleftrightarrow{AD}$. Now $\triangle PA'CE$ and $\triangle RB'BD$ are asymptotic triangles with $\overline{CA'} \simeq \overline{BB'}$ and $\angle A'CE \simeq \angle B'BD$. Hence $\angle CA'P \simeq \angle BB'R$ by Theorem 8.1.8.

$P \notin \overline{A'D'}$ and $R \notin \overline{A'D'}$ (Why?). Thus $\angle PA'D'$ and $\angle RB'D'$ exist and are congruent by the Angle Subtraction Theorem. Since $\overrightarrow{A'Q} | \overleftrightarrow{AD}$ and $\overrightarrow{B'R} | \overleftrightarrow{AD}$, we have $\overrightarrow{A'Q} | \overrightarrow{B'R}$. By Theorem 8.2.1, $\angle QA'D' < \angle RB'D'$. Since $\angle PA'D' \simeq \angle RB'D'$, $\angle QA'D' < \angle PA'D'$. Looking at the complements we see that $\angle AA'Q > \angle AA'P$ and $P \in \text{int}(\angle AA'Q)$. Because $\overrightarrow{A'Q} | \overleftrightarrow{AD}$, $\overrightarrow{A'P}$ intersects $\overleftrightarrow{AD}$ at a point F.

Now A' and F are on the same side of $\overleftrightarrow{CE}$, while A' and A are on opposite sides of $\overleftrightarrow{CE}$. Thus A and F are on opposite sides of $\overleftrightarrow{CE}$ and $\overline{AF}$ intersects $\overleftrightarrow{CE}$ at a point G. G is on $\overleftrightarrow{CE}$ since G is on the same side of $\overleftrightarrow{AA'}$ as F is. Thus $\overleftrightarrow{CE}$ intersects $\overline{AB}$ at a point G as claimed.

Step 2. We now construct the Saccheri quadrilateral using the point G. Let H be the unique point on $\overline{BD}$ with $\overline{CG} \simeq \overline{BH}$ as in Figure 8-25. Let G' and H' be the feet of the perpendiculars from G and H to l'. $\triangle A'CG \simeq \triangle B'BH$ by SAS so that $\overline{A'G} \simeq \overline{B'H}$ and $\angle CA'G \simeq \angle BB'H$. Thus $\angle GA'G' \simeq \angle HB'H'$ and $\triangle GA'G' \simeq \triangle HB'H'$ by HA. Hence $\overline{GG'} \simeq \overline{HH'}$ and $\square G'GHH'$ is a Saccheri quadrilateral.

Step 3. We now have a Saccheri quadrilateral with its lower base in l' and its upper base in l. By Problem A6 of Section 7.2, the line through the mid-

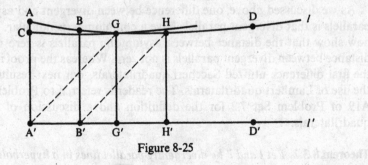

Figure 8-25

points of the bases of a Saccheri quadrilateral is perpendicular to both bases. Hence there is a line perpendicular to both l and l'.   □

Thus, divergent parallels have a common perpendicular while asymptotic parallels do not. Another property that distinguishes between the two types of parallels is the distance between them, which we define after the notion of the greatest lower bound (glb). The glb of a set of numbers is the largest number which is less than or equal to each number in the set.

Definition. If $\mathscr{B}$ is a set of real numbers, then $s \in \mathbb{R}$ is a **greatest lower bound** of $\mathscr{B}$ (written $s = \text{glb } \mathscr{B}$) if

 (i) $s \leq b$ for all $b \in \mathscr{B}$; and
(ii) if $r > s$ then there is an element $b_r \in \mathscr{B}$ with $b_r < r$.

The concept of the greatest lower bound is analogous to the least upper bound as defined in Section 7.3. If $\mathscr{B}$ is non-empty and if there is a number M with $M \leq b$ for all $b \in \mathscr{B}$ then $\mathscr{B}$ has a unique greatest lower bound.

Definition. In a metric geometry, the **distance from a point P to a line l** is

$$d(P, l) = \text{glb}\{d(P, Q) | Q \in l\}.$$

The **distance between the lines l' and l** is

$$d(l', l) = \text{glb}\{d(P, Q) | P \in l' \text{ and } Q \in l\}.$$

Both of the numbers defined above exist because the sets are non-empty and consist just of positive numbers. We know from Section 6.4 that in a neutral geometry $d(P, l) = PQ$ where Q is the (unique) foot of the perpendicular from P to l. It is not hard to show that in a metric geometry

$$d(l', l) = \text{glb}\{d(P, l) | P \in l'\} = \text{glb}\{d(Q, l') | Q \in l\}. \tag{3-1}$$

The distance between two lines l and l' may be thought of as the distance between two "closest" points. Care must be exercised, however, with this interpretation because there may not be points $P \in l$ and $Q \in l'$ with $d(l', l) = d(P, Q)$. For example, if $l = {}_{-1}L_1$ and $l' = {}_0L$ then $d(l, l') = 0$ but $l \cap l' = \varnothing$.

As we discussed above, one difference between divergent and asymptotic parallels is that divergent parallels have a common perpendicular. We shall now show that the distance between asymptotic parallels is zero while the distance between divergent parallels is not zero. Whereas the proof involving the first difference utilized Saccheri quadrilaterals, our next results require the use of Lambert quadrilaterals. The reader is referred to Problems A15–A19 of Problem Set 7.2 for the definition and a discussion of Lambert quadrilaterals.

Theorem 8.3.2. *Let l and l' be divergently parallel lines in a hyperbolic geometry. If $A \in l$ and $A' \in l'$ are points such that $\overline{AA'}$ is perpendicular to both l and*

l', *then*

$$d(l, l') = d(A, A').$$

Furthermore, if A—B—C *then* $d(B, l') < d(C, l')$.

PROOF. By Theorem 6.4.2, $d(A, l') = d(A, A')$. Let B be another point of l and let B' be the foot of the perpendicular from B to l'. Thus $\square B'A'AB$ has right angles at B', A', and A and so is a Lambert quadrilateral. By Problem A19 of Problem Set 7.2, $AA' \leq BB'$. Thus $d(A, l') = AA' \leq BB' = d(B, l')$. Hence by Equation (3-1),

$$d(l, l') = \text{glb}\{d(P, l') \mid P \in l\} = d(A, l') = d(A, A').$$

Now suppose that A—B—C and let C' be the foot of the perpendicular from C to l'. By Problem A10 of Problem Set 8.2, $\angle ABB'$ is acute so that $\angle CBB'$ is obtuse. We must show that $BB' < CC'$. See Figure 8-26. $\angle BCC' = \angle ACC'$ is acute by Problem A10 of Problem Set 8.2. Hence $\angle CBB' > \angle BCC'$ so that $BB' < CC'$ by Corollary 7.2.12. $\square$

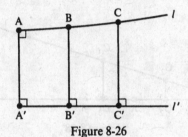

Figure 8-26

Let l and l' be divergently parallel lines in a hyperbolic geometry. We will write $\overline{AA'}$ for their common perpendicular which is guaranteed by Theorem 8.3.1 and assume $A \in l$ and $A' \in l'$. Theorem 8.3.2 tells us that the distance between l and l' is the length of the common perpendicular $\overline{AA'}$. Furthermore, the farther the point $C \in l$ is from A, the greater the distance from C to l'. Thus the lines l and l' get farther and farther apart at their "ends." How far apart can they get? We shall see in Theorem 8.3.5 that the answer is arbitrarily far, but first two preliminary results are necessary.

Theorem 8.3.3. *In a neutral geometry let* $\triangle ABC$ *have a right angle at* B. *Let* C' *be the point such that* A—C—C' *and* $\overline{AC} \simeq \overline{CC'}$. *Let* B' *be the foot of the perpendicular from* C' *to* $\overrightarrow{AB}$. *Then* $B'C' \geq 2BC$ *and* $AB' \leq 2AB$.

PROOF. The situation is sketched in Figure 8-27. Let D be the foot of the perpendicular from C' to $\overleftrightarrow{BC}$. Then $\angle ACB \simeq \angle C'CD$ and so $\triangle ACB \simeq \triangle C'CD$ by HA. Hence $\overline{CB} \simeq \overline{CD}$ and $\overline{AB} \simeq \overline{C'D}$.

$\square B'BDC$ is a Lambert quadrilateral. By Corollary 7.2.12, $B'C' \geq BD$ and $DC' \geq BB'$. Thus $B'C' \geq BD = BC + CD = 2BC$ and $AB' = AB + BB' \leq AB + DC' = 2AB$. $\square$

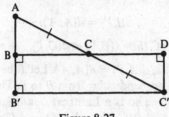

Figure 8-27

Theorem 8.3.4 (Aristotle's Theorem). *If $\angle ABC$ is an angle in a neutral geometry and if $r > 0$ then there is a point $E \in \overleftrightarrow{BC}$ such that $d(E, \overleftrightarrow{AB}) > r$.*

PROOF. If $\angle ABC$ is a right angle the result is trivial since $d(E, \overleftrightarrow{AB}) = d(E, B)$ in that case. If $\angle ABC$ is obtuse let A'—B—A. Then $\angle A'BC$ is acute and $\overleftrightarrow{A'B} = \overleftrightarrow{AB}$. Hence it suffices to consider the case where $\angle ABC$ is acute as in Figure 8-28.

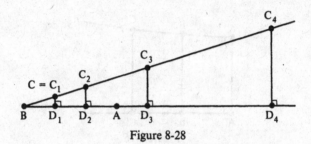

Figure 8-28

Choose points $C_1 = C, C_2, C_3 \cdots$ so that B—C_1—C_2—C_3—$\cdots$ and $\overline{BC_1} \simeq \overline{C_1C_2}$, $\overline{BC_2} \simeq \overline{C_2C_3}, \ldots$. Let D_i be the foot of the perpendicular from C_i to $\overleftrightarrow{AB}$.

By the first part of Theorem 8.3.3, $C_{n+1}D_{n+1} \geq 2C_nD_n$. By induction $C_{n+1}D_{n+1} \geq 2^nC_1D_1$ if $n \geq 1$. If n is large enough ($n > \log_2(r/C_1D_1)$) then $2^nC_1D_1 > r$ so that $C_{n+1}D_{n+1} > r$. We may let $E = C_{n+1}$ to obtain

$$d(E, \overleftrightarrow{AB}) = d(C_{n+1}, D_{n+1}) = C_{n+1}D_{n+1} > r. \qquad \square$$

We now prove that divergently parallel lines in a hyperbolic geometry get arbitrarily far apart.

Theorem 8.3.5. *In a hyperbolic geometry if l is divergently parallel to l' and $r > 0$, then there is a point $P \in l$ such that $d(P, l') > r$.*

PROOF. Let $\overleftrightarrow{AA'}$ be a common perpendicular to l and l' with $A \in l$ and $A' \in l'$. Let B be another point of l and let B' be the foot of the perpendicular from B to l'. Let $\overrightarrow{AE}$ be the unique ray through A with $\overrightarrow{AE}|\overrightarrow{A'B'}$. $E \in \text{int}(\angle A'AB)$ (Why?) so that $\angle BAE$ is acute. By Theorem 8.3.4 there is a point $P \in \overleftrightarrow{AB}$

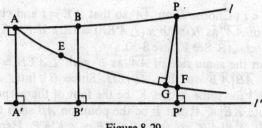

Figure 8-29

with $d(P, \overleftrightarrow{AE}) > r$. See Figure 8-29. Let G be the foot of the perpendicular from P to $\overleftrightarrow{AE}$ and let P' be the foot of the perpendicular from P to l'.

Applying the Crossbar Theorem to $\angle A'AP$ we see that $\overrightarrow{AE} \cap \overline{A'P} \neq \emptyset$. By Pasch's Theorem applied to $\triangle A'PP'$, $\overrightarrow{AE} \cap \overline{PP'} \neq \emptyset$. (Note $\overrightarrow{AE} \cap \overline{A'P'} = \emptyset$ because $\overleftrightarrow{AE} | \overleftrightarrow{A'B'}$.) In fact, $\overrightarrow{AE} \cap \overline{PP'} \neq \emptyset$ since $\overline{PP'}$ is on the same side of $\overleftrightarrow{AA'}$ as E. Hence $\overrightarrow{AE} \cap \overline{PP'} = \{F\}$ for some F with $P-F-P'$. Thus

$$d(P, l') = d(P, P') > d(P, F) > d(P, G) = d(P, \overleftrightarrow{AE}) > r.$$

Hence divergently parallel lines actually diverge! □

Since divergently parallel lines diverge, we might expect that asymptotic lines converge in the sense that the distance between them is zero. This is true as Corollary 8.3.8 will show.

Theorem 8.3.6. *If* $\overleftrightarrow{AB}$ *is strictly asymptotic to* $\overleftrightarrow{CD}$ *in a hyperbolic geometry then* $d(A, \overleftrightarrow{CD}) > d(B, \overleftrightarrow{CD})$.

PROOF. Let A' and B' be the feet of the perpendiculars from A and B to $\overleftrightarrow{CD}$. Choose $E \in \overrightarrow{B'B}$ with $\overline{AA'} \simeq \overline{EB'}$ so that $\square A'AEB'$ is a Saccheri quadrilateral. $\overrightarrow{AE}$ is divergently parallel to $\overrightarrow{A'B'} = \overleftrightarrow{CD}$ by Theorem 8.3.1. Since $\overleftrightarrow{AB} | \overleftrightarrow{A'B'}$, $\angle A'AB < \angle A'AE$. Hence $B'-B-E$ and $AA' = EB' > BB'$. Thus

$$d(A, \overleftrightarrow{CD}) = AA' > BB' = d(B, \overleftrightarrow{CD}). \qquad \square$$

Thus the distance between asymptotic rays gets smaller and smaller the farther out on the rays you go. Does it actually approach zero? This is the essence of the next result (which we restate in terms of distance in Corollary 8.3.8).

Theorem 8.3.7. *In a hyperbolic geometry, if* $\overleftrightarrow{AB} | \overleftrightarrow{CD}$ *and if* $t > 0$ *then there is a point* $P \in \overrightarrow{AB}$ *such that* $d(P, \overleftrightarrow{CD}) \leq t$.

PROOF. Let A' and B' be the feet of the perpendiculars from A and B to $\overleftrightarrow{CD}$. We may assume that $t < AA'$. (If not, let $0 < t^* < AA'$ and find $P \in \overrightarrow{AB}$ with

$d(P, \overline{CD}) \le t^* < t$.) Choose E on $\overrightarrow{AA'}$ so that $A'E = t$ and choose F on the opposite side of $\overleftrightarrow{AA'}$ as B' with $m(\angle A'EF) = \Pi(t)$. If $F - E - G$, we claim that $\overrightarrow{EG}$ intersects $\overrightarrow{AB}$. See Figure 8-30.

Let H be on the same side of $\overleftrightarrow{AA'}$ as B' with $\angle A'EH \simeq \angle A'EF$. Then $\overrightarrow{EH} | \overleftrightarrow{A'B'}$ and $\overrightarrow{AB} | \overleftrightarrow{A'B'}$ so that $\overrightarrow{AB} | \overrightarrow{EH}$. Since $G \in \text{int}(\angle AEH)$ (Why?), $\overrightarrow{EG} \cap \overrightarrow{AB} = \{K\}$ for some K. Let K' be the foot of the perpendicular from K to $\overleftrightarrow{A'B'}$. Note $\overrightarrow{KE} | \overrightarrow{K'A'}$. Let P be the point on $\overrightarrow{AB}$ such that $A - K - P$ and $\overrightarrow{KE} \simeq \overrightarrow{KP}$. $\overrightarrow{KP} | \overleftrightarrow{A'B'}$ so that $\angle K'KE \simeq \angle K'KP$. Hence $\triangle K'KE \simeq \triangle K'KP$ by SAS. Let P' be the foot of the perpendicular from P to $\overleftrightarrow{A'B'}$. $\overline{EK'} \simeq \overline{PK'}$ and $\angle EK'A' \simeq \angle PK'P'$ (Why?). Hence $\triangle EK'A' \simeq \triangle PK'P'$ by HA and $\overline{EA'} \simeq \overline{PP'}$. Hence

$$d(P, \overline{CD}) = d(P, P') = d(E, A') = t. \qquad \square$$

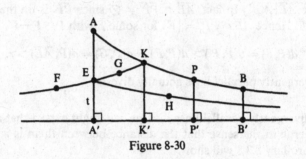

Figure 8-30

Corollary 8.3.8. *In a hyperbolic geometry, the distance between asymptotic rays is zero.*

Thus asymptotic rays actually do converge, and hence so do asymptotic lines at the end at which they are asymptotic. In Problem A5 you will show that asymptotic lines diverge at the end at which they are not asymptotic. The situation in hyperbolic geometry is quite different from that in Euclidean geometry. In hyperbolic geometry, two parallel lines either diverge or converge. In Euclidean geometry, parallel lines are equidistant.

We end this chapter with a brief discussion of the distance scale for a hyperbolic geometry and the idea of an isometry. The distance scale may be omitted since it is not used elsewhere. Isometries will be studied in detail in Chapter 11.

We know from Problem B16 of Section 2.2 that if $t > 0$ and if $\{\mathscr{S}, \mathscr{L}, d\}$ is a metric geometry so is $\{\mathscr{S}, \mathscr{L}, d'\}$ where $d'(A, B) = td(A, B)$. The metric geometries $\{\mathscr{S}, \mathscr{L}, d\}$ and $\{\mathscr{S}, \mathscr{L}, d'\}$ have the same segments and rays (Proof?). If one satisfies PSA so does the other. If $\{\mathscr{S}, \mathscr{L}, d, m\}$ is a protractor geometry, so is $\{\mathscr{S}, \mathscr{L}, d', m\}$. Finally, if $\{\mathscr{S}, \mathscr{L}, d, m\}$ satisfies SAS so does $\{\mathscr{S}, \mathscr{L}, d', m\}$. Switching from d to d' is called a *change of scale* and is like changing from inches to meters. In a Euclidean geometry, there isn't much to be gained by making such a change as we will see when we discuss the theory of similar triangles in Chapter 9. In particular, there is no "best" choice of a scale factor t for a Euclidean geometry.

Consider, for example, what happens to the formulas of trigonometry if a change of scale is introduced. The sine of an angle is the ratio of two lengths. Thus if the scale is changed by a factor of t each of the lengths is multiplied by a factor of t, but their ratio is unchanged. This does *not* happen in a hyperbolic geometry. Indeed there is a constant (depending on the scale) which appears throughout hyperbolic trigonometry. One way to fix the scale factor t is to insist that a certain distance have 45 as its angle of parallelism.

Definition. The **distance scale** of a hyperbolic geometry is the unique number s such that $\Pi(s) = 45$.

The distance scale is sometimes referred to as the absolute unit of length. The existence of such an object caused problems in the early development of hyperbolic geometry because it was so unlike the Euclidean situation where the choice of unit was totally a matter of taste and one choice was as good as another. Since Bolyai and Lobachevsky were developing their new geometry from what we would call the synthetic viewpoint, the distance scale entered into their work as an arbitrary constant whose value could not be determined. In particular, it meant that the theory of hyperbolic trigonometry was continually clouded with this constant whose value could not be determined. You can make the distance scale *seem* to disappear in the hyperbolic case by choosing $s = \ln(1 + \sqrt{2})$. This has the net result of making the constant in the hyperbolic trigonometric formulas become a factor of 1. For a detailed development of the theory of hyperbolic trigonometry and other aspects of hyperbolic geometry, see Martin [1975].

If s is the distance scale for the hyperbolic geometry $\{\mathscr{S}, \mathscr{L}, d, m\}$, then the distance scale for $\{\mathscr{S}, \mathscr{L}, d/s, m\}$ is 1. That means that for $\{\mathscr{S}, \mathscr{L}, d/s, m\}$ we have as our "standard" unit of distance, the length whose associated critical angle is 45. This would seem to be the most natural choice of scale for $\Pi(1) = 45$ here. However, for reasons in the field of differential geometry and hyperbolic trigonometry, it is better to choose the scale so that $\Pi(\ln(1 + \sqrt{2})) = 45$. That is, the distance scale of $\ln(1 + \sqrt{2})$ is the best choice. Note that $\ln(1 + \sqrt{2}) = \sinh^{-1}(1)$.

Definition. Let $\{\mathscr{S}, \mathscr{L}, d, m\}$ and $\{\mathscr{S}', \mathscr{L}', d', m'\}$ be two protractor geometries. A function $f: \mathscr{S} \to \mathscr{S}'$ is an **isometry** if

(i) f is a bijection;
(ii) $f(l) \in \mathscr{L}'$ if $l \in \mathscr{L}$;
(iii) $d'(f(A), f(B)) = d(A, B)$ for all $A, B \in \mathscr{S}$;
(iv) $m'(\angle f(A)f(B)f(C)) = m(\angle ABC)$ for every angle $\angle ABC$ in $\mathscr{S}$.

If there is an isometry between two geometries, we say that the two geometries are **isometric**.

If $f: \mathscr{S} \to \mathscr{S}'$ is an isometry of protractor geometries then there are no essential differences between the two geometries. Any theorem in one geometry is true in the other. In some sense all f does is change the names of the parts.

Example 8.3.9. Let $f: \mathbb{R}^2 \to \mathbb{R}^2$ by $f(a, b) = (a + 1, b - 3)$. Then f is an isometry between $\{\mathbb{R}^2, \mathscr{L}, d_E, m_E\}$ and itself.

We will show in Chapter 11 that there is an isometry between any two models of a Euclidean geometry. Hence there is essentially only one Euclidean geometry. A similar statement is not true in hyperbolic geometry. There is an isometry between two given models of a hyperbolic geometry if and only if they have the same distance scale. Hyperbolic geometries with different distance scales have a definite metric difference. However, from the synthetic view there is no essential difference. It is possible to find a bijection between hyperbolic geometries that preserves lines, betweenness and congruence.

Problem Set 8.3

Part A.

1. Let $l = {}_0L$ and $l' = {}_1L$ be lines in the Poincaré Plane. Show that $d(l, l') = 0$ from the definition of distance. (Hint: Consider $d(P, Q)$ where the y coordinate of both P and Q is large.)

2. If l and l' are lines in a metric geometry which are not parallel prove that $d(l, l') = 0$.

3. Prove Equation (3-1).

4. Prove that two distinct lines l and l' are parallel in a Euclidean geometry if and only if l and l' are equidistant (i.e., $d(P, l')$ is constant, independent of $P \in l$.)

*5. Suppose that $\overrightarrow{AB}$ is strictly asymptotic to $\overrightarrow{CD}$ in a hyperbolic geometry. Prove that $\overrightarrow{BA}$ and $\overrightarrow{DC}$ diverge; that is, if $r > 0$ there is a point $P \in \overrightarrow{BA}$ such that $d(P, \overrightarrow{CD}) > r$. (Thus asymptotic lines diverge at the ends at which they are not asymptotic.)

6. Prove that two distinct lines in a hyperbolic geometry cannot be equidistant.

7. Prove that equality holds in Theorem 8.3.3 if and only if the geometry is Euclidean.

8. Find the distance scale for the Poincaré Plane.

9. Prove that f as given in Example 8.3.9 really is an isometry.

10. Let $f: \mathbb{H} \to \mathbb{H}$ by $f(a, b) = (a + 2, b)$. Prove that f is an isometry.

11. Prove that the relation "is isometric to" is an equivalence relation on the set of protractor geometries.

12. Let $\{\mathscr{S},\mathscr{L},d,m\}$ and $\{\mathscr{S}',\mathscr{L}',d',m'\}$ be isometric protractor geometries. Prove that if one is a neutral geometry so is the other and if one is a hyperbolic geometry so is the other.

13. If $f:\mathscr{S}\to\mathscr{S}'$ and $g:\mathscr{S}'\to\mathscr{S}''$ are isometries, prove that $g\circ f:\mathscr{S}\to\mathscr{S}''$ is also an isometry.

Part B. "Prove" may mean "find a counterexample".

14. Let $\overline{AB}$ be a segment in the Poincaré Plane and let it be parametrized by $x = f(t)$, $y = g(t)$ for $a \le t \le b$. In differential geometry the hyperbolic length of $\overline{AB}$ is defined by the integral

$$\int_a^b \frac{\sqrt{\dot{x}^2 + \dot{y}^2}}{y}\,dt$$

where $\dot{x} = dx/dt$ and $\dot{y} = dy/dt$. Use this formula to derive the distance function for $\mathscr{H}$.

15. Let $\{\mathscr{S}, \mathscr{L}, d, m\}$ be a Euclidean geometry and $\mathscr{S}' = \{\mathscr{S}, \mathscr{L}, d', m\}$ a change of scale. Prove that $\mathscr{S}'$ is a Euclidean geometry.

16. Repeat Problem B15 for a hyperbolic geometry.

CHAPTER 9
Euclidean Geometry

9.1 Equivalent Forms of EPP

In the previous chapter we discussed properties possessed by hyperbolic geometries. Now we turn our attention to Euclidean geometries. In this first section we will present several equivalent formulations of the Euclidean Parallel Property. The proofs that a Euclidean geometry has certain properties are generally straightforward. However, the converse results that a neutral geometry with a certain property must be Euclidean strongly depend on the All or None Theorem and Chapter 8. If these converses are omitted, this chapter may be read right after Section 7.1.

In the second section we will be concerned with the theory of similar triangles and proportion. The third section will cover certain classical results of Euclidean geometry, including the Euler Line, the Nine Point Circle, and Morley's Theorem.

As we have noted before, considerable effort was spent trying to prove that EPP was a theorem in neutral geometry. This is really not so surprising when you consider the various equivalent formulations of EPP, some of which we give in this section and in the problems. Just the sheer weight of numbers was enough to "convince" many people that one of these forms must follow from the axioms of a neutral geometry, and hence so must EPP. Of course, we now know that this is incorrect. In fact, the All or None Theorem tells us that in a neutral geometry exactly one of EPP and HPP is satisfied. This will be the basis for many of the results in this section.

Recall that the defect of a triangle is 180 minus the sum of the measures of its three angles.

Theorem 9.1.1. *In a neutral geometry, EPP is satisfied if and only if there is a triangle with defect zero. Furthermore, if one triangle has defect zero then so does every triangle.*

PROOF. First suppose that EPP is satisfied. By Theorem 7.2.9 we know that the defect of any triangle is zero.

On the other hand, suppose that the defect of one triangle is zero. Then HPP cannot be satisfied because Theorem 8.2.4 says that the defect of any triangle is positive in a hyperbolic geometry. Since HPP is not satisfied, the All or None Theorem (Theorem 7.3.10) says that EPP must be satisfied.   □

Theorem 9.1.2. *In a neutral geometry, EPP is satisfied if and only if whenever a pair of parallel lines l and l' have a transversal t, then each pair of alternate interior angles are congruent.*

PROOF. First suppose that EPP is satisfied. Then by Problem A11 of Section 7.1 we know that a pair of alternate angles must be congruent.

On the other hand suppose that for every pair of parallel lines $l \| l'$ and transversal t, a pair of alternate interior angles are congruent. If HPP is satisfied, then by Theorem 8.1.4 there are lines l and l' which are strictly asymptotic (and hence parallel). Let $Q \in l$ and let t be the perpendicular from Q to l'. By Theorem 8.1.9, t is not perpendicular to l. Hence alternate interior angles are not congruent. Since this contradicts the assumption of HPP, it must be that EPP is satisfied.   □

Recall that a rectangle is a quadrilateral with four right angles.

Theorem 9.1.3. *In a neutral geometry, EPP is satisfied if and only if there exists a rectangle.*

PROOF. First suppose that EPP is satisfied. Let $\text{S}ABCD$ be a Saccheri quadrilateral. (We shall show that $\text{S}ABCD$ is actually a rectangle.) Then $\overline{BC} \| \overline{AD}$. Since $\overline{AB} \perp \overline{AD}$ and $\overline{AB}$ is transversal to $\overline{AD}$ and $\overline{BC}$, a pair of alternate interior angles must be congruent and so $\overline{AB} \perp \overline{BC}$. Thus $\angle A \simeq \angle B$. Since $\angle B \simeq \angle C$, we have $\angle A$, $\angle B$, $\angle C$ and $\angle D$ right angles. Hence $\square ABCD$ is a rectangle.

Now suppose that $\square ABCD$ is a rectangle. By Corollary 7.2.12 $\overline{AB} \simeq \overline{DC}$ so that $\square ABCD$ is also a Saccheri quadrilateral. If HPP is satisfied, then $\angle B$ is acute by Theorem 8.2.3. Since this is false, the All or None Theorem implies that EPP is satisfied.   □

Corollary 9.1.4. *A neutral geometry satisfies EPP if and only if every Saccheri quadrilateral is a rectangle.*

The next difference between EPP and HPP involves the perpendicular bisectors of the sides of a triangle. Theorems 9.1.5 and 9.1.6 will show that EPP is satisfied if and only if for every triangle, the perpendicular bisectors of the sides intersect at a common point.

Definition. A set of lines is **concurrent** if there is a point P that belongs to each of the lines. In this case we say that the lines **concur** at P.

Theorem 9.1.5. *In a Euclidean geometry, the perpendicular bisectors of the sides of* $\triangle ABC$ *are concurrent.*

PROOF. Let l be the perpendicular bisector of $\overline{AB}$ and let l' be the perpendicular bisector of $\overline{BC}$. See Figure 9-1. If $l\|l'$, then by Problem A5, $\overleftrightarrow{AB}\|\overleftrightarrow{BC}$. But this is impossible because $\overleftrightarrow{AB} \cap \overleftrightarrow{BC} = \{B\}$. Hence l intersects l' at some point O. By Theorem 6.4.6, $\overline{AO} \simeq \overline{BO}$ and $\overline{BO} \simeq \overline{CO}$. Hence $\overline{AO} \simeq \overline{CO}$ and so, by the same theorem, O lies on the perpendicular bisector of $\overline{AC}$. Hence the perpendicular bisectors of the sides of $\triangle ABC$ concur at O. □

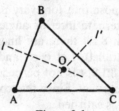

Figure 9-1

In the next theorem we will show that the result analogous to Theorem 9.1.5 is false in a hyperbolic geometry. We will actually construct a triangle such that two of the perpendicular bisectors are parallel. This will be done by exploiting the critical function to create asymptotic rays. The asymptotic rays then yield parallel lines. The trick involved here will also be used in Theorem 9.1.8. You are asked to construct a specific example in $\mathcal{H}$ in Problem A19.

Theorem 9.1.6. *In any hyperbolic geometry, there is a triangle such that the perpendicular bisectors of two of the sides are parallel.*

PROOF. We shall actually show that such a triangle can always be constructed in such a manner that it is an isosceles triangle with one angle prescribed. To this end, let $\angle ABC$ be any given angle and let $\overrightarrow{BD}$ be its angle bisector (so that $\angle ABD$ is acute). Since $0 < m(\angle ABD) < 90$ and the image of the critical function Π is the interval $(0, 90)$ (Theorem 8.2.8) there is a number t with $\Pi(t) = m(\angle ABD)$. We will construct an isosceles triangle $\triangle JBK$ whose congruent sides have length $2t$ and $\angle JBK = \angle ABC$.

Choose $E \in \overrightarrow{BA}$ and $F \in \overrightarrow{BC}$ with $BE = BF = t$ as in Figure 9-2. Let G and H be in the interior of $\angle ABC$ with $\overleftrightarrow{EG} \perp \overleftrightarrow{AB}$ and $\overleftrightarrow{FH} \perp BC$. ($\overleftrightarrow{EG}$ and $\overleftrightarrow{FH}$ will be the perpendicular bisectors of the two of the sides of $\triangle JBK$.) Since $m(\angle EBD) = \Pi(EB) = \Pi(t)$ and $\overrightarrow{EB} \perp \overrightarrow{EG}$, the definition of the critical function Π shows that $\overrightarrow{BD}|\overrightarrow{EG}$. Similarly $\overrightarrow{BD}|\overrightarrow{FH}$. Thus $\overrightarrow{EG}|\overrightarrow{FH}$ and in particular $\overleftrightarrow{EG} \| \overleftrightarrow{FH}$. Finally, let J and K be such that $J—E—B$, $B—F—K$, and $\overline{JE} \simeq \overline{EB} \simeq \overline{BF} \simeq \overline{FK}$. Then $\overleftrightarrow{EG}$ and $\overleftrightarrow{FH}$ are perpendicular bisectors of two of the sides of $\triangle JBK$ and are parallel. ☐

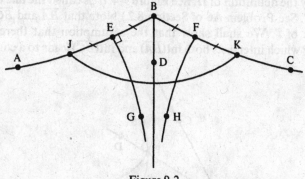

Figure 9-2

Corollary 9.1.7. *A neutral geometry satisfies* EPP *if and only if for every triangle, the perpendicular bisectors of the sides are concurrent.*

The next difference between hyperbolic and Euclidean geometries which we explore concerns lines through a point in the interior of an acute angle. Consider, for example, $\angle ABC$ in the Poincaré Plane $\mathcal{H}$ as pictured in Figure 9-3. For the point $P \in \text{int}(\angle ABC)$ there is no line (except $\overleftrightarrow{PB}$) which passes through both sides of $\angle ABC$. This situation cannot happen in a Euclidean geometry as the next result shows.

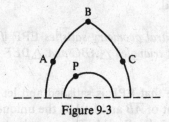

Figure 9-3

Theorem 9.1.8. *A neutral geometry satisfies* EPP *if and only if for every acute angle* $\angle ABC$ *and every point* $P \in \text{int}(\angle ABC)$ *there is a line l through P that intersects both* $\text{int}(\overline{BA})$ *and* $\text{int}(\overline{BC})$.

PROOF. First assume that EPP is satisfied. We shall show that the perpendicular from P to $\overleftrightarrow{AB}$ satisfies the theorem. Since $\angle ABC$ is acute so is $\angle ABP$ so that by Problem A1 of Section 7.1, the foot of the perpendicular from P to $\overleftrightarrow{AB}$ is in $\text{int}(\overline{BA})$. Let l be this perpendicular. Clearly $B \notin l$. By Euclid's Fifth Postulate l intersects $\overrightarrow{BC}$. Hence l is the desired line.

Now suppose that for every $P \in \text{int}(\angle ABC)$ there is a line l through P that intersects both $\text{int}(\overline{BA})$ and $\text{int}(\overline{BC})$. If EPP is not satisfied, then HPP is. We shall assume HPP and search for a contradiction.

Choose D so that $\overline{BD}$ bisects $\angle ABC$. As before we know that there is a positive number, t, with $\Pi(t) = m(\angle ABD)$. Choose $P \in \overline{BD}$ so that $BP = t$. Then the line l' through P perpendicular to $\overrightarrow{BP}$ is asymptotic to both $\overrightarrow{BA}$ and $\overrightarrow{BC}$ by the definition of Π. See Figure 9-4. (l' is called the **line of enclosure** of $\angle ABC$. See Problem A6 of Section 8.2.) Note that $\overrightarrow{BA}$ and $\overrightarrow{BC}$ lie on the same side of l'. We shall show that the assumption that there is a line l through P which intersects both int($\overrightarrow{BA}$) and int($\overrightarrow{BC}$) leads to a contradiction.

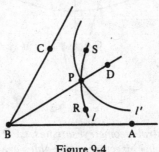

Figure 9-4

Since l' does not intersect $\overrightarrow{BA}$ and l does, $l' \neq l$. Hence there is a point $R \in l$ which is on the same side of l' as A. Choose S on l with R—P—S. Then S is on the opposite side of l' as $\overrightarrow{BA}$ (and $\overrightarrow{BC}$). Hence $\overrightarrow{PS}$ does not intersect either $\overrightarrow{BA}$ or $\overrightarrow{BC}$. On the other hand, int($\overrightarrow{PR}$) lies on one side of $\overrightarrow{BD}$ so that by Theorem 4.4.3, it cannot intersect both $\overrightarrow{BA}$ and $\overrightarrow{BC}$. Thus $l = \overrightarrow{PR} = \overrightarrow{PS}$ does not intersect both $\overrightarrow{BA}$ and $\overrightarrow{BC}$, which is a contradiction. Hence EPP is satisfied.   □

Theorem 9.1.9. *A neutral geometry satisfies* EPP *if and only if there are a pair of non-congruent triangles* $\triangle ABC$ *and* $\triangle DEF$ *with* $\angle A \simeq \angle D$, $\angle B \simeq \angle E$, *and* $\angle C \simeq \angle F$.

PROOF. First suppose that EPP is satisfied and let $\triangle ABC$ be any triangle. Let E be the midpoint of $\overline{AB}$ and let l be the unique line through E parallel to $\overrightarrow{BC}$. See Figure 9-5. By Pasch's Theorem, l intersects $\overline{AC}$ at a point F. $\angle AEF \simeq \angle ABC$ by Theorem 9.1.2 and Problem A2 of Section 7.1. Similarly, $\angle AFE \simeq \angle ACB$. Then $\triangle ABC$ and $\triangle AEF$ are the desired noncongruent triangles.

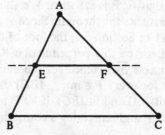

Figure 9-5

Now suppose there is a pair of noncongruent triangles $\triangle ABC$ and $\triangle DEF$ with corresponding angles congruent. If HPP is satisfied, then Theorem 8.2.6 implies that $\triangle ABC \simeq \triangle DEF$, which is false in our case. Hence EPP must be satisfied.   □

In several of the above proofs we used some powerful results, especially the fact that if HPP holds then the image of the critical function is $(0, 90)$. You may need similar methods for some of the exercises. However, it is possible to show all of these equivalences of EPP without the use of the critical function. For example Martin [1975] does this based on the following equivalent form of the All or None Theorem: In a neutral geometry the upper base angles of a Saccheri quadrilateral are either always right angles (EPP) or always acute angles (HPP).

PROBLEM SET 9.1
Part A.

1. In a neutral geometry, prove that EPP is satisfied if and only if $\|$ is a transitive relation.

2. In a neutral geometry, let $l \| l'$. Prove that EPP is satisfied if and only if any line perpendicular to l is also perpendicular to l'.

*3. In a neutral geometry let $l \| l'$. Prove that EPP is satisfied if and only if whenever a line (other than l) intersects l, it also intersects l'.

4. In a neutral geometry, prove that EPP is satisfied if and only if $\|$ is an equivalence relation.

5. In a Euclidean geometry prove that if $l \| l', r \perp l, r' \perp l'$, then $r \| r'$.

6. In a neutral geometry prove that EPP is satisfied if and only if for each $\triangle ABC$ there is a circle $\mathscr{C}$ (called the **circumcircle** or **circumscribed circle**) with $A, B, C \in \mathscr{C}$.

7. In a neutral geometry prove that EPP is satisfied if and only if for any three non-collinear points A, B, C there is a unique point O equidistant from $A, B,$ and C.

8. In a neutral geometry prove that EPP is satisfied if and only if the two angles of any open triangle are supplementary.

9. In a neutral geometry, prove that EPP is satisfied if and only if the measure of an exterior angle of any triangle equals the sum of the measures of the remote interior angles.

10. In a neutral geometry, prove that EPP is satisfied if and only if for every acute angle $\angle ABC$ the perpendicular to $\overrightarrow{BA}$ at $D \in \text{int}(\overrightarrow{BA})$ intersects $\overrightarrow{BC}$.

11. Let $\triangle ABC$ be given in a neutral geometry with B a point on the circle with diameter $\overline{AC}$. Prove that EPP is satisfied if and only if $\angle B$ is a right angle.

12. Let $\triangle ABC$ be given in a neutral geometry with $\angle B$ a right angle. Prove that EPP is satisfied if and only if B lies on the circle with diameter $\overline{AC}$.

13. In a neutral geometry, prove that EPP is satisfied if and only if whenever $l \perp r$, $r \perp s$, and $s \perp m$, then $l \cap m \neq \emptyset$.

14. In a neutral geometry, prove that EPP is satisfied if and only if there exists a pair of distinct equidistant lines.

*15. In a neutral geometry, prove that EPP is satisfied if and only if parallel lines are equidistant.

16. In a neutral geometry, prove that EPP is satisfied if and only if every Lambert quadrilateral is a rectangle.

17. Prove Corollary 9.1.7.

18. Modify the proof of Theorem 9.1.8 so that it works for any angle $\angle ABC$, not just an acute angle.

19. Give an example in $\mathscr{H}$ of a triangle such that the perpendicular bisectors of its sides are not concurrent.

Part B. "Prove" may mean "find a counterexample".

20. Prove that a neutral geometry satisfies HPP if and only if for every acute angle $\angle ABC$ and point $P \in \text{int}(\angle ABC)$ there is no line through P which intersects both $\text{int}(\overrightarrow{BA})$ and $\text{int}(\overrightarrow{BC})$. (Compare with Theorem 9.1.8.)

21. Let $\mathscr{C}$ be a circle with center O in a neutral geometry. Prove that EPP is satisfied if and only if for every acute angle $\angle ABC$ with $A, B, C \in \mathscr{C}$, $m(\angle ABC) = \frac{1}{2}m(\angle AOC)$.

22. Let $\mathscr{C}$ be a circle with center O in a neutral geometry. Let $\overline{AB}$ be a chord of $\mathscr{C}$ which is not a diameter, let l be tangent to $\mathscr{C}$ at A, and let C be a point on l on the same side of $\overleftrightarrow{AO}$ as B. Prove that EPP is satisfied if and only if $m(\angle CAB) = \frac{1}{2}m(\angle AOB)$.

23. State at least 12 equivalent forms of HPP.

9.2 Similarity Theory

This section deals with the idea of similarity. We shall define similar triangles to be triangles with corresponding angles congruent. Note that in a hyperbolic geometry this would mean that the triangles are congruent by Theorem 8.2.6. Thus in a hyperbolic geometry, the concept of similarity is identical to the concept of congruence.

The basic similarity result for a Euclidean geometry relates similarity to the equality of the ratio of corresponding sides of two triangles. This key result is Theorem 9.2.5 and, like many truly important results, it requires a great deal of work to prove. This work is contained in Theorems 9.2.1 through 9.2.5, especially Theorem 9.2.3. The basic similarity result will then be used to prove the Pythagorean Theorem and a proposition which will be crucial to the Euclidean area theory of the next chapter.

Theorem 9.2.1. *In a Euclidean geometry, let l_1, l_2, and l_3 be distinct parallel lines. Let t_1 intersect l_1, l_2, and l_3 at A, B, and C (respectively) and let t_2 intersect l_1, l_2, and l_3 at D, E, and F (respectively). If $\overline{AB} \simeq \overline{BC}$ then $\overline{DE} \simeq \overline{EF}$.*

PROOF. We shall assume that neither t_1 nor t_2 is a common perpendicular of l_1, l_2, l_3. (The case where either is a common perpendicular is left to Problem A1.)

Since $AB = BC$ and $l_1 \neq l_3$, then $A \neq C$ and A—B—C as in Figure 9-6. Let P be the foot of the perpendicular from A to l_2, Q be the foot of the perpendicular from B to l_3, R the foot of the perpendicular from D to l_2, and S the foot of the perpendicular from E to l_3.

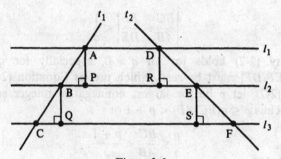

Figure 9-6

Since $\angle ABP \simeq \angle BCQ$, $\triangle ABP \simeq \triangle BCQ$ by HA. Thus $\overline{AP} \simeq \overline{BQ}$. By Problem A15 of Section 9.1, $\overline{AP} \simeq \overline{DR}$ and $\overline{BQ} \simeq \overline{ES}$. Hence $\overline{DR} \simeq \overline{ES}$. Since $\angle DER \simeq \angle EFS$, $\triangle DRE \simeq \triangle ESF$ by SAA. Thus $\overline{DE} \simeq \overline{EF}$.   ☐

The proof of the next result is left as Problem A2.

Theorem 9.2.2. *Let l_1, l_2, l_3 be distinct parallel lines in a Pasch geometry. Let t_1 intersect l_1, l_2, and l_3 at A, B, and C (respectively) and let t_2 intersect l_1, l_2, and l_3 at D, E, and F (respectively). If A—B—C then D—E—F.*

The next result is the key theorem from which we will derive the relationship between the corresponding sides of similar triangles.

Theorem 9.2.3. *Let l_1, l_2, l_3 be distinct parallel lines in a Euclidean geometry. Let t_1 and t_2 be two transversals which intersect l_1, l_2, l_3 at A, B, C and D, E, F, with A—B—C, as in Figure 9-7. Then*

$$\frac{BC}{AB} = \frac{EF}{DE}. \tag{2-1}$$

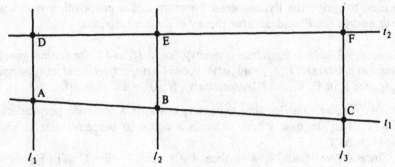

Figure 9-7

PROOF. Let q be any positive integer. We will show that

$$\left| \frac{BC}{AB} - \frac{EF}{DE} \right| < \frac{1}{q}. \tag{2-2}$$

If Inequality (2-2) holds for all $q > 0$, especially for q very large, $|BC/AB - EF/DE|$ must be zero, which proves Equation (2-1). To prove Inequality (2-2), let p be the largest nonnegative integer such that $p \le q(BC/AB)$. Thus $p \le q(BC/AB) < p + 1$ or

$$\frac{p}{q} \le \frac{BC}{AB} < \frac{p+1}{q}. \tag{2-3}$$

We shall break the segment $\overline{AB}$ into q segments each of length AB/q and then lay off $p + 1$ segments of this same length along $\overline{BC}$. Theorem 9.2.1 will then be applied to the resulting configuration which is illustrated in Figure 9-8 with $p = 7$ and $q = 5$.

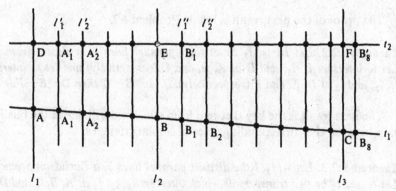

Figure 9-8

Let $A_1, A_2, \ldots, A_{q-1}$ be points of $\overline{AB}$ with

$$A - A_1 - A_2 - \cdots - A_{q-1} - B \quad \text{and} \quad \overline{AA_1} \simeq \overline{A_1 A_2} \simeq \cdots \simeq \overline{A_{q-1}B}. \tag{2-4}$$

Note that this means each segment $\overline{A_iA_{i+1}}$ has length AB/q. Similarly choose points $B_1, B_2, \ldots, B_{p+1}$ on $\overline{BC}$ so that the distance between neighboring points is AB/q. That is,

$$B—B_1—B_2—\cdots—B_{p+1} \quad \text{and} \quad BB_1 = B_1B_2 = \cdots = B_pB_{p+1} = \frac{AB}{q}. \quad (2\text{-}5)$$

We shall now show that $C \in \overline{B_pB_{p+1}}$. By Inequality (2-3) we have

$$BB_p = p \cdot BB_1 = \frac{p}{q} \cdot AB \leq BC$$

so that $B_p \in \overline{BC}$. Similarly

$$BB_{p+1} = (p+1) \cdot BB_1 = \frac{p+1}{q} \cdot AB > BC$$

so that $B_{p+1} \notin \overline{BC}$. Thus $B—C—B_{p+1}$. This means either $C = B_p$ or $B_p—C—B_{p+1}$. Hence $C \in \overline{B_pB_{p+1}}$ and $C \neq B_{p+1}$.

Let l_i' be the line through A_i parallel to $l_1 = \overline{AD}$. This line intersects t_2 at a point A_i'. Similarly let l_j'' be the line through B_j parallel to $l_1 = \overline{AD}$. This line intersects t_2 at a point B_j'. See Figure 9-8. By Theorem 9.2.2 we have

$$D—A_1'—A_2'—\cdots—A_{q-1}'—E$$
$$E—B_1'—B_2'—\cdots—B_p'—B_{p+1}'$$
$$E—F—B_{p+1}' \quad \text{and} \quad B_p' \in \overline{EF}.$$

We may now apply Theorem 9.2.1 to compare the distances between the A_i' and between the B_j'. Together with Condition (2-4), this theorem implies

$$\overline{DA_1'} \simeq \overline{A_1'A_2'} \simeq \cdots \simeq \overline{A_{q-1}'E}.$$

Hence

$$DE = q \cdot DA_1'. \quad (2\text{-}6)$$

Similarly from Condition (2-5) we have

$$\overline{EB_1'} \simeq \overline{B_1'B_2'} \simeq \cdots \simeq \overline{B_p'B_{p+1}'}.$$

Thus since $B_p' \in \overline{EF}$, $E—F—B_{p+1}$, and $EB_j' = j \cdot EB_1'$, we have

$$p \cdot EB_1' = EB_p' \leq EF < EB_{p+1}' = (p+1) \cdot EB_1'. \quad (2\text{-}7)$$

Since $\overline{EB_1'} \simeq \overline{A_{q-1}'E} \simeq \overline{DA_1'}$, Inequality (2-7) becomes

$$p \cdot DA_1' \leq EF < (p+1)DA_1'.$$

We may divide this inequality by DE to obtain

$$p \cdot \frac{DA_1'}{DE} \leq \frac{EF}{DE} < (p+1) \cdot \frac{DA_1'}{DE}$$

or, since $DE = q \cdot DA_1'$ by Equation (2-6),.

$$\frac{p}{q} \le \frac{EF}{DE} < \frac{p+1}{q}.$$

We may subtract this from Inequality (2-3) to get

$$-\frac{1}{q} = \frac{p}{q} - \frac{p+1}{q} < \frac{BC}{AB} - \frac{EF}{DE} < \frac{p+1}{q} - \frac{p}{q} = \frac{1}{q}.$$

Hence $|BC/AB - EF/DE| < 1/q$ and Inequality (2-2) is proved.   □

Corollary 9.2.4. *Let $\triangle DEF$ be a triangle in a Euclidean geometry. If D—G—E, D—H—F and $\overline{GH} \parallel \overline{EF}$, then*

$$\frac{DG}{DE} = \frac{DH}{DF}.$$ (2-8)

PROOF. Let $l_1 = \overleftrightarrow{EF}$, $l_2 = \overleftrightarrow{GH}$, and l_3 be the unique line through D parallel to l_2. See Figure 9-9. By Theorem 9.2.3 with $t_1 = \overleftrightarrow{DE}$ and $t_2 = \overleftrightarrow{DF}$, we have

$$\frac{GE}{DG} = \frac{HF}{DH}.$$

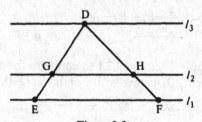

Figure 9-9

Since $DE = DG + GE$ and $DF = DH + HF$,

$$\frac{DE}{DG} = \frac{DG + GE}{DG} = 1 + \frac{GE}{DG} = 1 + \frac{HF}{DH} = \frac{DH + HF}{DH} = \frac{DF}{DH}$$

so that

$$\frac{DE}{DG} = \frac{DF}{DH} \quad \text{or} \quad \frac{DG}{DE} = \frac{DH}{DF}.$$ □

Definition. Two triangles $\triangle ABC$ and $\triangle DEF$ in a protractor geometry are **similar** (written $\triangle ABC \sim \triangle DEF$) if $\angle A \simeq \angle D$, $\angle B \simeq \angle E$, $\angle C \simeq \angle F$.

Three remarks are in order. First, to show that two triangles in a Euclidean geometry are similar it is sufficient to show that *two* of the pairs of corresponding angles are congruent. This is because the angle sum theorem guarantees that the remaining pair of angles are then also congruent. Second, as mentioned earlier, in a hyperbolic geometry $\triangle ABC \sim \triangle DEF$ if and only if $\triangle ABC \simeq \triangle DEF$ by Theorem 9.1.9. Third, as in the case of congruence, the notation for similarity includes the correspondence: $\triangle ABC \sim \triangle DEF$ means that A and D correspond, as do B and E, and also C and F.

The basic results of similarity theory follow from Corollary 9.2.4. The proofs generally involve constructing a line parallel to one side of a triangle and then invoking the corollary. Our first result gives the relationship between corresponding sides of similar triangles.

Theorem 9.2.5. *In a Euclidean geometry, the ratio of the lengths of corresponding sides of similar triangles is constant; that is, if $\triangle ABC \sim \triangle DEF$ then*

$$\frac{AB}{DE} = \frac{BC}{EF} = \frac{AC}{DF}. \tag{2-9}$$

PROOF. If $\overline{AB} \simeq \overline{DE}$ then $\triangle ABC \simeq \triangle DEF$ by ASA and each quotient in Equation (2-9) is 1. Thus we may assume that $AB \neq DE$. Suppose that $AB < DE$. (The case $AB > DE$ is similar.) Let G be the point on $\overline{DE}$ with $\overline{AB} \simeq \overline{DG}$. See Figure 9-10.

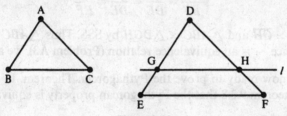

Figure 9-10

Let l be the unique line through G parallel to $\overleftrightarrow{EF}$. By Pasch's Theorem, l intersects $\overline{DF}$ at a point H. $\angle DGH \simeq \angle DEF \simeq \angle ABC$ (Why?). Thus $\triangle ABC \simeq \triangle DGH$ by ASA. By Corollary 9.2.4

$$\frac{DG}{DE} = \frac{DH}{DF}.$$

Since $\overline{DG} \simeq \overline{AB}$ and $\overline{DH} \simeq \overline{AC}$, this last equation becomes

$$\frac{AB}{DE} = \frac{AC}{DF}.$$

Similarly $AB/DE = BC/EF$.   □

Theorem 9.2.5 says that Equation (2-9) is a necessary condition for similarity. We now show that it is also sufficient.

Theorem 9.2.6 (SSS Similarity Theorem). *In a Euclidean geometry*, $\triangle ABC \sim$ $\triangle DEF$ *if and only if*

$$\frac{AB}{DE} = \frac{BC}{EF} = \frac{AC}{DF}. \tag{2-10}$$

PROOF. Because of Theorem 9.2.5 we need only show that Equation (2-10) implies that $\triangle ABC \sim \triangle DEF$. If $\overline{AB} \simeq \overline{DE}$, then $\triangle ABC \simeq \triangle DEF$ by the SSS Congruence Theorem. Thus $\triangle ABC \sim \triangle DEF$ in this case.

Assume that $AB < DE$. (The case $AB > DE$ is similar.) Let G be the point on $\overline{DE}$ with $\overline{AB} \simeq \overline{DG}$. By Pasch's Theorem, the line through G parallel to $\overleftrightarrow{EF}$ intersects $\overline{DF}$ at a point H. See Figure 9-10. By Theorem 9.2.4

$$\frac{DG}{DE} = \frac{DH}{DF}.$$

Thus

$$\frac{AC}{DF} = \frac{AB}{DE} = \frac{DG}{DE} = \frac{DH}{DF}$$

so that $\overline{AC} \simeq \overline{DH}$.

Note that $\angle DGH \simeq \angle E$ and $\angle DHG \simeq \angle F$ since $\overleftrightarrow{GH} \| \overleftrightarrow{EF}$ and the geometry is a Euclidean geometry. Thus $\triangle DGH \sim \triangle DEF$ so that

$$\frac{GH}{EF} = \frac{DG}{DE} = \frac{AB}{DE} = \frac{BC}{EF}.$$

Hence $\overline{BC} \simeq \overline{GH}$ and $\triangle ABC \simeq \triangle DGH$ by SSS. Thus $\triangle ABC \sim \triangle DGH \sim$ $\triangle DEF$. Since $\sim$ is an equivalence relation (Problem A3), we are done. $\square$

We are now ready to prove the Pythagorean Theorem. In fact, we shall show in Theorem 9.2.8 that the Pythagorean property is equivalent to EPP.

Theorem 9.2.7 (Pythagoras). *In a Euclidean geometry*, $\triangle ABC$ *has a right angle at B if and only if*

$$(AB)^2 + (BC)^2 = (AC)^2. \tag{2-11}$$

PROOF. First suppose that $\angle B$ is a right angle of $\triangle ABC$ and that D is the foot of the altitude from B. By Problem A4, $\triangle ADB \sim \triangle ABC \sim \triangle BDC$. Thus by Theorem 9.2.5

$$\frac{AB}{AD} = \frac{AC}{AB} \quad \text{and} \quad \frac{BC}{DC} = \frac{AC}{BC}.$$

Hence $(AB)^2 = (AC)(AD)$ and $(BC)^2 = (AC)(DC)$. Now $AC = AD + DC$ so that

$$(AC)^2 = (AC)(AD + DC)$$
$$= (AC)(AD) + (AC)(DC)$$
$$= (AB)^2 + (BC)^2.$$

Thus Equation (2-11) is valid if $\angle B$ is a right angle.

Now suppose that $\triangle ABC$ satifies Equation (2-11). We must show that $\angle B$ is a right angle. Let $\triangle PQR$ be a right triangle with right angle at Q and $\overline{PQ} \simeq \overline{AB}$, $\overline{QR} \simeq \overline{BC}$. Since $\angle Q$ is right we may apply Equation (2-11) to $\triangle PQR$ to obtain

$$(PR)^2 = (PQ)^2 + (QR)^2 = (AB)^2 + (BC)^2 = (AC)^2.$$

Hence $\overline{PR} \simeq \overline{AC}$ and $\triangle PQR \simeq \triangle ABC$ by SSS. But this means that $\angle B \simeq \angle Q$ is a right angle.   □

The next result says that EPP and the Pythagorean property are equivalent. Although Problem A5 of Section 5.3 shows that Pythagoras is false in $\mathcal{H}$, the proof given below (from Reyes [1897]) is interesting because it is a direct proof which does not rely on a model for a counterexample.

Theorem 9.2.8. *A neutral geometry satisfies* EPP *if and only if for every right triangle $\triangle ABC$ with right angle at B the equation*

$$(AC)^2 = (AB)^2 + (BC)^2 \tag{2-12}$$

is true.

PROOF. We need only show that Equation (2-12) for right triangles implies EPP. Let $\triangle ABC$ be a right triangle with right angle at B and $\overline{AB} \simeq \overline{BC}$. We shall show that the defect of $\triangle ABC$ is zero. (In fact, we will show that $m(\angle A) = m(\angle C) = 45$.) By Theorem 9.1.1 this will imply that EPP is satisfied.

Let D be the midpoint of $\overline{AC}$. Then $\triangle BAD \simeq \triangle BCD$ by SAS so that $\angle BDA \simeq \angle BDC$, as in Figure 9-11. Hence $\angle BDA$ is a right angle. We apply Equation (2-12) to $\triangle ABC$ to obtain

$$2(AB)^2 = (AB)^2 + (BC)^2 = (AC)^2 = (2 \cdot AD)^2 = 4(AD)^2$$

so that $(AB)^2 = 2(AD)^2$. We may also apply Equation (2-12) to $\triangle ADB$ to obtain

$$(AB)^2 = (AD)^2 + (DB)^2$$

so that

$$2(AD)^2 = (AB)^2 = (AD)^2 + (DB)^2$$

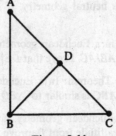

Figure 9-11

or

$$AD = BD.$$

Thus $\triangle ADB$ is isosceles as is $\triangle CDB$ and

$$\angle DBA \simeq \angle DAB \simeq \angle DCB \simeq \angle DBC.$$

Since $m(\angle ABC) = 90$, $m(\angle DBA) = 45$. Thus $\delta(\triangle ABC) = 180 - (45 + 90 + 45) = 0$ and EPP holds by Theorem 9.1.1. $\square$

The last result of this section will be useful in the study of area in the next chapter.

Theorem 9.2.9. *Let $\triangle ABC$ be a triangle in a Euclidean geometry. Let D be the foot of the altitude from A and let E be the foot of the altitude from B. Then*

$$(AD)(BC) = (BE)(AC).$$

PROOF. If $\angle C$ is a right angle then $E = C = D$ and the result is trivial. If $\angle C$ is not a right angle then $E \neq C$ and $D \neq C$. Since $\triangle BEC \sim \triangle ADC$ (Why?) we have

$$\frac{BC}{AC} = \frac{BE}{AD}$$

so that $(AD)(BC) = (BE)(AC)$. $\square$

PROBLEM SET 9.2

Part A.

1. Complete the proof of Theorem 9.2.1 in the cases where one or both of t_1 and t_2 are perpendicular to l_1, l_2, l_3.

2. Prove Theorem 9.2.2.

3. Prove that $\sim$ is an equivalence relation.

4. Let $\triangle ABC$ be a right triangle in a Euclidean geometry with right angle at B. If D is the foot of the altitude from B to $\overline{AC}$ prove that $\triangle ADB \sim \triangle ABC \sim \triangle BDC$.

5. Let $\triangle ABC \sim \triangle PQR$ in a neutral geometry. If $\overline{AB} \simeq \overline{PQ}$, prove that $\triangle ABC \simeq \triangle PQR$.

6. Let $\triangle ABC$ be a triangle in a Euclidean geometry. Suppose that $A—D—B$ and $A—E—C$ with $AD/AB = AE/AC$. Prove that $\overrightarrow{DE} \parallel \overrightarrow{BC}$.

*7. Prove the SAS Similarity Theorem: In a Euclidean geometry if $\angle B \simeq \angle Q$ and $AB/PQ = BC/QR$ then $\triangle ABC$ is similar to $\triangle PQR$.

8. Let $\triangle ABC \sim \triangle DEF$ in a Euclidean geometry. If G is the foot of the altitude from A and H is the foot of the altitude from D, prove that $AG/DH = AB/DE$.

9. In a Euclidean geometry let $\overline{AB}$ and $\overline{CD}$ be two chords of a circle $\mathscr{C}$ such that $\overline{AB}$ and $\overline{CD}$ intersect at a point E between A and B. Prove that $(AE)(EB) = (CE)(ED)$.

10. In a Euclidean geometry let $\mathscr{C}$ be a circle and let $B \in \mathrm{ext}(\mathscr{C})$. If $A, C, D \in \mathscr{C}$ with $\overrightarrow{AB}$ tangent to $\mathscr{C}$ and $B\text{—}D\text{—}C$, prove that $(AB)^2 = (BD)(BC)$.

11. In a Euclidean geometry let $\mathscr{C}$ be a circle with $C \in \mathrm{ext}(\mathscr{C})$. If $A, B, D, E \in \mathscr{C}$ with $A\text{—}B\text{—}C$ and $E\text{—}D\text{—}C$, prove that $(CA)(CB) = (CE)(CD)$.

12. In a Euclidean geometry let $\mathscr{C}$ be a circle with diameter $\overline{AB}$. Let l be the tangent to $\mathscr{C}$ at B and let C be any point on l. Prove that $\overline{AC}$ intersects $\mathscr{C}$ at a point $D \neq A$ and that $(AD)(AC)$ is a constant that does not depend on the choice of $C \in l$.

13. In a Euclidean geometry let $\mathscr{C}$ be a circle with diameter $\overline{AB}$. Let P be any point in $\mathrm{int}(\overline{AB})$ and let C, D, E be distinct points of $\mathscr{C}$ all on the same side of $\overleftrightarrow{AB}$ such that $\overline{DP} \perp \overline{AB}$ and $\angle CPD \simeq \angle DPE$. Prove that $(PD)^2 = (PC)(PE)$.

Part B. "Prove" may mean "find a counterexample".

14. Let $\triangle ABC$ be a triangle in a Euclidean geometry. If $\overrightarrow{AD}$ is the bisector of $\angle A$ with $B\text{—}D\text{—}C$ prove that $DB/DC = AB/AC$.

15. Let $\triangle ABC$ be a triangle in a Euclidean geometry. If $D\text{—}C\text{—}B$ and if the bisector of $\angle DCA$ intersects $\overrightarrow{BA}$ at E, prove that $EA/EB = CA/CB$.

9.3 Some Classical Theorems of Euclidean Geometry

One of the most fascinating aspects of mathematics is the discovery that concepts which we would not expect to be related are in fact related. In this section we will prove some of the more beautiful results of classical Euclidean geometry. We find these theorems so attractive because they contain unanticipated relationships.

For example, we will show that the three angle bisectors of a triangle all meet at one point (the incenter), as do the medians (at the centroid), the perpendicular bisectors of the sides (at the circumcenter), and the lines containing the altitudes (at the orthocenter). In addition, we shall prove the astounding result that the centroid, circumcenter, and orthocenter are collinear! The final result will be an equally surprising result regarding the trisectors of the angles of a triangle.

The first result of this section was given in Problem A12 of Section 6.4. We include it here for completeness. Recall that a collection of lines is called concurrent if there is a point, Q, which is on all of the lines. The point Q is called the **point of concurrence**.

Theorem 9.3.1. *In a neutral geometry, the angle bisectors of any triangle* $\triangle ABC$ *are concurrent. The point of concurrence, I, is called the* **incenter** *of* $\triangle ABC$.

PROOF. By the Crossbar Theorem, the bisector of $\angle A$ intersects $\overline{BC}$ at a point D. Likewise, the bisector of $\angle B$ intersects $\overline{AD}$ at a point I. See Figure 9-12. We will show that I belongs to each angle bisector. I is in the interior of $\angle A$ and the interior of $\angle B$. Hence it is in the interior of $\angle C$. (Why?)

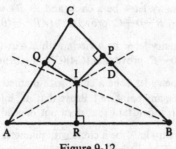

Figure 9-12

Let P, Q, and R be the feet of the perpendiculars from I to $\overleftrightarrow{BC}$, $\overleftrightarrow{AC}$, and $\overleftrightarrow{AB}$. $\triangle IQA \simeq \triangle IRA$ by HA so that $\overline{IQ} \simeq \overline{IR}$. Likewise $\triangle IRB \simeq \triangle IPB$ by HA so that $\overline{IR} \simeq \overline{IP}$. Hence $\overline{IQ} \simeq \overline{IP}$ so that $\triangle IQC \simeq \triangle IPC$ by HL. Thus $\angle ICQ \simeq \angle ICP$ and I lies on the bisector of $\angle C$. That is, the three angle bisectors concur at I.   □

Definition. A **median** of a triangle is a line segment joining a vertex to the midpoint of the opposite side.

Theorem 9.3.2. *In a Euclidean geometry, the medians of any triangle $\triangle ABC$ are concurrent. The point of concurrence, G, is called the* **centroid** *of $\triangle ABC$.*

PROOF. Let $\overline{AP}$, $\overline{BQ}$, and $\overline{CR}$ be the medians of $\triangle ABC$, as in Figure 9-13. The Crossbar Theorem implies that $\overline{AP} \cap \overline{BQ} \neq \emptyset$ and $\overline{BQ} \cap \overline{AP} \neq \emptyset$ so that $\overline{AP}$ intersects $\overline{BQ}$ at a point G. We must show that $G \in \overline{CR}$.

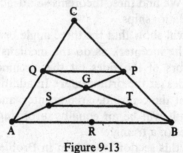

Figure 9-13

Now $\triangle QCP \sim \triangle ACB$ by the SAS Similarity Theorem (Problem A7 of Section 9.2). Hence $\angle CQP \simeq \angle CAB$ so that $\overleftrightarrow{QP} \| \overline{AB}$. Furthermore,

$$\frac{QP}{AB} = \frac{CQ}{CA} = \frac{1}{2}. \tag{3-1}$$

Similarly, if S and T are the midpoints of $\overline{AG}$ and $\overline{BG}$ then

$$\frac{ST}{AB} = \frac{1}{2}$$

so that Equation (3-1) implies that $\overline{QP} \simeq \overline{ST}$.

Since $\overleftrightarrow{QP} \| \overleftrightarrow{AB} \| \overleftrightarrow{ST}$, $\angle PQG \simeq \angle STG$. Furthermore, $\angle QGP \simeq \angle TGS$. Since $\overline{QP} \simeq \overline{ST}$, $\triangle QGP \simeq \angle TGS$ by SAA. Hence the point G has the property that it belongs to $\overline{AP}$ and

$$PG = GS = \tfrac{1}{2} AG. \tag{3-2}$$

A similar proof shows that $\overline{AP}$ and $\overline{CR}$ intersect at a point $G' \in \overline{AP}$ with $PG' = \tfrac{1}{2}AG'$. But there is only one point $X \in \overline{AP}$ with $PX = \tfrac{1}{2}AX$ so that $G = G'$. Hence the medians are concurrent at G.   □

Two things should be noted about this theorem. First, implicit in the proof is the fact that the medians intersect at a point G which is two-thirds of the way from a vertex to the opposite midpoint (Problem A3). Second, the theorem was proved only for a Euclidean geometry (unlike Theorem 9.3.1). This does not mean that it is false in hyperbolic geometry, only that the proof given does not work. In fact, it is true in hyperbolic geometry, but is more difficult to prove. See Greenberg [1980], Chapter 7, especially Problem K-19.

The following result is Theorem 9.1.5, and is mentioned here for completeness.

Theorem 9.3.3. *In a Euclidean geometry, the perpendicular bisectors of the sides of $\triangle ABC$ are concurrent. The point-of concurrence, O, is called the* **circumcenter** *of $\triangle ABC$.*

Theorem 9.3.3 also holds in a hyperbolic geometry, provided at least two of the perpendicular bisectors intersect (Problem A13 of Section 6.4).

Theorem 9.3.4. *In a Euclidean geometry, the lines containing the altitudes of $\triangle ABC$ are concurrent. The point of concurrence, H, is called the* **orthocenter** *of $\triangle ABC$.*

PROOF. This proof will use a different technique than that of Theorems 9.3.1, 9.3.2, and 9.3.3. We will construct another triangle $\triangle PQR$ such that a line containing a perpendicular bisector of a side of $\triangle PQR$ also contains an altitude of $\triangle ABC$. Since the perpendicular bisectors of the new triangle are concurrent by Theorem 9.3.3 so are the lines containing the altitudes of $\triangle ABC$. We shall now proceed with the construction of $\triangle PQR$.

Let l_1 be the unique line through A parallel to $\overleftrightarrow{BC}$, l_2 the line through B parallel to $\overleftrightarrow{AC}$, and l_3 the line through C parallel to $\overleftrightarrow{AB}$ as in Figure 9-14. Since $l_1 \| \overleftrightarrow{BC}$ and $l_2 \cap \overleftrightarrow{BC} \neq \varnothing$, $l_1 \cap l_2 \neq \varnothing$ by Problem A3 of Section 9.1. Likewise $l_2 \cap l_3 \neq \varnothing$ and $l_1 \cap l_3 \neq \varnothing$. Let the points of intersection be P, Q, and R as in Figure 9-14.

By using the fact that alternate interior angles are congruent a number of times, we see that

$$\triangle CAR \simeq \triangle ACB \simeq \triangle QBC$$

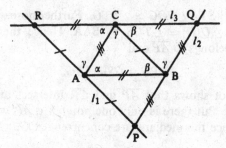

Figure 9-14

by ASA. Hence $\overline{CR} \simeq \overline{QC}$ and C is the midpoint of $\overline{RQ}$. Since $\overleftrightarrow{RQ} \| \overleftrightarrow{AB}$, the altitude from C to $\overline{AB}$ is perpendicular to $\overleftrightarrow{RQ}$. Hence the line containing this altitude is the perpendicular bisector of $\overline{RQ}$. Likewise the lines containing the other altitudes of $\triangle ABC$ are the perpendicular bisectors of the remaining sides of $\triangle PQR$. By Theorem 9.3.3 these lines are concurrent at a point H. Hence the lines containing the altitudes of $\triangle ABC$ are concurrent. □

Theorem 9.3.4 is also true in a hyperbolic geometry, provided at least two of the lines containing the altitudes intersect. See Greenberg [1980], Chapter 7.

The results on the concurrence of the perpendicular bisectors, altitudes, medians, and angle bisectors are surprising. The next result is truly astounding. The orthocenter, centroid, and circumcenter of a triangle would not seem to be related. However, the next theorem says that they are.

Theorem 9.3.5. *In a Euclidean geometry, the orthocenter, centroid, and circumcenter of $\triangle ABC$ are collinear.*

PROOF. Let H be the orthocenter, G be the centroid, and O the circumcenter of $\triangle ABC$. The proof will break into three cases depending on the shape of $\triangle ABC$.

If $\triangle ABC$ is equilateral, $H = G = O = I$ (the incenter) and the result is trivial. The case where $\triangle ABC$ is isosceles but not equilateral is left to Problem A6.

We assume, therefore, that the triangle is scalene. We first show that $O \neq G$ in this case. By Problem A3 of Section 6.4 the median from A to $\overline{BC}$ is not perpendicular to $\overline{BC}$. Hence G cannot lie on the perpendicular bisector of $\overline{BC}$ since G is not the midpoint of BC. (Why?) Thus $G \neq O$.

Let $l = \overleftrightarrow{OG}$. Let H' be the unique point on l with $O—G—H'$ and $GH' = 2 \cdot GO$. We shall prove H' is the orthocenter, H. See Figure 9-15.

Let P be the midpoint of $\overline{BC}$ so that $G \in \overline{AP}$. We will first show that $\overleftrightarrow{AH'} \| \overleftrightarrow{OP}$. Now $GA = 2 \cdot GP$ by Equation (3-2) and $GH' = 2 \cdot GO$ by construction. Since the vertical angles $\angle AGH'$ and $\angle PGO$ of Figure 9-15 are congruent, $\triangle AGH' \sim \triangle PGO$ by the SAS Similarity Theorem (Problem A7 of Section 9.2). Hence the alternate interior angles $\angle GAH'$ and $\angle GPO$ are congruent so that $AH' \| OP$.

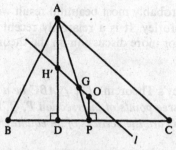

Figure 9-15

Since $\overline{OP} \perp \overline{BC}$, $\overline{AH'} \perp \overline{BC}$ also. Thus $\overline{AH'}$ contains the altitude from A to $\overline{BC}$. Thus the line containing the altitude from A goes through H'. Similarly the lines containing the other two altitudes also pass through H'. Hence H' is the orthocenter and $O, G, H = H'$ are collinear.   □

Note that this proof gives an alternative proof that the altitudes of a triangle are concurrent (at the point H').

Definition. The line containing the centroid, orthocenter, and circumcenter of a given nonequilateral triangle in a Euclidean geometry is called the **Euler line** of the triangle. The three points which are the midpoints of the segments joining the orthocenter of $\triangle ABC$ to its vertices are called the **Euler points** of $\triangle ABC$.

The Euler points appear in the next result whose proof is left to Problem B13.

Theorem 9.3.6 (Nine Point Circle). *Let $\triangle ABC$ be a triangle in a Euclidean geometry. Then the midpoints of the sides of $\triangle ABC$, the feet of the altitudes, and the Euler points all lie on the same circle. (See Figure 9-16.)*

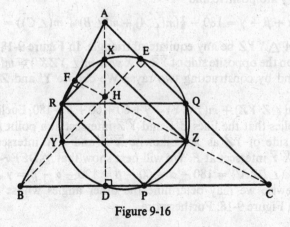

Figure 9-16

As our final and probably most beautiful result we present the following theorem due to F. Morley. It is a relatively recent theorem, having been discovered in 1899. For more discussion on the theorem and a converse see Kleven [1978].

Theorem 9.3.7 (Morley's Theorem). *Let $\triangle ABC$ be a triangle in a Euclidean geometry. Then the three points of intersection P, Q, R of adjacent trisectors of the angles of $\triangle ABC$ are the vertices of an equilateral triangle. (See Figure 9-17.)*

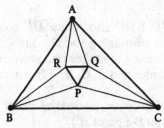

Figure 9-17

PROOF. The proof we give may be found in Coxeter [1961] along with references to other proofs. Rather than attacking the problem head on (which can be done) we instead start with an equilateral triangle $\triangle XYZ$ and around it construct a triangle which is similar to $\triangle ABC$. This construction will be carried out in such a manner that X, Y, Z will be the points of intersection of the angle trisectors of the new triangle.

First we define three numbers α, β, γ by

$$\alpha = 60 - \tfrac{1}{3}m(\angle A)$$
$$\beta = 60 - \tfrac{1}{3}m(\angle B)$$
$$\gamma = 60 - \tfrac{1}{3}m(\angle C).$$

Then α, β, γ are positive and

$$\alpha + \beta + \gamma = 180 - \tfrac{1}{3}(m(\angle A) + m(\angle B) + m(\angle C)) = 120. \qquad (3\text{-}3)$$

Now let $\triangle XYZ$ be any equilateral triangle. In Figure 9-18, we let X' be the point on the opposite side of $\overleftrightarrow{YZ}$ as X with $m(\angle YZX') = m(\angle ZYX') = \alpha$. ($X'$ is found by constructing two rays.) We choose Y' and Z' in a similar fashion.

Since $m(\angle Z'YZ) + m(Y'ZY) = \gamma + 60 + 60 + \beta > 180$, Euclid's Fifth Postulate implies that the lines $\overleftrightarrow{Z'Y}$ and $\overleftrightarrow{Y'Z}$ intersect at a point D which is on the same side of $\overleftrightarrow{YZ}$ as X'. Likewise $\overleftrightarrow{X'Z}$ and $\overleftrightarrow{Z'X}$ intersect at E while $\overleftrightarrow{Y'X}$ and $\overleftrightarrow{X'Y}$ intersect at F. We will next show that $\triangle DEF \sim \triangle ABC$.

Now $m(\angle DZX') = 180 - \alpha - 60 - \beta = 120 - \alpha - \beta = \gamma$ by Equation (3-3). Likewise we may determine the other angles whose measures are marked in Figure 9-18. Furthermore,

$$m(\angle YDZ) = 180 - 2\alpha - \beta - \gamma = 60 - \alpha$$
$$m(\angle ZEX) = 60 - \beta$$
$$m(\angle XFY) = 60 - \gamma.$$

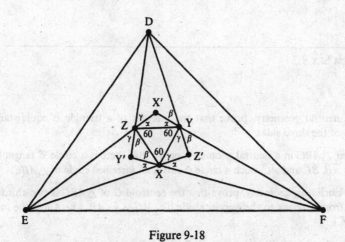

Figure 9-18

Now $\overline{X'Z} \simeq \overline{X'Y}$ by the converse of *Pons asinorum* (Theorem 6.2.2) and $\overline{ZX} \simeq \overline{YX}$ by construction. Hence $\triangle XZX' \simeq \triangle XYX'$ by SSS so that $\overline{X'X}$ bisects $\angle ZX'Y = \angle EX'F$. Furthermore

$$m(\angle EXF) = 180 - \alpha$$
$$= 90 + (90 - \alpha)$$
$$= 90 + \tfrac{1}{2}(180 - 2\alpha)$$
$$= 90 + \tfrac{1}{2}m(\angle EX'F).$$

By Problem A10, X is the incenter of $\triangle EX'F$ so that $\angle X'EX \simeq \angle FEX$. Likewise Z is the incenter of $\triangle DZ'E$ so that $\angle DEX' \simeq \angle X'EX$. Hence $\overline{EZ} = \overline{EX'}$ and $\overline{EX}$ are trisectors of $\angle DEF$. Thus

$$m(\angle DEF) = 3 \cdot m(\angle ZEX) = 3(60 - \beta) = m(\angle B).$$

In a similar manner, $m(\angle EFD) = m(\angle C)$ and $m(\angle FDE) = m(\angle A)$. Hence $\triangle DEF \sim \triangle ABC$. Furthermore $\triangle DEZ \sim \triangle ABR$, $\triangle EFX \sim \triangle BCP$, and $\triangle FDY \sim \triangle CAQ$. Thus by Theorem 9.2.5

$$\frac{PB}{XE} = \frac{CB}{FE} = \frac{AB}{DE} = \frac{RB}{ZE}$$

so that $\triangle XEZ \sim \triangle PBR$ by the SAS Similarity Theorem. Similarly, $\triangle ZDY \sim \triangle RAQ$ and $\triangle YFX \sim \triangle QCP$. Finally

$$\frac{PQ}{XY} = \frac{PC}{XF} = \frac{PB}{XE} = \frac{RB}{ZE} = \frac{RP}{ZX} = \frac{RP}{XY}$$

so that $PQ = RP$. Similarly $RP = QR$ so that $\triangle PQR$ is equilateral. $\square$

PROBLEM SET 9.3

Part A.

1. In a neutral geometry, prove that the incenter of a triangle is equidistant from each of the three sides.

2. Given $\triangle ABC$ in a neutral geometry, prove that there is a circle $\mathscr{C}$ tangent to the lines $\overleftrightarrow{AB}$, $\overleftrightarrow{BC}$ and $\overleftrightarrow{AC}$. Such a circle is called an **inscribed circle** of $\triangle ABC$.

3. In a Euclidean geometry, prove that the centroid G of $\triangle ABC$ is two-thirds of the way from a vertex to the opposite side (i.e., $AG = \frac{2}{3} \cdot AP$, where P is the midpoint of $\overline{BC}$).

4. Given three noncollinear points A, B, C in a Euclidean geometry, prove that there is a circle $\mathscr{C}$ with A, B, $C \in \mathscr{C}$. Such a circle is called a **circumscribed circle**.

5. In the Poincaré Plane $\mathscr{H}$ show that there are three noncollinear points which do not all lie on the same circle.

6. Complete the proof of Theorem 9.3.5 in the case of an isosceles triangle.

7. In a Euclidean geometry, prove that the incenter of an isosceles triangle which is not equilateral lies on the Euler line.

8. In a Euclidean geometry, prove that the circumcenter of a right triangle is the midpoint of the hypotenuse.

9. In a Euclidean geometry, prove that the Euler line of a right triangle is the line containing the median to the hypotenuse.

10. In a Euclidean geometry suppose that I is a point on the bisector of $\angle UVW$. Prove that I is the incenter of $\triangle UVW$ if and only if $m(\angle UIW) = 90 + \frac{1}{2}m(\angle UVW)$ and $I \in \mathrm{int}(\triangle UVW)$.

11. In the proof of Morley's Theorem show that $\overleftrightarrow{XX'}$, $\overleftrightarrow{YY'}$, and $\overleftrightarrow{ZZ'}$ are concurrent.

Part B. "Prove" may mean "find a counterexample".

12. In a Euclidean geometry prove that the circumcenter O of $\triangle ABC$ lies in the interior of $\triangle ABC$.

13. Prove Theorem 9.3.6. Hint: Look at Figure 9-16 carefully.

14. In a Euclidean geometry let P be a point on the circumcircle of $\triangle ABC$ and let X, Y, Z be the feet of the perpendiculars from P to the sides of $\triangle ABC$. Prove that X, Y, Z are collinear. The line $\overleftrightarrow{XY}$ is called the **Simson line** of P.

15. In a Euclidean geometry let P be a point such that the feet of the perpendiculars from P to the sides of $\triangle ABC$ are collinear. Prove that P lies on the circumcircle of $\triangle ABC$.

16. Prove that the center of the nine point circle of a nonequilateral triangle lies on the Euler line.

17. Let H be the orthocenter of $\triangle ABC$. Prove that the nine point circle of $\triangle ABH$ is the same as that of $\triangle ABC$.

Part C. Expository exercises.

18. Prepare a lecture for a high school class which describes your favorite of these classical theorems of Euclidean geometry. You should pay attention to whether the students would understand the content or both the content and the proof. In your description of the lecture explain why the theorem that you are quoting has a strong appeal to you.

19. Write an essay on Euler and his contributions to mathematics.

CHAPTER 10
Area

10.1 The Area Function

In this chapter we shall be interested in the concept of area in a neutral geometry. We shall start off with the definition of an area function and an investigation of the properties of a Euclidean area function. In Sections 10.2 and 10.3 we will prove the existence of area functions for Euclidean and hyperbolic geometries respectively. In the last section we will consider a beautiful theorem due to J. Bolyai which says that if two polygonal regions have the same area then one may be cut into a finite number of pieces and rearranged to form the other.

Informally, we are accustomed to making statements such as "the area of a triangle is $bh/2$" or "the area of a circle is πr^2." Such language is imprecise, of course. What we really mean is that "the area of the region bounded by the triangle is $bh/2$." Thus we must first define what we mean by a region in a neutral geometry. We shall adopt the view here that a region is a polygon together with its interior. After first defining polygons and polygonal regions, we will define area as a certain real valued function whose domain is the set of regions.

Definition. A subset $\bar{P}$ of a metric geometry is a **polygon of degree** $n \geq 3$ (or **n-gon**) if there are n distinct points $P_1, P_2, \ldots, P_n$ (called the **vertices** of $\bar{P}$) such that

(i) $\bar{P} = \overline{P_1 P_2} \cup \overline{P_2 P_3} \cup \cdots \cup \overline{P_{n-1} P_n} \cup \overline{P_n P_1}$ and $\qquad$ (1-1)
(ii) the interiors of the segments in Equation (1-1) are pairwise disjoint.

In this case we write $\bar{P} = \bigtriangleup P_1 P_2 \cdots P_n$. The segments in Equation (1-1) are called the **sides** (or **edges**) of $\bar{P}$. Two vertices P_i and P_j are **consecutive** if $\overline{P_i P_j}$ is a side of P.

It should be noted that this definition is just the generalization of the idea of a quadrilateral (4-gon). The notion of convexity also extends to the concept of n-gons.

Definition. A polygon $\bar{P}$ in a Pasch geometry is a **convex polygon** if for every pair of consecutive vertices P and Q, all other vertices of $\bar{P}$ lie in single halfplane H_{PQ} of the line $\overrightarrow{PQ}$. H_{PQ} is called the **half plane determined by the vertices** P and Q.

The **interior** of a convex polygon $\bar{P} = \triangle P_1 P_2 \cdots P_n$ is the intersection of the half planes determined by consecutive vertices:

$$\text{int}(\bar{P}) = H_{P_1 P_2} \cap H_{P_2 P_3} \cap \cdots \cap H_{P_n P_1}.$$

Example 10.1.1. In Figure 10-1, (a) represents a convex polygon while (b) represents a nonconvex polygon. (c) and (d) are not polygons at all.

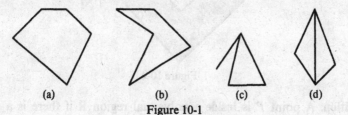

| (a) | (b) | (c) | (d) |

Figure 10-1

Definition. A **triangular region** T in a Pasch geometry is a set which is the union of a triangle and its interior. The triangular region determined by $\triangle ABC$ will be denoted $\blacktriangle ABC$ so that $\blacktriangle ABC = \triangle ABC \cup \text{int}(\triangle ABC)$.

A **polygonal region** R in a Pasch geometry is a set such that there are triangular regions $T_1, T_2, \ldots, T_l$ which satisfy

(i) $R = T_1 \cup T_2 \cup \cdots \cup T_l$
(ii) If $i \neq j$, $T_i \cap T_j$ is either empty, consists of a common vertex, or consists of a common edge of T_i and T_j.

The set of all polygonal regions of a Pasch geometry, $\{\mathscr{S}, \mathscr{L}, d\}$, is denoted $\mathscr{R}(\mathscr{S})$ or $\mathscr{R}$.

Example 10.1.2. In Figure 10-2 we have a polygonal region made up of 5 triangular regions T_1, T_2, T_3, T_4, T_5. As Figure 10-3 indicates, the triangular regions that make up a polygonal region are not unique. Note also that Figure 10-4 also illustrates a polygonal region.

You should note that we did not define a polygonal region as the union of a polygon and its interior. This is because the interior of a polygon has only been defined for a convex polygon.

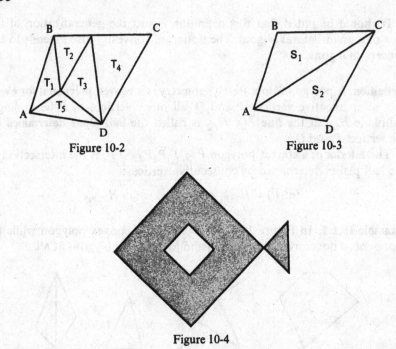

Figure 10-2   Figure 10-3

Figure 10-4

Definition. A point P is **inside** a polygonal region **R** if there is a triangle $\triangle ABC$ such that $\blacktriangle ABC \subset \mathbf{R}$ and $P \in \text{int}(\triangle ABC)$. If $P \in \mathbf{R}$ but P is not inside **R**, then P is a **boundary point** of **R**. The inside of **R**, ins(**R**), is the set of all points inside **R**. The **boundary** of **R**, bd(**R**) is the set of all boundary points.

Thus a point P is inside **R** if we can find some (small) triangle $\triangle ABC$ with $P \in \text{int}(\triangle ABC)$ such that the triangle fits inside the region. As we might expect, if $\bar{P} = \bigtriangleup P_1 P_2 \cdots P_n$ is a convex polygon then $\mathbf{P} = \bar{P} \cup \text{int}(\bar{P})$ is a polygonal region and $\text{int}(\bar{P}) = \text{ins}(\mathbf{P})$.

Theorem 10.1.3. *Let* $\bar{P} = \bigtriangleup P_1 \cdots P_n$ *be a convex polygon and let*

$$\mathbf{P} = \blacktriangle P_1 \cdots P_n = \bigtriangleup P_1 \cdots P_n \cup \text{int}(\bigtriangleup P_1 \cdots P_n).$$

Then

(i) **P** *is a polygonal region*
(ii) ins(**P**) = int($\bar{P}$)
(iii) bd(**P**) = $\bar{P}$.

PROOF. The actual details of the proof are left to Problem A4. The basic idea in part (i) is to use induction on n to show that

$$\blacktriangle P_1 \cdots P_n = \blacktriangle P_1 P_2 P_3 \cup \blacktriangle P_1 P_3 P_4 \cup \cdots \cup \blacktriangle P_1 P_{n-1} P_n.$$

See Figure 10-5.   □

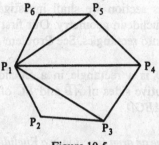

Figure 10-5

Definition. If $\bar{P} = \bigtriangleup P_1 \cdots P_n$ is a convex polygon, then $\mathbf{P} = \blacktriangle P_1 \cdots P_n$ as defined in Theorem 10.1.3 is called the **convex polygonal region** determined by $\bar{P}$.

Because of Theorem 10.1.3 we abuse our notation and refer to the interior of a convex polygonal region $\mathbf{R}$, int($\mathbf{R}$), instead of the inside of $\mathbf{R}$, ins($\mathbf{R}$).

We are now in a position to define area. Certainly "area" must assign to each polygonal region a positive number such that if two regions intersect only along an edge or at a vertex (or, equivalently, along their boundary), then the area of their union is the sum of their areas. This last statement is the same as saying that the whole area is the sum of its parts and that edges have "zero area." We would also want two triangular regions determined by "identical" (i.e., congruent) triangles to have the same area. In addition, we prefer that the area function be normalized so that a square of side a has area a^2. More precisely, the definition is as follows.

Definition. In a neutral geometry an **area function** is a function $\sigma : \mathcal{R} \to \mathbb{R}$ such that

  (i) $\sigma(\mathbf{R}) > 0$ for every region $\mathbf{R} \in \mathcal{R}$.
 (ii) If $\triangle ABC \simeq \triangle DEF$ then $\sigma(\blacktriangle ABC) = \sigma(\blacktriangle DEF)$.
(iii) If $\mathbf{R}_1$ and $\mathbf{R}_2$ are two polygonal regions whose intersection contains only boundary points of $\mathbf{R}_1$ and $\mathbf{R}_2$ then

$$\sigma(\mathbf{R}_1 \cup \mathbf{R}_2) = \sigma(\mathbf{R}_1) + \sigma(\mathbf{R}_2).$$

(iv) If $\mathbf{R}$ is the convex polygonal region determined by a square whose sides have length a, then

$$\sigma(\mathbf{R}) = a^2.$$

It would be natural at this point to prove that every neutral geometry has an area function. We shall assert this fact now, but not prove it until Section 10.2 (for a Euclidean geometry) and Section 10.3 (for a hyperbolic geometry). This is because both theorems are technically quite involved.

Theorem 10.1.4. *For any neutral geometry there is an area function σ.*

For the rest of this section we shall investigate the consequences of Theorem 10.1.4 in a Euclidean geometry. Our first result can be proved by breaking a square up into rectangles. See Problem A5.

Definition. If $\square ABCD$ is a rectangle in a Euclidean geometry then the lengths of two consecutive sides of $\overline{AB}$ and $\overline{BC}$ of $\square ABCD$ are called the **length** and **width** of $\square ABCD$.

Theorem 10.1.5. *Let σ be an area function in a Euclidean geometry. If $\square ABCD$ is a rectangle of length a and width b then*

$$\sigma(\blacksquare ABCD) = ab.$$

We might note that Theorem 10.1.5 is also true in a hyperbolic geometry, precisely because there are no rectangles! The remaining results of the section are false in hyperbolic geometry. Our proofs depend upon the existence of rectangles which, of course, invalidates them in a hyperbolic geometry.

Theorem 10.1.6. *In a Euclidean geometry, if $\triangle ABC$ has a right angle at B then*

$$\sigma(\blacktriangle ABC) = \tfrac{1}{2}(AB)(BC).$$

PROOF. Let D be the point on the same side of $\overline{BC}$ as A such that $\overline{CD} \simeq \overline{AB}$ and $\overline{CD} \perp \overline{BC}$, as in Figure 10-6. Since the geometry is Euclidean, $\square ABCD$ is a rectangle with length AB and width BC. Since $\triangle ABC \simeq \triangle CDA$, the result follows from Theorem 10.1.5.   ☐

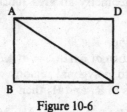

Figure 10-6

The proofs of the next three results are left as exercises.

Corollary 10.1.7. *In a Euclidean geometry if $\triangle ABC$ has base $\overline{BC}$ of length b and altitude $\overline{AD}$ of length h, then $\sigma(\blacktriangle ABC) = \tfrac{1}{2}bh$.*

Definition. $\square ABCD$ is a **trapezoid** if $\overline{AD} \parallel \overline{BC}$. In this case we say that $\overline{AD}$ is the **lower base** and $\overline{BC}$ is the **upper base**.

Theorem 10.1.8. *A trapezoid is a convex polygon.*

Theorem 10.1.9. *In a Euclidean geometry, if* $\square ABCD$ *is a trapezoid then*

$$\sigma(\blacksquare ABCD) = \tfrac{1}{2}(AD + BC) \cdot h$$

where h is the distance from $\overleftrightarrow{AD}$ *to* $\overleftrightarrow{BC}$. *In particular, if* $\square ABCD$ *is a parallelogram then* $\sigma(\blacksquare ABCD) = (AD) \cdot h$.

The area function can be used to prove the basic theorem of similarity theory (Theorem 9.2.3). In fact, this is what Euclid did. (See Problem A12 for the proof.) Although such an approach appears to be simpler, it depends on knowing the existence of an area function. As we shall see in the next section, the existence of an area function in Euclidean geometry can be proved using similarity theory and in particular Theorem 9.2.3. Thus if you wish to avoid circular reasoning, the simpler proof of Theorem 9.2.3 can be obtained only at the expense of assuming, as an axiom, that an area function exists.

As the last result in this section, we present Euclid's proof of the Pythagorean Theorem. Although this proof is more complicated than the one we presented earlier (Theorem 9.2.7), it is important because it shows a basic difference between the view of Euclid and our own view. We have taken a metric view of geometry and so may use the full power of the ruler postulate and our knowledge of the real numbers. However, real numbers were not available to Euclid. In fact, Euclid was unable to measure the length of the hypotenuse of an isosceles right triangle with each leg of unit length because the length of the hypotenuse is irrational: $\sqrt{2}$.

Note that in the following restatement of the Pythagorean Theorem there is *no* mention of the lengths of the sides of the triangle. Also, the term "equal" means "equal in area." We cannot resist quoting the theorem as Euclid did.

Theorem 10.1.10 (Pythagoras). *In a Euclidean geometry, if* $\triangle ABC$ *is a right triangle then the square on the hypotenuse is equal to the sum of the squares on the legs.*

PROOF. Let $\triangle ABC$ have a right angle at B. Let $\square ABDE$ be a square with $\overline{DE}$ on the opposite side of $\overleftrightarrow{AB}$ as C. Similarly, let $\square BCFG$ and $\square ACIH$ be squares constructed as in Figure 10-7. Euclid's statement of the theorem is equivalent to

$$\sigma(\blacksquare ACIH) = \sigma(\blacksquare ABDE) + \sigma(\blacksquare BCFG). \tag{1-2}$$

Let J be the foot of the altitude from B to $\overline{AC}$. Since $\overleftrightarrow{AC} \| \overleftrightarrow{HI}$, $\overleftrightarrow{BJ}$ is perpendicular to $\overleftrightarrow{HI}$ at a point K. Furthermore, $A—J—C$ and $H—K—I$ (Why?). We shall verify Equation (1-2) by showing that $\sigma(\blacksquare ABDE) = \sigma(\blacksquare AJKH)$ and $\sigma(\blacksquare BCFG) = \sigma(\blacksquare JCIK)$ and then using the addition law for area:

$$\sigma(\blacksquare ACIH) = \sigma(\blacksquare AJKH) + \sigma(\blacksquare JCIK). \tag{1-3}$$

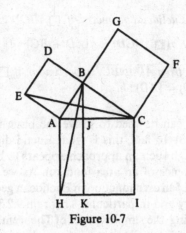

Figure 10-7

Note that the right angles $\angle EAB$ and $\angle HAC$ are congruent and that $\angle BAC = \angle CAB$. Now $B \in \text{int}(\angle EAC)$ and $C \in \text{int}(\angle HAB)$ so that by Angle Addition $\angle EAC \simeq \angle HAB \simeq \angle BAH$. By construction $\overline{AE} \simeq \overline{AB}$ and $\overline{AC} \simeq \overline{AH}$ so that $\triangle EAC \simeq \triangle BAH$ by SAS. Hence

$$\sigma(\blacktriangle EAC) = \sigma(\blacktriangle BAH).$$

If we view $\overline{EA}$ as the base of $\triangle EAC$, then the height of $\triangle EAC$ is BA. This is because the desired height is the distance between the parallel lines $\overleftrightarrow{BC}$ and $\overleftrightarrow{EA}$. Thus

$$\sigma(\blacktriangle EAC) = \tfrac{1}{2}(EA)(BA) = \tfrac{1}{2}\sigma(\blacksquare ABDE).$$

Likewise,

$$\sigma(\blacktriangle BAH) = \tfrac{1}{2}(AH)(AJ) = \tfrac{1}{2}\sigma(\blacksquare AJKH)$$

so that

$$\sigma(\blacksquare ABDE) = \sigma(\blacksquare AJKH). \qquad (1\text{-}4)$$

Similarly,

$$\sigma(\blacksquare BCFG) = \sigma(\blacksquare JCIK). \qquad (1\text{-}5)$$

Combining Equations (1-3), (1-4), and (1-5) we obtain Equation (1-2) as desired. □

Other area proofs of the Pythagorean Theorem (including one due to a former President of the United States!) are sketched in Problems A10 and A11.

PROBLEM SET 10.1

Part A.

1. Prove that the interior of a convex polygon is non-empty.

2. Prove that every 3-gon is a convex polygon.

3. Prove that a convex polygonal region is a convex set.

4. Prove Theorem 10.1.3.

5. Prove Theorem 10.1.5. Hint: See Figure 10-8.

Figure 10-8

6. Prove Corollary 10.1.7.

7. Prove Theorem 10.1.8.

8. Prove Theorem 10.1.9

*9. Prove that in a Euclidean geometry if two triangles have the same height, then the ratio of their areas equals the ratio of their bases.

10. Use Figure 10-9 to give a proof of the Pythagorean Theorem using area.

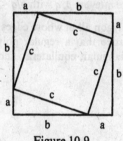

Figure 10-9

11. Use Figure 10-10 to give a proof of the Pythagorean Theorem due to James Garfield, the twentieth President of the United States.

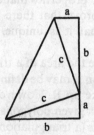

Figure 10-10

12. Use the following outline to prove the basic similarlity theorem (Theorem 9.2.3): In a Euclidean geometry, if parallel lines l_1, l_2, l_3 are cut by transversals t_1, t_2 at points A, B, C and D, E, F respectively with $A—B—C$ and $D—E—F$ then $AB/BC = DE/EF$.

    a. Prove $\sigma(\blacktriangle CBE) = \sigma(\blacktriangle FEB)$ and $\sigma(\blacktriangle ABE) = \sigma(\blacktriangle DEB)$.

    b. Prove $AB/BC = \sigma(\blacktriangle ABE)/\sigma(\blacktriangle CBE)$.

    c. Prove $DE/EF = AB/BC$.

Part B. "Prove" may mean "find a counterexample".

13. Prove that the boundary of a polygonal region is a polygon.

14. Let $\mathbf{R}_1$ and $\mathbf{R}_2$ be polygonal regions in a Euclidean geometry. If there is a point $P \in \text{ins}(\mathbf{R}_1) \cap \text{ins}(\mathbf{R}_2)$, then prove $\mathbf{R}_1 \cap \mathbf{R}_2$ is also a polygonal region.

15. Let $\tau : \mathscr{R} \to \mathbb{R}$ be a function defined on the set of polygonal regions in a Euclidean geometry such that it satisfies axioms (i), (ii), and (iii) of an area function. Suppose further that if $\square ABCD$ is a square whose side has length 1 then $\tau(\blacksquare ABCD) = 1$. Prove that τ actually is an area function by the following outline.

    a. If q is a positive integer and $\square ABCD$ is a square whose side has length $1/q$, prove $\tau(\blacksquare ABCD) = 1/q^2$.

    b. If p and q are positive integers and $\square ABCD$ is a square whose side has length p/q, then prove that $\tau(\blacksquare ABCD) = p^2/q^2$.

    c. If $\square ABCD$ is a square whose side has length a, prove that $\tau(\blacksquare ABCD) = a^2$ by considering positive integers p, q with $p/q \le a < (p + 1)/q$.

16. A **regular polygon** is a polygon all of whose edges are congruent and all of whose angles are congruent. Prove that a regular polygon is convex. (Note that for polygons, the terms equiangular, equilateral, and regular are different, which is not the case for triangles.)

10.2　The Existence of Euclidean Area

In the previous section we defined what is meant by an area function and discovered some of the basic properties that a Euclidean area function must possess. Now we want to prove that there really is an area function in a Euclidean geometry and that it is unique. The proof will involve some technicalities.

    The basic idea is to define the area of a triangular region first. From that the area of a polygonal region $\mathbf{R}$ may be defined as the sum of the areas of the triangular regions that make up $\mathbf{R}$. However, there is a technical problem with this approach because a given polygonal region can be subdivided into a union of triangular regions (a triangulation) in many ways. We must show that the sum of the areas of the triangular regions does not depend upon the actual choice of subdivision. This is where the technical difficulties become severe.

    Because we do not know at the beginning of this section that the function we define actually is an area function, we should not use the word "area" in the definition. Thus we will define the "size" of a polygonal region in a

Euclidean geometry (and will hope that "size" is an area function, a fact that we will eventually prove).

In keeping with the approach outlined above we will first define the size of a triangular region and then the size of a polygonal region with respect to a particular triangulation. The primary theorem will be that the size of a region does not depend upon the particular triangulation used.

Definition. The **size of a triangular region** in a Euclidean geometry is one-half the length of one side multiplied by the length of the altitude to that side:

$$s(\blacktriangle ABC) = \tfrac{1}{2}(BC)(AD)$$

where D is the foot of the altitude from A.

By Theorem 9.2.9, it doesn't matter which side we choose in the definition of size of a triangular region: the product (base)(height) is independent of the choice of the base. Recall that this result was a consequence of similarity theory so that our development of the existence of a Euclidean area function depends on similarity theory.

Definition. A **triangulation** τ of a polygonal regional **R** is a set $\tau = \{T_1, T_2, \ldots, T_n\}$ of triangular regions whose union is **R** such that any two elements of τ are either disjoint or intersect only along a common edge or at a common vertex. $P \in \mathbf{R}$ is a **vertex** of τ if P is a vertex of one (or more) of the members of τ. An **edge** of τ is an edge of a member of τ.

From the very definition of a region, every polygonal region has a triangulation. However, as Figures 10-2 and 10-3 showed, a region may have more than one triangulation.

Definition. Let $\tau = \{T_1, T_2, \ldots, T_n\}$ be a triangulation of a polygonal region **R** in a Euclidean geometry. The **size of R with respect to** τ is the sum of the sizes of the members of τ:

$$s_\tau(\mathbf{R}) = s(T_1) + \cdots + s(T_n).$$

As stated above, the key technical result of this section will be that $s_\tau(\mathbf{R}) = s_{\tau'}(\mathbf{R})$ for any two triangulations τ and τ' of **R**. Our basic method for proving this result is to find a third triangulation $\bar\tau$ which is in a sense "smaller" than τ and τ' such that $s_\tau(\mathbf{R}) = s_{\bar\tau}(\mathbf{R}) = s_{\tau'}(\mathbf{R})$. This "smaller" triangulation will be formed by cutting the given triangular regions into triangles and trapezoids, and then further cutting up these regions into triangles in a special way.

Definition. A **base triangulation** of a triangular region $\mathbf{R} = \blacktriangle ABC$ is a triangulation all of whose vertices except for one lie on a single side of $\triangle ABC$. (See Figure 10-11.)

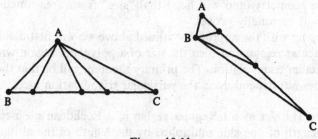

Figure 10-11

The proof of the following result is not difficult and is left to Problem A1.

Lemma 10.2.1. *In a Euclidean geometry, the size of triangular region* $\blacktriangle ABC$ *with respect to any base triangulation* τ *of* $\blacktriangle ABC$ *equals the size of* $\blacktriangle ABC$:

$$s_\tau(\blacktriangle ABC) = s(\blacktriangle ABC) = \tfrac{1}{2}(base)(height).$$

Definition. A base triangulation of a trapezoidal region $\blacksquare ABCD$ is a triangulation τ such that each vertex of τ lies either on $\overline{BC}$ or on $\overline{AD}$. (See Figure 10-12.)

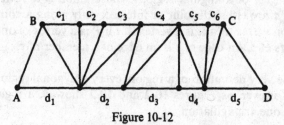

Figure 10-12

Lemma 10.2.2. *In a Euclidean geometry, the size of a trapezoidal region* $\blacksquare ABCD$ *with respect to a base triangulation* τ *is the product of the height of* $\square ABCD$ *with the average of the lengths of its bases:*

$$s_\tau(\blacksquare ABCD) = \tfrac{1}{2}h(b_1 + b_2)$$

where $b_1 = BC$ *and* $b_2 = AD$. *In particular, the size is independent of the specific base triangulation used.*

PROOF. For each triangular region in τ choose as its base the edge parallel to $\overline{AD}$. Then each triangular region has height h = height of $\square ABCD$. Let the lengths of the bases of the triangular regions be denoted $c_1, c_2, \ldots, c_k$ and $d_1, d_2, \ldots, d_l$ as in Figure 10-12. Then since each of the triangular regions has height h

$$s_\tau(\blacksquare ABCD) = \tfrac{1}{2}hc_1 + \cdots + \tfrac{1}{2}hc_k + \tfrac{1}{2}hd_1 + \cdots + \tfrac{1}{2}hd_l$$
$$= \tfrac{1}{2}hb_1 + \tfrac{1}{2}hb_2 = \tfrac{1}{2}h(b_1 + b_2). \qquad \square$$

> **Notation.** If $\square ABCD$ is a trapezoid then $s(\blacksquare ABCD)$ is the size of $\blacksquare ABCD$ with respect to any base triangulation. Similarly, $s(\blacktriangle ABC)$ is the size of $\triangle ABC$ with respect to any base triangulation of $\blacktriangle ABC$. This notation is well defined by Lemmas 10.2.1 and 10.2.2.

Next we must investigate what happens when we decompose a triangle into a triangle and a trapezoid.

Lemma 10.2.3. *Let* $\triangle ABC$ *be given in a Euclidean geometry with* $A\!-\!D\!-\!B$ *and* $A\!-\!E\!-\!C$ *where* $\overline{DE} \parallel \overline{BC}$. *Then*

$$s(\blacktriangle ABC) = s(\blacktriangle ADE) + s(\blacksquare BDEC)$$

PROOF. Consider $\overline{DE}$ as the base of $\triangle ADE$ and $\overline{BC}$ as the base of $\triangle ABC$. See Figure 10-13. The height h of $\triangle ABC$, the height h_1 of $\triangle ADE$, and the height h_2 of the trapezoid $\square BDEC$ are related by

$$h = h_1 + h_2.$$

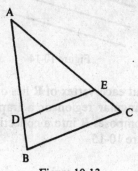

Figure 10-13

If $b_1 = DE$ and $b_2 = BC$, then by Theorem 9.2.3

$$\frac{h_1}{h} = \frac{b_1}{b_2}$$

so that $b_2 h_1 = b_1 h = b_1 h_1 + b_1 h_2$ or $b_1 h_1 = b_2 h_1 - b_1 h_2$. Thus

$$
\begin{aligned}
s(\blacktriangle ADE) + s(\blacksquare BDEC) &= \tfrac{1}{2} b_1 h_1 + \tfrac{1}{2} h_2 (b_1 + b_2) \\
&= \tfrac{1}{2}(b_2 h_1 + b_2 h_2) \\
&= \tfrac{1}{2} b_2 (h_1 + h_2) = s(\blacktriangle ABC).
\end{aligned}
$$
$\square$

Definition. A finite set of lines, $\mathscr{F}$, is a **family of parallel lines** in a Euclidean geometry if for any $l, l' \in \mathscr{F}$, $l \parallel l'$.

Suppose that **R** is a polygonal region in a Euclidean geometry and that $\mathscr{F}$ is a family of parallel lines such that each vertex of **R** lies on a line of $\mathscr{F}$. The lines of $\mathscr{F}$ can be named $l_1, l_2, \ldots, l_k$ so that if l is a common transversal of the lines in $\mathscr{F}$ and $l \cap l_i = A_i$, then

$$A_1 - A_2 - A_3 - \cdots - A_k.$$

We may then say that l_i and l_{i+1} are consecutive lines in the family. For each i let H_i^+ be the half plane of l_i that contains l_{i+1} and let H_i^- be the half plane of l_i that contains l_{i-1}. (H_1^- and H_k^+ are not defined but could be.) For each i with $1 \le i \le k-1$, we define the **strip** between l_i and l_{i+1} to be:

$$\mathbf{B}_i = l_i \cup l_{i+1} \cup (H_i^+ \cap H_{i+1}^-).$$

See Figure 10-14.

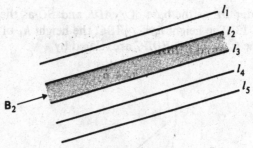

Figure 10-14

Since we assumed that each vertex of **R** lies on a line of $\mathscr{F}$ then for each i, $\mathbf{R} \cap \mathbf{B}_i$ is either a triangular region or a trapezoidal region (or empty). Thus the family $\mathscr{F}$ decomposes **R** into a collection of triangular and trapezoidal regions. See Figure 10-15.

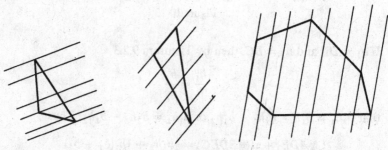

Figure 10-15

Definition. If $\mathscr{F}$ is a family of parallel lines in a Euclidean geometry such that every vertex of a polygonal region **R** lies on a line of $\mathscr{F}$, then the collection of triangular and trapezoidal regions described above is called the **parallel decomposition** of **R** induced by $\mathscr{F}$.

The proof of the next result is left to Problem A2. Be sure to consider two cases depending on whether or not the family $\mathscr{F}$ is parallel to a side of the triangular region.

Lemma 10.2.4. *Let* $\mathbf{R} = \triangle ABC$ *be a triangular region and let* $\mathscr{F}$ *be a finite family of parallel lines in a Euclidean geometry such that A, B, and C lie on lines of* $\mathscr{F}$. *Then the size of* $\mathbf{R}$ *is the sum of the sizes of the triangular and trapezoidal regions which are the members of the parallel decomposition of* $\mathbf{R}$ *induced by* $\mathscr{F}$.

We are now ready for the major technical result that the size of a polygonal region does not depend upon the choice of triangulation.

Theorem 10.2.5. *In a Euclidean geometry if* τ *and* τ' *are triangulations of the polygonal region* $\mathbf{R}$ *then* $s_\tau(\mathbf{R}) = s_{\tau'}(\mathbf{R})$.

PROOF. The method of proof is to subdivide the regions in τ and τ' to get a new triangulation $\bar{\tau}$ with $s_\tau(\mathbf{R}) = s_{\bar{\tau}}(\mathbf{R}) = s_{\tau'}(\mathbf{R})$. We start this procedure by choosing a family of parallel lines $\mathscr{F}$ and inducing from that family a parallel decomposition of the triangular regions of τ and τ'.

Two edges of a triangulation are either disjoint, identical (and intersect in a segment), or intersect at a single point (a vertex). Similarly, if e is an edge of one triangulation of $\mathbf{R}$ and f is an edge of another triangulation of $\mathbf{R}$, then e and f are either disjoint, intersect in a segment (which may be only part of e or f), or intersect in a single point (which may not be a vertex of either triangulation). We let $\mathscr{V}$ be the set of all points which are intersections of two edges:

$$\mathscr{V} = \{P | \{P\} = e \cap f, \text{ where } e \text{ is an edge of either}$$
$$\tau \text{ or } \tau' \text{ and } f \text{ is an edge of either } \tau \text{ or } \tau'\}.$$

Note that $\mathscr{V}$ contains all the vertices of τ and all the vertices of τ' as well as some additional points. Let l' be any line and define

$$\mathscr{F} = \mathscr{F}(\tau, \tau', l') = \{l \mid l \| l' \text{ and } V \in l \text{ for some } V \in \mathscr{V}\}.$$

See Figure 10-16 where the points of $\mathscr{V}$ are marked on both triangulations and $\mathscr{F}$ is marked on τ. In this example $\mathscr{V}$ contains exactly one point which is not a vertex of either τ or τ'.

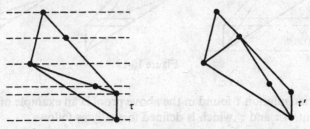

Figure 10-16

$\mathscr{F}$ induces a parallel decomposition of each triangular region in τ. This gives a decomposition ρ of **R** into triangles and trapezoids. By definition, the size of **R** with respect to τ is the sum of the sizes of the triangular regions in τ. By Lemma 10.2.4 the size of each of these regions is the sum of the sizes of the regions in the induced parallel decomposition. Thus we may write $s_\tau(\mathbf{R}) = s_\rho(\mathbf{R})$.

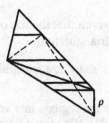

Figure 10-17

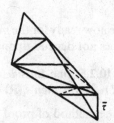

Figure 10-18

Finally each edge of τ' creates base decompositions of the regions in ρ into triangles and trapezoids. See Figure 10-17. Choose a diagonal of any remaining trapezoid to give a base decomposition. See Figure 10-18. The net result is we have a new triangulation $\bar\tau$ of **R**.

By Lemmas 10.2.1 and 10.2.2 we have

$$s_{\bar\tau}(\mathbf{R}) = s_\rho(\mathbf{R}) = s_\tau(\mathbf{R}).$$

We may carry out the same procedure starting with τ' and using the same family $\mathscr{F}$. The crucial point is that *the decomposition ρ' so obtained is the same as ρ.* This is because each is determined by the lines of $\mathscr{F}$, the edges of τ and the edges of τ'. See Figure 10-19. In particular $\bar\tau' = \bar\tau$ if we choose the same diagonals of the remaining trapezoids so that

$$s_{\bar\tau'}(\mathbf{R}) = s_{\rho'}(\mathbf{R}) = s_{\bar\tau}(\mathbf{R}) = s_\tau(\mathbf{R}). \qquad \square$$

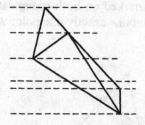

Figure 10-19

The triangulation $\bar\tau$ found in the above proof is an example of a common refinement of τ and τ' which is defined formally as follows.

Definition. Let τ and $\bar{\tau}$ be two triangulations of a polygonal region **R**. If every triangular region of $\bar{\tau}$ is contained in a triangular region of τ, then $\bar{\tau}$ is a **refinement** of τ.

Theorem 10.2.6. *In a Euclidean geometry let* $\sigma: \mathcal{R} \to \mathbb{R}$ *be defined by*

$$\sigma(\mathbf{R}) = s_\tau(\mathbf{R})$$

where τ *is any triangulation of the polygonal region* **R***. Then* σ *is an area function.*

PROOF. Because of Theorem 10.2.5, σ is well defined; that is, it does not depend on the choice of triangulation τ. Thus all that is necessary is to verify the four axioms of area are satisfied. This is left to Problem A5. Be careful with the third axiom: If **R** is the union of $\mathbf{R}_1$ and $\mathbf{R}_2$ and τ_i is a triangulation of $\mathbf{R}_i$, then $\tau_1 \cup \tau_2$ need not be a triangulation of **R**. See Figure 10-20 and Problem A6. □

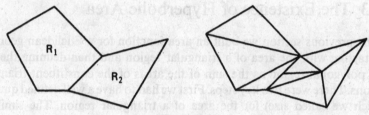

Figure 10-20

Theorem 10.2.7. *In a Euclidean geometry there is exactly one area function.*

PROOF. By Theorem 10.2.6 we know that σ is an area function, so there is at least one. We must therefore show that if α is an area function then $\alpha = \sigma$.

Let α be any area function for a Euclidean geometry. By the third axiom of area, the area of a polygonal region is the sum of the areas of the triangular regions in any triangulation. Hence if α agrees with σ on triangular regions we must have $\alpha = \sigma$. But this follows from Corollary 10.1.7 because there we showed that $\alpha(\triangle ABC) = \frac{1}{2}(\text{base})(\text{height})$ and this latter expression is the definition of σ. Thus there is only one area function in a Euclidean geometry. □

PROBLEM SET 10.2

Part A.

1. Prove Lemma 10.2.1.

2. Prove Lemma 10.2.4.

3. Give an example of a point of a polygonal region **R** which is a vertex in one triangulation but not in another.

4. Carry out the details of the proof of Theorem 10.2.5 for the triangulation τ_1 and τ_2 of $\triangle ABC$ and the line l' pictured in Figure 10-21.

5. Prove Theorem 10.2.6.

6. Explain why $\tau_1 \cup \tau_2$ does not give a triangulation of $\mathbf{R}_1 \cup \mathbf{R}_2$ in Figure 10-20.

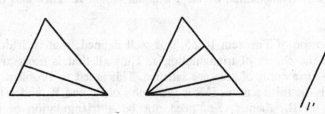

Figure 10-21

10.3 The Existence of Hyperbolic Area

In the previous section we built an area function for a Euclidean geometry by starting with the area of a triangular region and then defining the area of a polygonal region as the sum of the areas of the constituent triangular regions. There were two key steps. First we had to have a well defined quantity (which we called size) for the area of a triangular region. The similarity theorem which said that (base)(height) was independent of the choice of base was crucial for this step. Second, we had to show that if we had two different triangulations τ and τ' of a region **R** then each gave the same size for **R** so that we had a well defined area for any polygonal region. We did this by finding a common refinement $\overline{\tau}$ by means of a parallel decomposition.

For hyperbolic geometry we will follow a similar course, but with different proofs because of two difficulties. The first difficulty is that since (base)(height) is not independent of the choice of base in a hyperbolic geometry we will need a different definition for the area of a triangular region. See Problem A1. Since parallel lines are neither unique nor equidistant in the hyperbolic case we will also need a replacement for the idea of a parallel decomposition.

This section may be omitted if the reader is willing to accept the existence of a hyperbolic area function. The area function constructed here is based on the concept of the defect of a triangle. The area of a polygonal region will be defined to be the sum of the defects of the triangles of any triangulation of the region. In Section 10.4 we will show that this is the only choice for an area function, up to a constant multiple.

To start, we need a function which assigns to each triangular region a positive number in such a way that congruent triangles are assigned the same number. We already have such a function: the defect.

Definition. The **defect of a triangular region R** is the defect of the triangle that determines **R**:

$$\delta(\blacktriangle ABC) = \delta(\triangle ABC).$$

Our initial goal is to show that if τ and τ' are two triangulations of a polygonal region **R**, then the sum of the defects of the triangular regions in τ and τ' are the same.

Definition. Let **R** be the convex polygonal region determined by $\bar{P} = \bigwedge P_1 \cdots P_n$ and let $P \in \text{int}(\bar{P})$. The triangulation of **R** by triangular regions whose vertices are P together with a pair of consecutive vertices of $\bar{P}$ is called a **star triangulation** of **R** with respect to P. (See Figure 10-22.) Thus

$$\tau^*(\mathbf{R}, P) = \{\blacktriangle PP_1P_2, \ldots, \blacktriangle PP_{n-1}P_n, \blacktriangle PP_nP_1\} \tag{3-1}$$

is the star triangulation of **R** with respect to P.

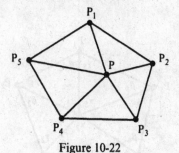

Figure 10-22

The proof that $\tau^*(\mathbf{R}, P)$ as defined by Equation (3-1) actually is a triangulation is left to Problem A2. The next piece of notation is useful because it avoids breaking statements into special cases like $\triangle PP_nP_1$, $\triangle PP_{i-1}P_i$, etc.

Notation. If $\bar{P} = \bigwedge P_1P_2 \cdots P_n$ is a polygon of degree n then P_0 and P_{n+1} are defined by

$$P_0 = P_n \quad \text{and} \quad P_{n+1} = P_1.$$

The angle at vertex P_i is

$$\angle P_i = \angle P_{i-1}P_iP_{i+1} \quad \text{for } 1 \le i \le n.$$

Lemma 10.3.1. *Let* $\mathbf{R} = \blacktriangle P_1 \cdots P_n$ *be a convex polygonal region in a neutral geometry and let* $P \in \text{int}(\mathbf{R})$. *Then the sum of the defects of the triangular regions in* $\tau^*(\mathbf{R}, P)$ *is*

$$\delta(\mathbf{R}, P) = 180(n - 2) - \sum_{i=1}^{n} m(\angle P_i). \tag{3-2}$$

In particular $\delta(\mathbf{R}, P)$ *does not depend on the choice of* $P \in \text{int}(\mathbf{R})$.

PROOF. For $1 \leq i \leq n$ define $\alpha_i, \beta_i, \gamma_i$ by

$$\alpha_i = m(\angle P_i P P_{i+1})$$
$$\beta_i = m(\angle P P_i P_{i+1})$$
$$\gamma_i = m(\angle P P_{i+1} P_i)$$

as marked in Figure 10-23.

We have the following relations

$$\delta(\blacktriangle P P_i P_{i+1}) = 180 - (\alpha_i + \beta_i + \gamma_i)$$

$$\sum_{i=1}^{n} \alpha_i = 360$$

$$m(\angle P_i) = \gamma_{i-1} + \beta_i \quad \text{since } P \in \text{int}(\angle P_i)$$

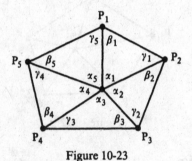

Figure 10-23

where $\gamma_0 = \gamma_n$. Thus we have

$$\delta(\mathbf{R}, P) = \sum_{i=1}^{n} (180 - (\alpha_i + \beta_i + \gamma_i))$$

$$= 180n - \left(\sum_{i=1}^{n} \alpha_i \right) - (\gamma_n + \beta_1 + \gamma_1 + \beta_2 + \cdots + \gamma_{n-1} + \beta_n)$$

$$= 180n - 360 - \sum_{i=1}^{n} m(\angle P_i)$$

$$= 180(n - 2) - \sum_{i=1}^{n} m(\angle P_i). \qquad \square$$

Because of this lemma we can make the following definition.

Definition. The **defect of a convex polygonal region R**, denoted $\delta(\mathbf{R})$, is the sum of the defects of a star triangulation of $\mathbf{R}$ with respect to any point $P \in \text{int}(\mathbf{R})$.

Following the notation of the previous proof, if $\mathbf{R} = \blacksquare P_1 \cdots P_n$ then

$$\delta(\mathbf{R}) = 180(n-2) - \sum_{i=1}^{n} m(\angle P_i).$$

Lemma 10.3.2. *Let $\mathbf{R}$ be a convex polygonal region in a neutral geometry and let l be a line which intersects the interior of $\mathbf{R}$. Then l decomposes $\mathbf{R}$ into two convex polygonal regions $\mathbf{R}_1$ and $\mathbf{R}_2$.*

PROOF. The details are left to Problem A5. Note that $\mathbf{R}_1$ and $\mathbf{R}_2$ can be defined in terms of the half planes determined by l. The important assertion is that $\mathbf{R}_1$ and $\mathbf{R}_2$ are convex polygonal regions. See Figure 10-24. $\square$

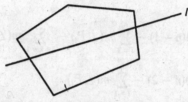

Figure 10-24

Lemma 10.3.3. *Let $\mathbf{R} = \blacktriangle P_1 \cdots P_n$ be a convex polygonal region and let l be a line that intersects the interior of $\mathbf{R}$. Let $\mathbf{R}_1$ and $\mathbf{R}_2$ be the two convex polygonal regions of Lemma 10.3.2. Then*

$$\delta(\mathbf{R}) = \delta(\mathbf{R}_1) + \delta(\mathbf{R}_2).$$

PROOF. There are three cases depending on whether l contains 0, 1, or 2 vertices of $\mathbf{R}$. (Why can't l contain more than two vertices?) We shall consider the case where l contains one vertex and leave the other cases to Problem A6. See Figure 10-25.

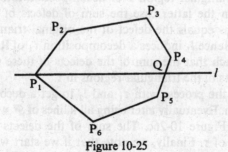

Figure 10-25

We may label our vertices so that $P_1 \in l$. l must also intersect a side $\overline{P_k P_{k+1}}$ (Why?) at a point Q with $2 \leq k \leq n-1$ and $P_k - Q - P_{k+1}$. Then $\mathbf{R}_1 = \blacktriangle P_1 P_2 \cdots P_k Q$ has $k + 1$ vertices and $\mathbf{R}_2 = \blacktriangle Q P_{k+1} \cdots P_n P_1$ has $n - k + 2$ vertices. By Lemma 10.3.1 and Equation (3-2)

$$\delta(\mathbf{R}_1) = 180(k + 1 - 2) - m(\angle QP_1P_2) - \left(\sum_{i=2}^{k} m(\angle P_i) \right) - m(\angle P_kQP_1)$$

$$\delta(\mathbf{R}_2) = 180(n - k + 2 - 2) - m(\angle P_1QP_{k+1})$$

$$- \left(\sum_{i=k+1}^{n} m(\angle P_i) \right) - m(\angle P_nP_1Q).$$

Since $Q \in \text{int}(\angle P_1)$ we have

$$m(\angle QP_1P_2) + m(\angle P_nP_1Q) = m(\angle P_1).$$

Since $P_k - Q - P_{k+1}$

$$m(\angle P_kQP_1) + m(\angle P_1QP_{k+1}) = 180.$$

Hence

$$\delta(\mathbf{R}_1) + \delta(\mathbf{R}_2) = 180(n - 1) - \sum_{i=2}^{k} m(\angle P_i) - \sum_{i=k+1}^{n} m(\angle P_i) - m(\angle P_1) - 180$$

$$= 180(n - 2) - \sum_{i=1}^{n} m(\angle P_i)$$

$$= \delta(\mathbf{R}). \qquad \square$$

Theorem 10.3.4. *Let τ and τ' be triangulations of the same polygonal region* **R**. *Then the sum of the defects of the triangular regions of τ equals the sum of the defects of the triangular regions of τ'.*

PROOF. Each edge of a triangular region in either τ or τ' determines a line. Let $\mathcal{F}$ be the set of all such lines. $\mathcal{F}$ contains a finite number of lines which can be named $l_1, l_2, \ldots, l_k$. Note that unlike the set of lines used in the previous section, this set $\mathcal{F}$ is *not* made up of parallel lines.

Consider the triangular regions of τ. l_1 either does not intersect the interior of a particular triangular region **T** or decomposes it into two convex polygonal regions. In the latter case the sum of defects of these two convex polygonal regions equals the defect of the original triangular region **T** by Lemma 10.3.3. Hence l_1 induces a decomposition τ_1 of **R** into convex polygonal regions such that the sum of the defects of these regions equals the sum of the defects of the triangular regions in τ.

We continue the process with τ_1 and l_2 to get a decomposition τ_2, and then l_3 and so on. Eventually after using all k lines of $\mathcal{F}$ we receive a decomposition $\bar{\tau}$. See Figure 10-26c. The sum of the defects associated with $\bar{\tau}$ is equal to that of τ. Finally, we note that if we start with τ' instead of τ, we get the same decomposition $\bar{\tau}$. See Figure 10-26d. Hence, the sum of the defects associated with τ' equals the sum of the defects associated with $\bar{\tau}$ and hence with τ. (If we take star triangulations of the convex polynomial regions of $\bar{\tau}$ then we will have a common refinement of τ and τ'.) $\qquad \square$

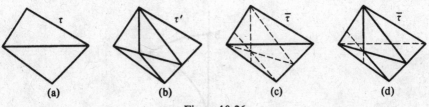

Figure 10-26

Definition. The **total defect** of a polygonal region **R**, $\delta(\mathbf{R})$, is the sum of the defects of the triangular regions of any triangulation τ of **R**.

This definition makes sense precisely because of Theorem 10.3.4.

Theorem 10.3.5. *In a hyperbolic geometry the total defect function* $\delta: \mathscr{R} \to \mathbb{R}$ *is an area function.*

PROOF. We leave most of the details to Problem A8 and consider only the fourth axiom. We must show that if $\square ABCD$ is a square whose side has length a then $\delta(\blacksquare ABCD) = a^2$. This is true *precisely because there are no squares* in a hyperbolic geometry, and so every square has defect a^2. In other words, the statement is true vacuously. (Of course, every square also has defect 1 or 7 or 22/7. In fact, every square has any defect we desire.) $\square$

We end this section with an alternative and optional description of the area function in the Poincaré Plane $\mathscr{H}$. This description is motivated by differential geometry and is defined in terms of an integral. See Millman and Parker [1977] for more details. It is included so that the reader can see an alternative tool for computing a hyperbolic area. Since Euclidean areas can be given by integrals, you might expect that the (hyperbolic) area of a region in the Poincaré Plane can also be given as an integral. This is actually the case.

Definition. The **hyperbolic area** of a polygonal region **R** in $\mathscr{H}$ is

$$\alpha(\mathbf{R}) = \frac{180}{\pi} \iint_{\mathbf{R}} \frac{1}{y^2}\, dy\, dx. \tag{3-3}$$

We shall verify that this formula actually gives the defect in the case of one specific triangle. Of course, this does not prove that it always gives the defect.

Example 10.3.6. Let **R** be the triangular region in $\mathscr{H}$ with vertices at $A = (0, 1)$, $B = (0, 5)$ and $C = (3, 4)$. Show that $\alpha(\mathbf{R}) = \delta(\mathbf{R})$. See Figure 10-27.

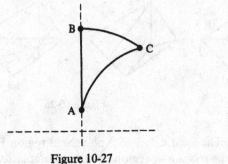

Figure 10-27

SOLUTION. The sides of **R** are given by

$$\overline{AB} = {}_0L, \qquad \overline{BC} = {}_0L_5, \qquad \overline{AC} = {}_4L_{\sqrt{17}}$$

$$m(\angle A) = \cos^{-1}\left(\frac{\langle(0,1),(1,4)\rangle}{1 \cdot \sqrt{17}}\right) = \cos^{-1}\left(\frac{4}{\sqrt{17}}\right)$$

$$m(\angle B) = 90$$

$$m(\angle C) = \cos^{-1}\left(\frac{\langle(-4,3),(-4,-1)\rangle}{5 \cdot \sqrt{17}}\right) = \cos^{-1}\left(\frac{13}{5\sqrt{17}}\right)$$

Hence

$$\delta(\mathbf{R}) = 90 - \cos^{-1}\left(\frac{4}{\sqrt{17}}\right) - \cos^{-1}\left(\frac{13}{5\sqrt{17}}\right) \doteq 25.0576$$

On the other hand

$$\alpha(\mathbf{R}) = \frac{180}{\pi} \int_0^3 \int_{\sqrt{17-(x-4)^2}}^{\sqrt{25-x^2}} \left(\frac{1}{y^2}\right) dy\, dx$$

$$= \frac{180}{\pi} \int_0^3 \left(-\frac{1}{\sqrt{25-x^2}} + \frac{1}{\sqrt{17-(x-4)^2}}\right) dx$$

$$= \frac{180}{\pi} \left(-\mathrm{Sin}^{-1}\frac{x}{5} + \mathrm{Sin}^{-1}\left(\frac{x-4}{\sqrt{17}}\right)\right)\Bigg|_0^3$$

(where $\mathrm{Sin}^{-1}(t)$ means measure in *radians* and $\sin^{-1}(t)$ means in *degrees*)

$$= -\sin^{-1}\left(\frac{3}{5}\right) + \sin^{-1}\left(\frac{-1}{\sqrt{17}}\right) + \sin^{-1}(0) - \sin^{-1}\left(\frac{-4}{\sqrt{17}}\right)$$

$$\doteq 25.0576.$$

Hence $\alpha(\mathbf{R}) = \delta(\mathbf{R})$.

We see in this computation why there is the factor of $180/\pi$ in the definition α: it is to change from radian measure to degree measure. □

PROBLEM SET 10.3

Part A.

1. Give an example of a triangle $\triangle ABC$ in the Poincaré Plane $\mathcal{H}$ such that

   (BC)(length of altitude from A) $\neq$ (AB)(length of altitude from C).

   Thus (base)(height) is not well defined for a triangle in a hyperbolic geometry.

2. If $\mathbf{R}$ is a convex polygonal region and $P \in \text{int}(\mathbf{R})$ in a neutral geometry, prove that $\tau^*(\mathbf{R}, P)$ is a triangulation of $\mathbf{R}$.

3. Show by example that $\tau^*(\mathbf{R}, P)$ as defined by Equation (3-1) need not be a triangulation if $\mathbf{R}$ is not convex.

4. In the proof of Lemma 10.3.1, prove the assertion that $\sum \alpha_i = 360$.

5. Prove Lemma 10.3.2.

6. Prove Lemma 10.3.3 for the two remaining cases.

7. Carry out the construction of Theorem 10.3.4 for the case which is pictured in Figure 10-28.

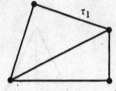 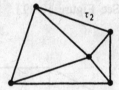

τ_1 τ_2

Figure 10-28

8. Prove Theorem 10.3.5.

9. Prove that the area of a Saccheri quadrilateral in a hyperbolic geometry is always less than 360.

10. Verify that Equation (3-3) gives the area of the Poincaré triangle with vertices at $(0, 5)$, $(0, 3)$ and $(2, \sqrt{21})$. (See Problem A2 of Section 5.1.)

11. Repeat Problem 10 for the Poincaré triangle with vertices at $(5, 1)$, $(8, 4)$, and $(1, 3)$.

Part B. "Prove" may mean "find a counterexample".

12. Prove that Equation (3-3) gives the area of any polygonal region in $\mathcal{H}$. (Hint: you need only consider triangles. Since integrals and the defect are both additive, you can assume that one side of the triangle is part of a type I line.)

13. Let $\delta: \mathcal{R} \to \mathbb{R}$ be the defect function for a Euclidean geometry. Prove that δ is not an area function. Which area axioms are not satisfied by δ?

14. In Example 10.3.6 we showed that the integral and the defect were both approximately equal to 25.0576. Use trigonometric identities to show that the integral is exactly equal to the defect.

10.4 Bolyai's Theorem

The two area functions we developed in Sections 10.2 and 10.3 seem to be quite different. The Euclidean area function was built using distance, and in particular the lengths of the base and altitude of a triangle. The hyperbolic area depended upon angle measurement, not distance. As we shall see in this chapter both of these area functions lead to a very beautiful and surprising theorem due to J. Bolyai, one of the founders of hyperbolic geometry. Roughly, Bolyai's Theorem states that if two polygonal regions have the same area, then one can be cut up into a finite number of pieces and reassembled to form the other. (This is the basis of many interesting puzzles and games for children.) The most difficult part of this proof will be showing it is true for triangular regions (Theorem 10.4.6). Before we attack that problem, we will define "equivalent by finite decomposition," which formalizes the concept of "cut up and reassemble."

Definition. Two polygonal regions $\mathbf{R}$ and $\mathbf{R}'$ of a neutral geometry are **equivalent by finite decomposition** ($\mathbf{R} \equiv \mathbf{R}'$) if there exist triangulations $\tau = \{\mathbf{T}_1, \mathbf{T}_2, \ldots, \mathbf{T}_k\}$ of $\mathbf{R}$ and $\tau' = \{\mathbf{T}'_1, \mathbf{T}'_2, \ldots, \mathbf{T}'_k\}$ of $\mathbf{R}'$ such that $\mathbf{T}_i \simeq \mathbf{T}'_i$ for $1 \le i \le k$. (See Figure 10-29.)

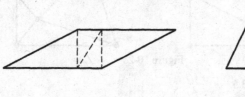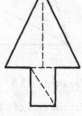

Figure 10-29

The proof of the following theorem is left to Problem A2.

Theorem 10.4.1. *If σ is an area function in a neutral geometry and $\mathbf{R}$ and $\mathbf{R}'$ are polygonal regions with $\mathbf{R} \equiv \mathbf{R}'$ then $\sigma(\mathbf{R}) = \sigma(\mathbf{R}')$.*

The goal of this chapter is to prove the converse of Theorem 10.4.1, namely

Bolyai's Theorem. *If $\mathbf{R}$ and $\mathbf{R}'$ are polygonal regions in a neutral geometry and σ is an area function with $\sigma(\mathbf{R}) = \sigma(\mathbf{R}')$ then $\mathbf{R}$ is equivalent by finite decomposition to $\mathbf{R}'$.*

We will prove Bolyai's Theorem by first verifying it for triangular regions. It will be shown that to each triangular region $\blacktriangle ABC$ we may associate a Saccheri quadrilateral $\square GBCH$ (Theorem 10.4.2) so that $\blacktriangle ABC \equiv$ $\blacksquare GBCH$ (Theorem 10.4.3). By showing that under certain conditions two Saccheri quadrilaterals are equivalent (Theorem 10.4.5) we shall eventually be able to verify Bolyai's Theorem for triangular regions (Theorem 10.4.6). The proof of Bolyai's Theorem for polygonal regions follows from the triangular case by essentially a proof by induction. Most of the hard work is contained in the triangular case.

Theorem 10.4.2. *Let $\triangle ABC$ be a triangle in a neutral geometry with D the midpoint of $\overline{AB}$ and E the midpoint of $\overline{AC}$. Let F, G, and H be the feet of the perpendiculars from A, B, and C to $\overleftrightarrow{DE}$ as in Figure 10-30. Then $\square GBCH$ is a Saccheri quadrilateral, called the Saccheri quadrilateral associated with side $\overline{BC}$ of $\triangle ABC$.*

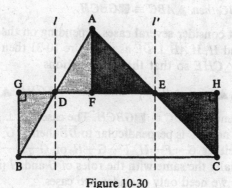

Figure 10-30

PROOF. We first suppose that neither $\overleftrightarrow{AB}$ nor $\overleftrightarrow{AC}$ is perpendicular to $\overleftrightarrow{DE}$. Let l be the line perpendicular to $\overleftrightarrow{DE}$ at D as in Figure 10-30. $l \neq \overleftrightarrow{AB}$ and A—D—B so that A and B lie on opposite sides of l. Since $\overleftrightarrow{GB} \parallel l \parallel \overleftrightarrow{AF}$ (Why?), G and F lie on opposite sides of l. Likewise, F and H are on opposite sides of the line l' perpendicular to $\overleftrightarrow{DE}$ at E. Hence

$$G—D—F \quad \text{and} \quad F—E—H. \qquad (4\text{-}1)$$

Since the vertical angles $\angle BDG$ and $\angle ADF$ are congruent, HA implies that

$$\triangle BDG \simeq \triangle ADF \quad \text{so that } \overline{BG} \simeq \overline{AF}. \qquad (4\text{-}2)$$

Likewise

$$\triangle CEH \simeq \triangle AEF \quad \text{so that } \overline{CH} \simeq \overline{AF}. \qquad (4\text{-}3)$$

Thus $\overline{BG} \simeq \overline{AF} \simeq \overline{CH}$ and $\square GBCH$ is a Saccheri quadrilateral.

The case where one of $\overleftrightarrow{AB}$ or $\overleftrightarrow{AC}$ is perpendicular to $\overleftrightarrow{BC}$ is left to Problem A3. See Figure 10-31 for the case $\overleftrightarrow{AB} \perp \overleftrightarrow{DE}$. $\qquad\qquad \square$

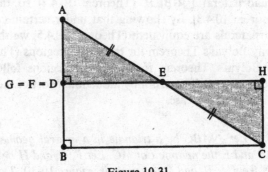

Figure 10-31

Theorem 10.4.3. *If* $\boxed{\text{S}}GBCH$ *is the Saccheri quadrilateral associated with side* $\overline{BC}$ *of* $\triangle ABC$ *then* $\blacktriangle ABC \equiv \blacksquare GBCH$.

PROOF. We must consider several cases depending on the relative positions of D, E, F, G, and H. If $\overleftrightarrow{AB} \perp \overleftrightarrow{DE}$ as in Figure 10-31 then it is easy to show that $\triangle ADE \simeq \triangle CHE$ so that the triangulations

$$\tau = \{\blacktriangle ADE, \blacktriangle DEB, \blacktriangle BEC\} \quad \text{and} \quad \tau' = \{\blacktriangle CHE, \blacktriangle DEB, \blacktriangle BEC\}$$

give the equivalence $\blacktriangle ABC \equiv \blacksquare GBCH$. The case $\overleftrightarrow{AC} \perp \overleftrightarrow{DE}$ is similar.

If neither $\overleftrightarrow{AB}$ nor $\overleftrightarrow{AC}$ is perpendicular to $\overleftrightarrow{DE}$ then F, G, and H are distinct points so that either G—F—H, F—G—H or G—H—F. The latter two cases are essentially the same with the roles of G and H (and also B and C) reversed. Hence we need only consider two cases.

Case 1. G—F—H. This situation was illustrated in Figure 10-30. $\blacktriangle ABC$ and $\blacksquare GBCH$ have triangulations

$$\tau = \{\blacktriangle ADF, \blacktriangle AEF, \blacktriangle BDF, \blacktriangle BFE, \blacktriangle BEC\}$$
$$\tau' = \{\blacktriangle BDG, \blacktriangle CEH, \blacktriangle BDF, \blacktriangle BFE, \blacktriangle BEC\}.$$

Because of Congruences (4-2) and (4-3), τ and τ' show that $\blacktriangle ABC \equiv \blacksquare GBCH$.

Case 2. F—G—H. This situation is illustrated in Figure 10-32. We shall first construct a quadrilateral $\blacksquare E_0BCE$ which is equivalent by finite decomposition to $\blacktriangle ABC$. Let E_0 be the point with E_0—D—E and $\overline{E_0D} \simeq \overline{DE}$. Since D is the midpoint of $\overline{AB}$, $\overline{DB} \simeq \overline{DA}$. The vertical angles $\angle E_0DB$ and $\angle EDA$ are congruent so that by SAS we have

$$\triangle E_0DB \simeq \triangle EDA. \tag{4-4}$$

The triangulations $\tau = \{\blacktriangle ADE, \blacktriangle DBE, \blacktriangle BEC\}$ and $\tau' = \{\blacktriangle BDE_0, \blacktriangle DBE, \blacktriangle BEC\}$ then show that

$$\blacktriangle ABC \equiv \blacksquare E_0BCE. \tag{4-5}$$

All that remains is to show that $\blacksquare E_0BCE \equiv \blacksquare GBCH$. Suppose that D—G—E as in Figure 10-32. Now $\overline{GB} \simeq \overline{HC}$ by Theorem 10.4.2. Since E is the midpoint of $\overline{AC}$, $\overline{AE} \simeq \overline{EC}$. Because $\triangle E_0DB \simeq \triangle EDA$, we have $\overline{E_0B} \simeq \overline{AE} \simeq \overline{EC}$. Hence $\triangle E_0GB \simeq \triangle EHC$ by HL. Because D—G—E, we have triangulations

$$\tau = \{\blacktriangle E_0GB, \blacktriangle GBE, \blacktriangle EBC\} \quad \text{and} \quad \tau' = \{\blacktriangle EHC, \blacktriangle GBE, \blacktriangle EBC\}$$

which imply that

$$\blacksquare E_0BCE \equiv \blacksquare GBCH.$$

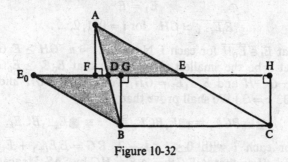

Figure 10-32

However, G may not be between D and E, as Figure 10-33 shows, so that τ and τ' are not triangulations of $\blacksquare E_0BCE$ and $\blacksquare GBCH$. Thus a different argument is needed. We will proceed by induction.

We first show that $E_0E = GH$. Since $F \neq G$ we have F—D—G and $\triangle ADF \simeq \triangle BDG$ by HA so that $\overline{FD} \simeq \overline{DG}$. (Note that F—D—G—H.)

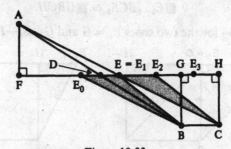

Figure 10-33

Likewise $F \neq H$ implies that F—E—H and $\overline{FE} \simeq \overline{EH}$. Thus

$$
\begin{aligned}
2 \cdot FE = FH &= FD + DH \\
&= DG + DH \\
&= DG + DG + GH \\
&= (2 \cdot DG) + GH \\
&= (2 \cdot FD) + GH.
\end{aligned}
$$

Hence

$$2 \cdot (FE - FD) = GH > 0$$

so that $FE > FD$ and F—D—E—H. Thus $FE - FD = DE$ and

$$E_0E = 2 \cdot DE = GH.$$

We next define a sequence of quadrilaterals. Choose points $E_1, E_2, E_3, \ldots$ on $\overrightarrow{DE}$ so that

$$E_0\text{—}E_1\text{—}E_2\text{—}E_3\text{—}\cdots$$
$$E_1 = E$$
$$\overline{E_iE_{i+1}} \simeq GH \quad \text{for } i = 0, 1, 2, \ldots.$$

Note that $E_i \in \overline{E_0H}$ for each i. Now $E_0E_n = n \cdot GH \geq E_0G$ if n is large enough. Let k be the smallest integer such that $E_0E_k \geq E_0G$. Then since E_0—E_{k-1}—G—H and $E_{k-1}E_k = GH$, we have $E_k \in \overline{GH}$ and $E_k \neq H$. (In Figure 10-33 $k = 3$.) We shall prove that

$$\blacksquare E_0BCE_1 \equiv \blacksquare E_1BCE_2 \equiv \cdots \equiv \blacksquare E_{k-1}BCE_k.$$

Now for each i with $0 \leq i \leq k - 2$, $E_iG = E_iE_{i+1} + E_{i+1}G = GH + E_{i+1}G' = E_{i+1}H$ so that $\triangle E_iGB \simeq \triangle E_{i+1}HC$ by SAS. Hence $\overline{E_iB} \simeq \overline{E_{i+1}C}$ and $\angle GE_iB \simeq \angle HE_{i+1}C$. Thus $\triangle E_{i+1}E_iB \simeq \triangle E_{i+2}E_{i+1}C$ by SAS. But this implies that $\blacksquare E_iBCE_{i+1} \equiv \blacksquare E_{i+1}BCE_{i+2}$. Hence

$$\blacksquare E_0BCE_1 \equiv \blacksquare E_1BCE_2 \equiv \cdots \equiv \blacksquare E_{k-1}BCE_k. \tag{4-6}$$

By the choice of k we have $E_k \in \overline{GH}$ and $E_k \neq H$. Thus $\triangle E_{k-1}GB \simeq \triangle E_kHC$ by HL and

$$\blacksquare E_{k-1}BCE_k \equiv \blacksquare GBCH. \tag{4-7}$$

See Figure 10-34 for the two cases $E_k = G$ and G—E_k—H.

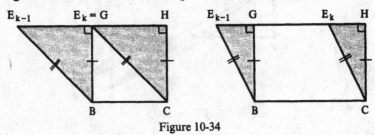

Figure 10-34

Since $\equiv$ is an equivalence relation (Problem A1) we may combine Equivalences (4-5), (4-6) and (4-7) to obtain

$$\blacktriangle ABC \equiv \blacksquare E_0BCE_1 \equiv \blacksquare E_{k-1}BCE_k \equiv \blacksquare GBCH. \qquad \square$$

Let $\triangle ABC$ be given. The next result says that we may construct another triangle with one side $\overline{BC}$ and another side of arbitrary length (greater than AB) such that the triangular regions are equivalent. In the Euclidean case the theorem is clear if we replace "equivalent" by "have the same area" (see Problem A4). As it is stated it requires the full force of Theorem 10.4.3.

Theorem 10.4.4. *Let $\triangle ABC$ be a triangle in a neutral geometry and let r be a number greater than AB. Then there is a point P with $BP = r$ and $\blacktriangle PBC \equiv \blacktriangle ABC$.*

PROOF. Let D, E, G, H be as in Theorem 10.4.2 so that $\blacktriangle ABC \equiv \blacksquare GBCH$. Now

$$\tfrac{1}{2}r > \tfrac{1}{2} \cdot AB = BD \geq BG$$

where the last inequality is a consequence of $\overline{BG} \perp \overline{DE}$. By Theorem 6.5.8 there is a point M on $\overleftrightarrow{GH}$ with $BM = \tfrac{1}{2}r$. See Figure 10-35. If P is the point with B—M—P and $\overline{BM} \simeq \overline{MP}$ then $BP = 2 \cdot BM = r$.

We must show that $\blacktriangle PBC \equiv \blacktriangle ABC$. The first step is to show that if N is the midpoint of $\overline{PC}$ then N is actually on $\overleftrightarrow{DE} = \overleftrightarrow{GH}$.

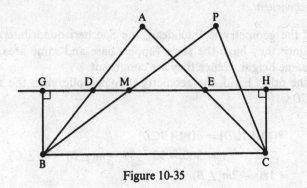

Figure 10-35

Since $\square GBCH$ is a Saccheri quadrilateral, the perpendicular bisector l of $\overline{BC}$ is also perpendicular to $\overleftrightarrow{GH} = \overleftrightarrow{DE}$ by Problem A6 of Section 7.2. If G' and H' are the feet of the perpendiculars from B to C to $\overleftrightarrow{MN}$, then Theorem 10.4.2 shows that $\square G'BCH'$ is a Saccheri quadrilateral. Thus the same problem (applied to $\boxed{S}G'BCH'$) shows that $l \perp \overleftrightarrow{MN}$. Since $M \in \overleftrightarrow{DE}$ and there is only one line through M perpendicular to l, $\overleftrightarrow{MN} = \overleftrightarrow{DE}$ and so $N \in \overleftrightarrow{DE}$.

This means that $G' = G$ and $H' = H$. Thus, invoking Theorem 10.4.3

$$\blacktriangle PBC \equiv \blacksquare G'BCH' = \blacksquare GBCH \equiv \blacktriangle ABC. \qquad \square$$

In the proofs that follow, we shall proceed by considering the two cases, Euclidean and hyperbolic. Because of Theorem 10.2.7, in the Euclidean case there is only one area function. However, in the hyperbolic case things are not the same. If α is any area function for a hyperbolic geometry and $t > 0$ then $\beta_t = t\alpha$ is also an area function. Philosophically this happens since there is no "normalization" because there are no squares (i.e., the fourth axiom for area is vacuous for a hyperbolic geometry—see the proof of Theorem 10.3.5). It will turn out that for a hyperbolic geometry, every area function is of the form $\beta_t = t\delta$ for some t (Theorem 10.4.9) so that the defect is essentially the only hyperbolic area function. Our problem is that we cannot prove the uniqueness result of Theorem 10.4.9 until we prove Bolyai's Theorem. For this reason, we need the following terminology.

Definition. The **special area function** for a neutral geometry is the Euclidean area function for a Euclidean geometry and the defect function for a hyperbolic geometry.

Theorem 10.4.5. *If two Saccheri quadrilaterals* ⑤ *ABCD and* ⑤ *PQRS in a neutral geometry have congruent upper bases and the same special area, then they are congruent.*

PROOF. If the geometry is Euclidean, the Saccheri quadrilaterals are rectangles. Since they have the same (upper) base and same area, they must have the same height. Hence they are congruent.

If on the other hand, the geometry is hyperbolic, then the special area of ■$ABCD$ is

$$\alpha(\blacksquare ABCD) = \delta(\blacksquare ABCD)$$
$$= 360 - 90 - 90 - m(\angle B) - m(\angle C)$$
$$= 180 - 2m(\angle B).$$

Likewise $\alpha(\blacksquare PQRS) = 180 - 2m(\angle Q)$. Since $\alpha(\blacksquare PQRS) = \alpha(\blacksquare ABCD)$, $\angle B \simeq \angle Q$. Since $\angle A \simeq \angle B$ and $\angle P \simeq \angle Q$, we see that corresponding angles of the two Saccheri quadrilaterals are congruent.

We shall now show that corresponding sides are congruent. If $\overline{AB}$ is not congruent to $\overline{PQ}$ then one of those segments is longer, say $\overline{PQ}$. We will show that this assumption leads to the existence of a rectangle, which is impossible in a hyperbolic geometry. See Figure 10-36.

Choose $E \in \overline{BA}$ with $\overline{BE} \simeq \overline{QP}$ and $F \in \overline{CD}$ with $\overline{CF} \simeq \overline{RS}$. Since the upper bases are congruent by assumption, $\triangle PQR \simeq \triangle EBC$ by SAS. Hence

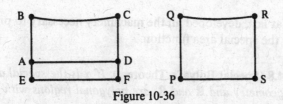

Figure 10-36

$\overline{PR} \simeq \overline{EC}$ and $\angle SRP \simeq \angle FCE$ by Angle Subtraction. Hence $\triangle SRP \simeq \triangle FCE$ by SAS. Thus $\angle CFE$ is a right angle. Likewise $\angle BEF$ is a right angle so that $\square EBCF$ is a Saccheri quadrilateral.

If $A \neq E$ and $D \neq F$ then $\square EADF$ is a rectangle, which is impossible in a hyperbolic geometry. Hence $A = E$ and $D = F$ so that $\overline{AB} = \overline{EB} \simeq \overline{PQ}$ and $\overline{AD} = \overline{EF} \simeq \overline{PS}$. Thus $\square ABCD \simeq \square PQRS$. $\square$

We are now ready to prove Bolyai's theorem in the special case in which both regions are triangles. This result will be derived by first showing that each region is equivalent by finite decomposition to a Saccheri quadrilateral and then showing that the two Saccheri quadrilaterals are congruent.

Theorem 10.4.6. *Let α be the special area function in a neutral geometry. If $\alpha(\triangle ABC) = \alpha(\triangle DEF)$ then $\blacktriangle ABC \equiv \blacktriangle DEF$.*

PROOF. If $\triangle ABC \simeq \triangle DEF$ then $\blacktriangle ABC \equiv \blacktriangle DEF$. Hence we consider the case where one of the triangles has a side which is longer than the corresponding side of the other triangle. Without loss of generality we shall assume that $\overline{DE} > \overline{AB}$. By Theorem 10.4.4, with $r = DE$, there is a point P with $\overline{PB} \simeq \overline{DE}$ and $\blacktriangle PBC \equiv \blacktriangle ABC$.

By Theorem 10.4.2 there is a Saccheri quadrilateral $\boxed{S} GPBH$ with $\blacksquare GPBH \equiv \blacktriangle PBC$. The same result applied to $\triangle DEF$ says that there is a Saccheri quadrilateral $\boxed{S} MDEN$ with $\blacksquare MDEN \equiv \blacktriangle DEF$. By Theorem 10.4.1

$$\alpha(\blacksquare MDEN) = \alpha(\blacktriangle DEF) = \alpha(\blacktriangle ABC)$$
$$= \alpha(\blacktriangle PBC) = \alpha(\blacksquare GPBH).$$

Since $\overline{DE} \simeq \overline{PB}$, Theorem 10.4.5 says that $\boxed{S} MDEN \simeq \boxed{S} GPBH$. Hence $\blacksquare MDEN \equiv \blacksquare GPBH$. Gathering these results together we have

$$\blacktriangle ABC \equiv \blacktriangle PBC \equiv \blacksquare GPBH \equiv \blacksquare MDEN \equiv \blacktriangle DEF. \qquad \square$$

The proof of the next result follows from Theorem 8.2.10 in the hyperbolic case. The Euclidean case is left as Problem A5.

Theorem 10.4.7. *Let α be the special area function in a neutral geometry and let $\triangle ABC$ be a triangle. If $0 < x < \alpha(\blacktriangle ABC)$ then there is a point $Q \in \overline{BC}$ with $\alpha(\blacktriangle ABQ) = x$.*

We have, at last, developed all the machinery necessary to prove Bolyai's Theorem for the special area function.

Theorem 10.4.8 (Special Bolyai's Theorem). *If α is the special area function in a neutral geometry and R and R' are polygonal regions with $\alpha(R) = \alpha(R')$ then $R \equiv R'$.*

PROOF. First we want to show that R and R' have triangulations

$$\bar{\tau} = \{\bar{T}_1, \bar{T}_2, \ldots, \bar{T}_k\}$$
$$\bar{\tau}' = \{\bar{S}_1, \bar{S}_2, \ldots, \bar{S}_k\}$$

where $\alpha(\bar{T}_i) = \alpha(\bar{S}_i)$ for each i.

Let $\tau = \{T_1, \ldots, T_m\}$ be any triangulation of R and let $\tau' = \{S_1, \ldots, S_n\}$ be any triangulation of R'. We will cut triangles off of the triangulations for R and R' in such a manner that the areas of the discarded triangles are the same. As we do this, we create new regions R_1 and R'_1 with $\alpha(R_1) = \alpha(R'_1)$ and triangulations τ_1 and τ'_1 of these regions where the total number of triangular regions in τ_1 and τ'_1 is strictly less than the total in τ and τ'.

If $\alpha(T_m) = \alpha(S_n)$ let R_1 be the region with triangulation $\tau_1 = \{T_1, \ldots, T_{m-1}\}$ and let R'_1 be the region with triangulation $\tau'_1 = \{S_1, \ldots, S_{n-1}\}$. Note that the total number of regions in τ_1 and τ'_1 is $m + n - 2 < m + n$.

If $\alpha(T_m) > \alpha(S_n)$ then let T_m be the triangular region $\triangle ABC$. By Theorem 10.4.7 with $x = \alpha(S_n)$ there is a point $Q \in \overline{BC}$ with $\alpha(\triangle ABQ) = \alpha(S_n)$. Hence we can break T_m into two parts $T'_m = \triangle AQC$ and $T''_m = \triangle ABQ$ with $\alpha(T''_m) = \alpha(S_n)$. Let R_1 be the region with triangulation $\tau_1 = \{T_1, \ldots, T_{m-1}, T'_m\}$ and R'_1 the region with triangulation $\tau'_1 = \{S_1, \ldots, S_{n-1}\}$. Note that the total number of regions in τ_1 and τ'_1 is $m + n - 1 < m + n$.

If $\alpha(T_m) < \alpha(S_n)$ we do a similar step breaking up S_n into two pieces, one of which has the same special area as T_m.

We may now repeat the process with R_1 and R'_1, cutting off a triangular region from both, again of equal area, and again so that the corresponding triangulations have fewer elements. Eventually we reach the stage where R_p and R'_p have triangulations consisting of a single triangular region each (so that R_p and R'_p are actually triangular regions) and these regions have equal special area. The desired triangulations $\bar{\tau}$ and $\bar{\tau}'$ are formed from R_p, R'_p and the triangular regions previously cut off. Since at each stage we cut a single triangle off each region, $\bar{\tau}$ and $\bar{\tau}'$ have the same number of elements.

Thus we have triangulations $\bar{\tau} = \{\bar{T}_1, \ldots, \bar{T}_k\}$ of R and $\bar{\tau}' = \{\bar{S}_1, \ldots, \bar{S}_k\}$ of R' with $\alpha(\bar{T}_i) = \alpha(\bar{S}_i)$ for each i. By Theorem 10.4.6, $\bar{T}_i \equiv \bar{S}_i$. Hence $R \equiv R'$. $\square$

The Special Bolyai's Theorem can be used to prove the uniqueness of the hyperbolic area function up to a scale factor.

Theorem 10.4.9. *If σ is an area function in a hyperbolic geometry, then* $\sigma = t \cdot \delta$ *for some* $t > 0$, *where* δ *is the defect function.*

PROOF. Since area functions are determined by their values on triangular regions we only need to show that there is a $t > 0$ with $\sigma(\blacktriangle ABC) = t \cdot \delta(\blacktriangle ABC)$ for every $\triangle ABC$. This is the same as proving that

$$\frac{\sigma(\blacktriangle ABC)}{\delta(\blacktriangle ABC)} = \frac{\sigma(\blacktriangle DEF)}{\delta(\blacktriangle DEF)} \quad \text{for all } \triangle ABC, \triangle DEF. \tag{4-8}$$

We first prove Equation (4-8) is true if the two triangles have the same defect. In this case the special Bolyai's Theorem shows that $\blacktriangle ABC \equiv \blacktriangle DEF$. By Theorem 10.4.1 we have $\sigma(\blacktriangle ABC) = \sigma(\blacktriangle DEF)$ so that Equation (4-8) holds in this case.

Now suppose that $\delta(\blacktriangle ABC) \neq \delta(\blacktriangle DEF)$. We may suppose that

$$\delta(\blacktriangle ABC) > \delta(\blacktriangle DEF). \tag{4-9}$$

If we let $x = \delta(\blacktriangle DEF)$ in Theorem 10.4.7 then there is a point $Q \in \overline{BC}$ with

$$\delta(\blacktriangle ABQ) = x = \delta(\blacktriangle DEF).$$

Hence by the special Bolyai's Theorem, $\blacktriangle ABQ \equiv \blacktriangle DEF$. But this means that $\sigma(\blacktriangle ABQ) = \sigma(\blacktriangle DEF)$ by Theorem 10.4.1 so that

$$\sigma(\blacktriangle ABC) > \sigma(\blacktriangle ABQ) = \sigma(\blacktriangle DEF). \tag{4-10}$$

The importance of Inequality (4-10) is that it says that if one triangle has larger δ-area than another, then it also has larger σ-area. Intuitively, this means that σ is "relative size preserving."

Our method of proving Equation (4-8) for the general case will be to show that the difference between

$$\frac{\sigma(\blacktriangle DEF)}{\sigma(\blacktriangle ABC)} \quad \text{and} \quad \frac{\delta(\blacktriangle DEF)}{\delta(\blacktriangle ABC)}$$

is less than $1/q$ for any positive integer q. Let $q > 0$ be an integer. By repeated use of Theorem 10.4.7, there are points $P_0, P_1, \ldots$ such that $B = P_0 - P_1 - P_2 - \cdots - P_q = C$ and $\delta(\blacktriangle AP_iP_{i+1}) = 1/q \cdot \delta(\triangle ABC)$. See Figure 10-37.

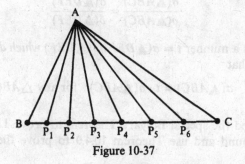

Figure 10-37

Since the triangular regions $\blacktriangle AP_iP_{i+1}$ all have the same defect, we have

$$\sigma(\blacktriangle ABP_i) = \frac{i}{q} \cdot \sigma(\blacktriangle ABC) \quad \text{for } i = 1, 2, \ldots, q. \qquad (4\text{-}11)$$

The unique point $Q \in \overline{BC}$ such that $\delta(\blacktriangle ABQ) = \delta(\blacktriangle DEF)$ lies in a unique segment $\overline{P_iP_{i+1}}$. More precisely, let p be the unique integer such that either

$$P_p = Q \quad \text{or} \quad P_p \!-\! Q \!-\! P_{p+1}.$$

For this value of p we have the following inequalities

$$\delta(\blacktriangle ABP_p) \le \delta(\blacktriangle ABQ) < \delta(\blacktriangle ABP_{p+1}) \qquad (4\text{-}12)$$

$$\frac{p}{q} \cdot \delta(\blacktriangle ABC) \le \delta(\blacktriangle ABQ) < \frac{p+1}{q} \cdot \delta(\blacktriangle ABC) \qquad (4\text{-}13)$$

$$\frac{p}{q} \le \frac{\delta(\blacktriangle ABQ)}{\delta(\blacktriangle ABC)} < \frac{p+1}{q}. \qquad (4\text{-}14)$$

Since σ is "relative size preserving" (Inequality (4-10)) we may replace δ by σ in Inequalities (4-11), (4-12), (4-13) and (4-14). In particular we have

$$\frac{p}{q} \le \frac{\sigma(\blacktriangle ABQ)}{\sigma(\blacktriangle ABC)} < \frac{p+1}{q}. \qquad (4\text{-}15)$$

As in the proof of Theorem 9.2.3, Inequalities (4-14) and (4-15) may be subtracted to yield

$$\left| \frac{\sigma(\blacktriangle ABQ)}{\sigma(\blacktriangle ABC)} - \frac{\delta(\blacktriangle ABQ)}{\delta(\blacktriangle ABC)} \right| < \frac{1}{q}.$$

Since q can be arbitrarily large, the expression on the left must be zero. Hence

$$\frac{\sigma(\blacktriangle DEF)}{\sigma(\blacktriangle ABC)} = \frac{\sigma(\blacktriangle ABQ)}{\sigma(\blacktriangle ABC)} = \frac{\delta(\blacktriangle ABQ)}{\delta(\blacktriangle ABC)} = \frac{\delta(\blacktriangle DEF)}{\delta(\blacktriangle ABC)}$$

so that we have Equation (4-8). But this means that

$$\frac{\sigma(\blacktriangle ABC)}{\delta(\blacktriangle ABC)} = \frac{\sigma(\blacktriangle DEF)}{\delta(\blacktriangle DEF)}.$$

Hence there is a number $t = \sigma(\blacktriangle DEF)/\delta(\blacktriangle DEF)$ *which does not depend on* $\triangle ABC$ such that

$$\sigma(\blacktriangle ABC) = t \cdot \delta(\blacktriangle ABC) \quad \text{for any } \triangle ABC. \qquad \square$$

Having used the special Bolyai's Theorem to prove Theorem 10.4.9, we now turn around and use Theorem 10.4.9 to prove the general Bolyai's Theorem.

Theorem 10.4.10 (Bolyai's Theorem). *If σ is an area function in a neutral geometry and if* **R**, **R**′ *are polygonal regions with* σ(**R**) = σ(**R**′) *then* **R** *is equivalent by finite decomposition to* **R**′.

PROOF. If the geometry is Euclidean then σ must be the unique Euclidean area and the result was proved in Theorem 10.4.8.

If the geometry is hyperbolic then there is a real number $t > 0$ with σ = $t \cdot \delta$. Hence if σ(**R**) = σ(**R**′) then δ(**R**) = δ(**R**′). By the special Bolyai's Theorem **R** ≡ **R**′.   □

In Problem A1 of Section 10.3 you gave an example to show that the product (base)(height) could not be used as an area function in hyperbolic geometry. Our final result shows that in fact, no function of height and base alone will give an area function. This result comes from the fact that the defect (or special area) of a triangle is bounded by 180.

Corollary 10.4.11. *The area of a triangle in a hyperbolic geometry is not determined by just the base and height.*

PROOF. Choose points P_0—P_1—P_2—$\cdots$—P_n—$\cdots$ with $\overline{P_0P_1} \simeq \overline{P_1P_2} \simeq$ $\cdots$. Let Q be a point with $\overline{QP_0} \perp \overrightarrow{P_0P_1}$ as in Figure 10-38. Note that each of the triangles, $\triangle QP_iP_{i+1}$, have the same base ($P_iP_{i+1} = P_0P_1$) and the same height (QP_0). Thus if the area of a triangle depends just on its height and base then σ($\blacktriangle QP_iP_{i+1}$) must be independent of i for any area function σ. But since σ = $t\delta$ for some $t > 0$ by Theorem 10.4.9, then there is an $r > 0$ (independent of i) for which

$$\delta(\blacktriangle QP_iP_{i+1}) = r \quad \text{for } i = 0, 1, \ldots .$$

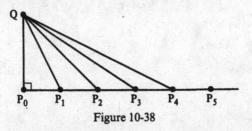

Figure 10-38

Thus $\delta(\blacktriangle QP_0P_i) = i\delta(\blacktriangle QP_0P_1) = ir$. If we take i to be large so that $i > 180/r$ then we conclude that

$$\delta(\blacktriangle QP_0P_i) > 180$$

which is a contradiction. Thus the triangles $\triangle QP_iP_{i+1}$ cannot all have the same hyperbolic area. Therefore, the area of a triangle cannot depend just on the base and height of the triangle.   □

Problem Set 10.4

Part A.

1. Prove that "equivalent by finite decomposition," $\equiv$, is an equivalence relation.

2. Prove Theorem 10.4.1.

3. Complete the proof of Theorem 10.4.2 in the cases where $\overleftrightarrow{AB}$ or $\overleftrightarrow{AC}$ is perpendicular to $\overleftrightarrow{DE}$.

4. Prove the area version of Theorem 10.4.4 for a Euclidean geometry without using Theorem 10.4.3 or Bolyai's Theorem: If $\triangle ABC$ is a triangle in a Euclidean geometry and if $r > AB$, then there is a point P with $PB = r$ such that $\alpha(\blacktriangle PBC) = \alpha(\blacktriangle ABC)$, where α is the area function.

5. Prove Theorem 10.4.7 for a Euclidean geometry.

6. Prove that in a Euclidean geometry any polygonal region $\mathbf{R}$ is equivalent by finite decomposition to an equilateral triangle.

7. Let $\mathbf{R}$ be any polygonal region in a hyperbolic geometry. If $\sigma = t\delta$ is a hyperbolic area function and $\sigma(\mathbf{R}) < 180t$ prove that $\mathbf{R}$ is equivalent by finite decomposition to a triangle.

8. In the proof of Theorem 10.4.9 if τ has m triangles and τ' has n triangles, what is the maximum number of triangles in the triangulation $\bar{\tau}$ which was constructed?

Part B. "Prove" may mean "find a counterexample".

9. Repeat Problem A6 for a hyperbolic geometry.

CHAPTER 11
The Theory of Isometries

11.1 Collineations and Isometries

In mathematics when we have a class of objects satisfying certain axioms (such as incidence geometries) it is natural to study functions that send one object to another. Such functions are most interesting when they preserve special properties of the objects. If $\mathcal{G} = \{\mathcal{S}, \mathcal{L}, d\}$ and $\mathcal{G}' = \{\mathcal{S}', \mathcal{L}', d'\}$ are metric geometries and if $\varphi: \mathcal{S} \to \mathcal{S}'$ is a function, what geometric properties could we reasonably require φ to have?

Because there are two basic concepts in a metric geometry, namely the ideas of lines and distance, there are two important types of functions between metric geometries. One type (a collineation) sends lines to lines. The other type (an isometry) preserves distance. In this section we will carefully define collineations and isometries. We shall see that an isometry of neutral geometries is a collineation and also preserves angle measure. We will end the section with a proof that the Euclidean Plane $\mathscr{E}$ is essentially the only model of a Euclidean geometry.

Recall that if $\varphi: X \to Y$, and if $Z \subset X$ then $\varphi(Z) = \{\varphi(z) | z \in Z\}$.

Definition. If $\mathcal{I} = \{\mathcal{S}, \mathcal{L}\}$ and $\mathcal{I}' = \{\mathcal{S}', \mathcal{L}'\}$ are incidence geometries, then $\varphi: \mathcal{S} \to \mathcal{S}'$ **preserves lines** if for every line l of $\mathcal{S}$, $\varphi(l)$ is a line of $\mathcal{S}'$; that is, $\varphi(l) \in \mathcal{L}'$ if $l \in \mathcal{L}$.

φ is a **collineation** if φ is a bijection which preserves lines.

Example 11.1.1. Let $\mathcal{I} = \mathcal{I}' = \{\mathbb{R}^2, \mathcal{L}_E\}$. Show that $\varphi: \mathbb{R}^2 \to \mathbb{R}^2$ by $\varphi(x, y) = (2x + y, y - x + 5)$ is a collineation.

SOLUTION. An inverse for φ is given by

$$\psi(x, y) = \left(\frac{x - y + 5}{3}, \frac{x + 2y - 10}{3}\right)$$

so that φ is a bijection. We must show that φ preserves lines. If $l = L_a$ then

$$\varphi(l) = \{(2a + y, y - a + 5) \,|\, y \in \mathbb{R}\} = \{(u, v) \,|\, v = u - 3a + 5\}$$
$$= L_{1,5 - 3a} \in \mathcal{L}_E.$$

If $l = L_{m,b}$ then

$$\varphi(l) = \{(2x + mx + b, mx + b - x + 5) \,|\, x \in \mathbb{R}\}$$
$$= \{((2 + m)x + b, (m - 1)x + b + 5) \,|\, x \in \mathbb{R}\}.$$

Thus $\varphi(l) = L_b$ if $m = -2$ and $\varphi(l) = L_{n,c}$ where $n = (m - 1)/(m + 2)$ and $c = 5 + 3b/(m + 2)$ if $m \neq -2$. Hence φ preserves lines and is a collineation.
□

Lemma 11.1.2. Let $\mathcal{I} = \{\mathcal{S}, \mathcal{L}\}$ and $\mathcal{I}' = \{\mathcal{S}', \mathcal{L}'\}$ be incidence geometries. Let $\varphi: \mathcal{S} \to \mathcal{S}'$ be a bijection such that if $l \in \mathcal{L}$ then $\varphi(l) \subset l'$ for some $l' \in \mathcal{L}'$ and if $t' \in \mathcal{L}'$ then $\varphi^{-1}(t') \subset t$ for some $t \in \mathcal{L}$. Then φ is a collineation.

PROOF. We must show that $\varphi(l)$ is a line, not just a subset of a line. Let $l = \overline{AB}$. Then $A' = \varphi(A)$ and $B' = \varphi(B)$ are distinct. Since $\varphi(l) \subset l'$ for some $l' \in \mathcal{L}'$ and $A', B' \in l'$, we must have $l' = \overline{A'B'}$.

On the other hand, $\varphi^{-1}(l') \subset t$ for some $t \in \mathcal{L}$, and $A, B \in \varphi^{-1}(l')$. Hence $\overline{AB} = t$. If $C' \in l'$ then $C = \varphi^{-1}(C') \in \overline{AB} = l$ and $C' = \varphi(C) \in \varphi(l)$. Hence $\varphi(l) = l'$ and φ preserves lines.
□

The importance of this result is that it is often easier to show that φ and φ^{-1} send lines *into* lines than it is to show that φ sends lines *onto* lines.

Example 11.1.3. Let $\mathcal{I} = \mathcal{I}' = \{\mathbb{H}, \mathcal{L}_H\}$ and let $\varphi: \mathbb{H} \to \mathbb{H}$ by

$$\varphi(x, y) = \left(\frac{-x}{x^2 + y^2}, \frac{y}{x^2 + y^2}\right). \tag{1-1}$$

Show that φ is a collineation.

SOLUTION. We first note geometrically what φ does to $\mathbb{H}$. If $r = \sqrt{x^2 + y^2}$ is the radius coordinate (for polar coordinates) then Equation (1-1) may be written as

$$\varphi(x, y) = \left(\frac{-x}{r^2}, \frac{y}{r^2}\right).$$

If we set

$$j(x, y) = \frac{1}{r^2}(x, y) \quad \text{and} \quad \rho(x, y) = (-x, y)$$

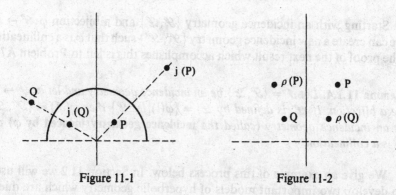

Figure 11-1   Figure 11-2

then $\varphi = \rho \circ j$. The function j is called inversion in the unit circle and is pictured in Figure 11-1. ρ is a (Euclidean) reflection across the y-axis as in Figure 11-2. φ is thus inversion in a circle followed by reflection across the y-axis.

We now show that φ is a collineation. An easy computation shows that $\varphi \circ \varphi(P) = P$ for all $P \in \mathbb{H}$. Hence φ is its own inverse and is a bijection. Because $\varphi = \varphi^{-1}$, we may use Lemma 11.1.2 to show that φ preserves lines by showing that for each $l \in \mathscr{L}_H$, $\varphi(l) \subset l'$ for some $l' \in \mathscr{L}_H$. In particular we do not need to show that $\varphi(l) = l'$. There are four cases to consider: $l = {}_0L$, $l = {}_aL$ with $a \neq 0$, $l = {}_cL_r$ with $c \neq \pm r$, and $l = {}_cL_r$ with $c = \pm r$. Check carefully all the assertions which are made below.

If $l = {}_0L$ and $P \in l$ then $P = (0, y)$ for some $y > 0$. Thus $\varphi(P) = (0, 1/y) \in {}_0L$ so that $\varphi(l) \subset l$.

If $l = {}_aL$ with $a \neq 0$ then for $P = (a, y) \in l$ we have

$$\varphi(P) = \left(\frac{-a}{a^2 + y^2}, \frac{y}{a^2 + y^2} \right) = (z, w).$$

A routine calculation shows that

$$\left(z + \frac{1}{2a} \right)^2 + w^2 = \frac{1}{4a^2}$$

so that $\varphi(l) \subset {}_dL_s$ with $d = -1/2a$ and $s = |1/2a|$.

If $l = {}_cL_r$ with $c \neq \pm r$ then

$$\varphi(l) \subset {}_dL_s \quad \text{with} \quad d = \frac{c}{r^2 - c^2} \quad \text{and} \quad s = \left| \frac{r}{r^2 - c^2} \right|. \qquad (1\text{-}2)$$

Finally

$$\varphi({}_{\pm r}L_r) \subset {}_{\pm a}L \quad \text{where} \quad a = \frac{-1}{2r}. \qquad (1\text{-}3)$$

Hence in all cases $\varphi(l) \subset l'$ for some $l' \in \mathscr{L}_H$.   □

Starting with an incidence geometry $\{\mathscr{S}, \mathscr{L}\}$ and a bijection $\varphi: \mathscr{S} \to \mathscr{S}'$, we can create a new incidence geometry $\{\mathscr{S}', \mathscr{L}'\}$ such that φ is a collineation. The proof of the next result which accomplishes this is left to Problem A7.

Lemma 11.1.4. *Let $\mathscr{I} = \{\mathscr{S}, \mathscr{L}\}$ be an incidence geometry and let $\varphi: \mathscr{S} \to \mathscr{S}'$ be a bijection. If $\mathscr{L}'$ is defined by $\mathscr{L}' = \{\varphi(l) | l \in \mathscr{L}\}$ then $\varphi(\mathscr{I}) = \{\mathscr{S}', \mathscr{L}'\}$ is an incidence geometry (called the* **incidence geometry induced by** *φ) and φ is a collineation.*

We give an example of this process below. In Section 11.2 we will use it to develop two important models of hyperbolic geometry which are due to F. Klein and H. Poincaré.

Example 11.1.5. If $\varphi: \mathbb{R}^2 \to \mathbb{R}^2$ by $\varphi(x, y) = (x, y^3)$ then φ gives a collineation from $\mathscr{E} = \{\mathbb{R}^2, \mathscr{L}_E\}$ to a new model $\varphi(\mathscr{E}) = \{\mathbb{R}^2, \mathscr{L}'\}$. Some of the lines of $\varphi(\mathscr{E})$ are pictured in Figure 11-3.

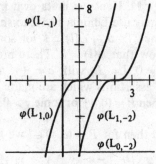

Figure 11-3

So far we have briefly discussed bijections which preserve lines. We now turn our attention to maps that preserve distance in a metric geometry. For convenience we adopt the following notation for the rest of the book.

Notation. If $\varphi: \mathscr{S} \to \mathscr{S}'$ is a function and $A \in \mathscr{S}$ then

$$\varphi A = \varphi(A).$$

Definition. Let $\mathscr{G} = \{\mathscr{S}, \mathscr{L}, d\}$ and $\mathscr{G}' = \{\mathscr{S}', \mathscr{L}', d'\}$ be metric geometries. An **isometry** from $\mathscr{G}$ to $\mathscr{G}'$ is a function $\varphi: \mathscr{S} \to \mathscr{S}'$ such that for all $A, B \in \mathscr{S}$

$$d'(\varphi A, \varphi B) = d(A, B). \tag{1-4}$$

A function φ satisfying Equation (1-4) is said to **preserve distance**.

Note that we have *not* assumed that an isometry is a bijection. It is an easy exercise to show that an isometry is injective (Problem A8). Hence an isometry is a bijection if and only if it is surjective. We shall see that an isometry of neutral geometries is surjective in Lemma 11.1.16.

Lemma 11.1.6. *An isometry of neutral geometries preserves betweenness. More precisely, if $\{\mathscr{S},\mathscr{L},d\}$ is a metric geometry, if $\{\mathscr{S}',\mathscr{L}',d',m'\}$ is a neutral geometry, if $\varphi:\mathscr{S} \to \mathscr{S}'$ is an isometry, and if A, B and C are points of $\mathscr{S}$ with A—B—C then φA—φB—φC. Furthermore if $l \in \mathscr{L}$ then $\varphi(l) \subset l'$ for some $l' \in \mathscr{L}'$.*

PROOF. If A—B—C in $\mathscr{S}$ then A, B, C are collinear and

$$d(A,B) + d(B,C) = d(A,C)$$

so that

$$d'(\varphi A, \varphi B) + d'(\varphi B, \varphi C) = d'(\varphi A, \varphi C).$$

Since the strict triangle inequality is true in $\mathscr{S}'$ (Theorem 6.3.8) φA, φB, and φC must be collinear so that φA—φB—φC.

Now let $l = \overleftrightarrow{AB}$ and $l' = \overleftrightarrow{\varphi A \varphi B}$. If $D \in l$ and $D \neq A$, $D \neq B$ then either D—A—B, A—D—B, or A—B—D. By the first part of the proof φD—φA—φB, φA—φD—φB, or φA—φB—φD. In any case $\varphi D \in l'$ and $\varphi(l) \subset l'$. □

We cannot prove that an isometry is surjective yet (although it is) but we can show that the image of a line is a line.

Lemma 11.1.7. *If $\{\mathscr{S},\mathscr{L},d\}$ is a metric geometry, if $\{\mathscr{S}',\mathscr{L}',d',m'\}$ is a neutral geometry, and if $\varphi:\mathscr{S} \to \mathscr{S}'$ is an isometry, then the image of a line of $\mathscr{S}$ under φ is a line of $\mathscr{S}'$.*

PROOF. Since $\overleftrightarrow{AB} = \overrightarrow{AB} \cup \overrightarrow{BA}$, it is sufficient to prove that $\varphi(\overrightarrow{AB}) = \overrightarrow{\varphi A \varphi B}$ for all $A \neq B$. We shall prove this by a judicious choice of rulers. Let f be a ruler for $\overrightarrow{AB}$ with origin A and B positive. Similarly let f' be a ruler for $\overrightarrow{\varphi A \varphi B}$ with origin φA and φB positive.

If $D' \in \overrightarrow{\varphi A \varphi B}$ then $f'(D') = s \geq 0$. There is a unique point $D \in \overrightarrow{AB}$ with $f(D) = s$. We claim $\varphi D = D'$. Now

$$d'(\varphi D, \varphi A) = d(D, A) = |f(D) - f(A)| = f(D) = s. \qquad (1\text{-}5)$$

On the other hand

$$d'(\varphi D, \varphi A) = |f'(\varphi D) - f'(\varphi A)| = |f'(\varphi D)|. \qquad (1\text{-}6)$$

Hence $f'(\varphi D) = \pm s$. If we can show that $f'(\varphi D) \geq 0$ then φD must be D'.

If $D = A$ or $D = B$ then $\varphi D = \varphi A$ or $\varphi D = \varphi B$. If A—D—B or A—B—D then by Lemma 11.1.6, φA—φD—φB or φA—φB—φD. In all cases $\varphi D \in$

$\overline{\varphi A \varphi B}$ so that $f'(\varphi D) \geq 0$. Hence $f'(\varphi D) = +s$ and $\varphi D = D'$. Thus $\varphi(\overline{AB}) = \overline{\varphi A \varphi B}$. $\qquad\qquad\qquad\qquad\qquad\qquad\qquad\qquad\qquad\qquad\square$

The proof of the next result uses the triangle inequality and is left to Problem A14.

Lemma 11.1.8. *If $\varphi : \mathscr{S} \to \mathscr{S}'$ is an isometry of neutral geometries and A, B, C are noncollinear points of $\mathscr{S}$, then φA, φB, φC are noncollinear points of $\mathscr{S}'$.*

We now turn our attention to the effect of an isometry on angle measure.

Definition. A function $\varphi : \mathscr{S} \to \mathscr{S}'$ of protractor geometries **preserves right angles** if $\angle \varphi A \varphi B \varphi C$ is a right angle in $\mathscr{S}'$ whenever $\angle ABC$ is a right angle in $\mathscr{S}$.

φ **preserves angle measure** if for any $\angle ABC$ in $\mathscr{S}$, $m'(\angle \varphi A \varphi B \varphi C) = m(\angle ABC)$ where m is the angle measure of $\mathscr{S}$ and m' is the angle measure of $\mathscr{S}'$.

Our goal is to show that isometries preserve angle measure. Note that this will be true only because we have taken the convention that all angle measures are degree measures (i.e., right angles have measure 90). If we consider the Euclidean Plane with both degree and radian measure then the identity function is an isometry that does not preserve angle measure.

Lemma 11.1.9. *If $\varphi : \mathscr{S} \to \mathscr{S}'$ is an isometry of neutral geometries then φ preserves right angles.*

PROOF. Let $\angle ABC$ be a right angle in $\mathscr{S}$. We must show that $\angle \varphi A \varphi B \varphi C$ is a right angle. Let D be the unique point such that $D\text{---}B\text{---}C$ and $\overline{DB} \simeq \overline{BC}$ as in Figure 11-4. Then $\triangle ABC \simeq \triangle ABD$ by SAS. Hence $\overline{AC} \simeq \overline{AD}$. Since φ preserves distance we have

$$\overline{\varphi A \varphi B} \simeq \overline{\varphi A \varphi B}, \qquad \overline{\varphi A \varphi C} \simeq \overline{\varphi A \varphi D}, \qquad \overline{\varphi B \varphi C} \simeq \overline{\varphi B \varphi D}$$

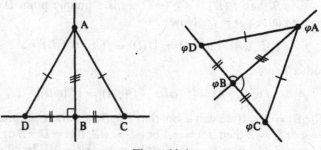

Figure 11-4

so that $\triangle \varphi A \varphi B \varphi C \simeq \triangle \varphi A \varphi B \varphi D$ by SSS. Thus $\angle \varphi A \varphi B \varphi C \simeq \angle \varphi A \varphi B \varphi D$. Since $\varphi D - \varphi B - \varphi C$, $\angle \varphi A \varphi B \varphi C$ and $\angle \varphi A \varphi B \varphi D$ form a linear pair of congruent angles. Hence each is a right angle. □

The next two lemmas tell us that isometries preserve the interiors of angles and angle bisectors. The proof of the first is left to Problem A16.

Lemma 11.1.10. *If $\varphi : \mathscr{S} \to \mathscr{S}'$ is an isometry of neutral geometries and $D \in$ int($\angle ABC$) then $\varphi D \in$ int($\angle \varphi A \varphi B \varphi C$).*

Lemma 11.1.11. *Let $\varphi : \mathscr{S} \to \mathscr{S}'$ be an isometry of neutral geometries. If $\overrightarrow{BD}$ is the bisector of $\angle ABC$ in $\mathscr{S}$ then $\overrightarrow{\varphi B \varphi D}$ is the bisector of $\angle \varphi A \varphi B \varphi C$.*

PROOF. We may assume that $\overline{BC} \simeq \overline{BA}$. By the Crossbar Theorem $\overrightarrow{BD}$ intersects $\overline{AC}$ at a point E. See Figure 11-5. Then $\triangle ABE \simeq \triangle CBE$ by SAS so that $\overline{AE} \simeq \overline{CE}$. Since φ preserves distance, $\triangle \varphi A \varphi B \varphi E \simeq \triangle \varphi C \varphi B \varphi E$ by SSS. Then $\angle \varphi A \varphi B \varphi E \simeq \angle \varphi C \varphi B \varphi E$. Since $\varphi C - \varphi E - \varphi A$, $\varphi E \in$ int($\angle \varphi A \varphi B \varphi C$) and $\overrightarrow{\varphi B \varphi E}$ bisects $\angle \varphi A \varphi B \varphi C$. Now $\overrightarrow{\varphi B \varphi D} = \varphi(\overrightarrow{BD}) = \varphi(\overrightarrow{BE}) = \overrightarrow{\varphi B \varphi E}$. Hence $\overrightarrow{\varphi B \varphi D}$ bisects $\angle \varphi A \varphi B \varphi C$. □

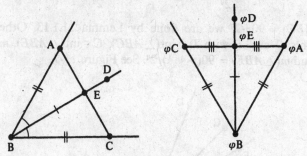

Figure 11-5

Lemma 11.1.12. *Let $\varphi : \mathscr{S} \to \mathscr{S}'$ be an isometry of neutral geometries. If $m(\angle ABC) = 90/2^q$ for some integer $q \geq 0$ then $m'(\angle \varphi A \varphi B \varphi C) = 90/2^q$ also.*

PROOF. Bisect a right angle q times and apply Lemma 11.1.11. See Figure 11-6 where $q = 3$. □

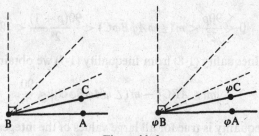

Figure 11-6

Lemma 11.1.13. *If* $\varphi:\mathscr{S} \to \mathscr{S}'$ *is an isometry of neutral geometries and* $m(\angle ABC) = 90p/2^q$ *where* p *and* q *are integers with* $0 < p < 2^{q+1}$ *then* $m'(\angle \varphi A \varphi B \varphi C) = 90p/2^q$ *also.*

PROOF. Choose points $D_0, D_1, \ldots, D_p$ so that $A = D_0 - D_1 - \cdots - D_p = C$ and $m(\angle D_i B D_{i+1}) = 90/2^q$ for $0 \leq i < p$. Then

$$m'(\angle \varphi D_i \varphi B \varphi D_{i+1}) = 90/2^q \quad \text{and} \quad \varphi A = \varphi D_0 - \varphi D_1 - \cdots \varphi D_p = \varphi C.$$

By Angle Addition

$$m'(\angle \varphi A \varphi B \varphi C) = \sum_{i=0}^{p-1} m'(\angle \varphi D_i \varphi B \varphi D_{i+1}) = \frac{90p}{2^q}. \qquad \square$$

Theorem 11.1.14. *An isometry* $\varphi:\mathscr{S} \to \mathscr{S}'$ *of neutral geometries preserves angle measure.*

PROOF. If q is a sufficiently large positive integer then we may find an integer p with $0 < p < 2^{q+1}$ such that

$$0 < \frac{90p}{2^q} \leq m(\angle ABC) < \frac{90(p + 1)}{2^q} < 180. \qquad (1\text{-}7)$$

If $m(\angle ABC) = 90p/2^q$ we are done by Lemma 11.1.13. Otherwise there exist points D and E with $D \in \text{int}(\angle ABC)$, $C \in \text{int}(\angle ABE)$, $m(\angle ABD) = 90p/2^q$, and $m(\angle ABE) = 90(p + 1)/2^q$. See Figure 11-7.

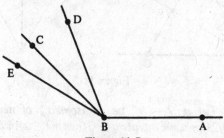

Figure 11-7

Then $C \in \text{int} \angle DBE$ so that $\varphi C \in \text{int}(\angle \varphi D \varphi B \varphi E)$ and

$$0 < \frac{90p}{2^q} < m'(\angle \varphi A \varphi B \varphi C) < \frac{90(p + 1)}{2^q} < 180. \qquad (1\text{-}8)$$

Subtracting Inequality (1-8) from Inequality (1-7) we obtain

$$\left| m(\angle ABC) - m'(\angle \varphi A \varphi B \varphi C) \right| < \frac{90}{2^q}.$$

Since this inequality is true for all large values of the integer q, we must have $m(\angle ABC) = m'(\angle \varphi A \varphi B \varphi C)$. $\qquad \square$

An amusing corollary to Theorem 11.1.14 is that if a Pasch geometry can be made into a neutral geometry by choosing an angle measure, then it can be done in only one way if degree measure is used.

Corollary 11.1.15. *If $\{\mathscr{S}, \mathscr{L}, d\}$ is a Pasch geometry then there is at most one degree measure m such that $\{\mathscr{S}, \mathscr{L}, d, m\}$ is a neutral geometry.*

PROOF. Suppose that both $\{\mathscr{S}, \mathscr{L}, d, m\}$ and $\{\mathscr{S}, \mathscr{L}, d, m'\}$ are neutral geometries. The function $\varphi : \mathscr{S} \to \mathscr{S}$ given by $\varphi P = P$ for all $P \in \mathscr{S}$ is an isometry. Hence $m(\angle ABC) = m'(\angle \varphi A \varphi B \varphi C) = m'(\angle ABC)$ for any $\angle ABC$. Hence $m = m'$.   □ ·

We are now able to prove that an isometry of neutral geometries is a collineation.

Lemma 11.1.16. *If $\varphi : \mathscr{S} \to \mathscr{S}'$ is an isometry of neutral geometries then φ is surjective.*

PROOF. Let $D' \in \mathscr{S}'$ and let A, B be two points of $\mathscr{S}$. If $D' \in \overrightarrow{\varphi A \varphi B}$ then $D' = \varphi D$ for some $D \in \overrightarrow{AD}$ by Lemma 11.1.7. Now assume $D' \notin \overrightarrow{\varphi A \varphi B}$ and choose P, Q on opposite sides of $\overleftrightarrow{AB}$ with $m(\angle PAB) = m(\angle QAB) = m'(\angle D' \varphi A \varphi B)$ and $d(P, A) = d(Q, A) = d'(D', \varphi A)$, as in Figure 11-8.

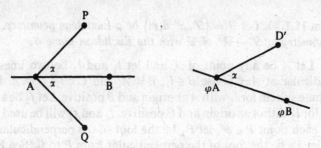

Figure 11-8

Then $m'(\angle \varphi P \varphi A \varphi B) = m(\angle PAB) = m'(\angle D' \varphi A \varphi B)$. Likewise

$$m'(\angle \varphi Q \varphi A \varphi B) = m'(\angle D' \varphi A \varphi B).$$

Thus φP and φQ are two points of $\mathscr{S}'$ whose distance from φA is $d'(D', \varphi A)$ and $m'(\angle \varphi P \varphi A \varphi B) = m'(\angle \varphi Q \varphi A \varphi B) = m'(\angle D' \varphi A \varphi B)$. Hence either $\varphi P = D'$ or $\varphi Q = D'$ (Why?) and $D' \in \text{image}(\varphi)$.   □

Theorem 11.1.17. *An isometry $\varphi : \mathscr{S} \to \mathscr{S}'$ of neutral geometries is a collineation.*

PROOF. By Problem A8 and Lemma 11.1.16, φ is a bijection. Thus by Lemma 11.1.7, φ is a collineation.   □

Corollary 11.1.18. *Let* $\varphi : \mathscr{S} \to \mathscr{S}$ *be an isometry of a neutral geometry with itself. Then* φ^{-1} *is an isometry,* φ *preserves angle measure, and* φ *is a collineation.*

The importance of isometries is that they preserve all geometric properties: distance, angle measure, congruence, betweenness, and incidence. In effect all an isometry does is rename the points. The internal structure of two isometric geometries is the same. This is illustrated in the next result.

Theorem 11.1.19. *If* $\varphi : \mathscr{S} \to \mathscr{S}'$ *is an isometry of neutral geometries then* $\mathscr{S}$ *satisfies EPP if and only if* $\mathscr{S}'$ *does.*

FIRST PROOF. Note that $l \| l'$ in $\mathscr{S}$ if and only if $\varphi(l) \| \varphi(l')$ in $\mathscr{S}'$ (Why?). Thus there is a unique line through $P \notin l$ parallel to l if and only if there is a unique line through φP parallel to $\varphi(l)$.

SECOND PROOF. Let $\triangle ABC$ be a triangle in $\mathscr{S}$. Since φ preserves angle measure, $\triangle ABC$ and $\triangle \varphi A \varphi B \varphi C$ have the same defect. Since $\mathscr{S}$ satisfies EPP if and only if some triangle has defect 0 and a similar statement holds for $\mathscr{S}'$, the result is immediate.   □

The last result of this section tells us that it is not misleading to refer to $\mathscr{E}$ as *the* Euclidean Plane because any Euclidean geometry is isometric to $\mathscr{E}$.

Theorem 11.1.20. *Let* $\mathscr{G} = \{\mathscr{S}, \mathscr{L}, d, m\}$ *be a Euclidean geometry. Then there is an isometry* $\varphi : \mathscr{S} \to \mathbb{R}^2$ *of* $\mathscr{G}$ *with the Euclidean plane* $\mathscr{E}$.

PROOF. Let A be any point of $\mathscr{S}$ and let l_1 and l_2 be two lines which are perpendicular at A. Choose $B \in l_1$, $B \neq A$, and $C \in l_2$, $C \neq A$. Let f_1 be a coordinate system for l_1 with A as origin and B positive. Let f_2 be a coordinate system for l_2 with A as origin and C positive. f_1 and f_2 will be used to define φ.

For each point $P \in \mathscr{S}$, let P_1 be the foot of the perpendicular from P to l_1 and let P_2 be the foot of the perpendicular from P to l_2. See Figure 11-9.

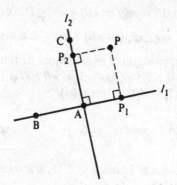

Figure 11-9

We define $\varphi : \mathscr{S} \to \mathbb{R}^2$ by

$$\varphi P = (f_1(P_1), f_2(P_2)).$$

In order to prove that φ is an isometry we must show that $d(P, Q) = d_E(\varphi P, \varphi Q)$. Suppose that $\overline{PQ}$ is not parallel to either l_1 or l_2. (The cases where this is not true are left to Problem A21.) Let m_1 be the line through P parallel to l_1 and let m_2 be the line through Q parallel to l_2. Since $l_1 \perp l_2$, we have $m_1 \perp m_2$ and $m_1 \cap m_2 = \{R\}$ for some R. See Figure 11-10.

$$d(P, R) = |f_1(P_1) - f_1(R_1)|$$

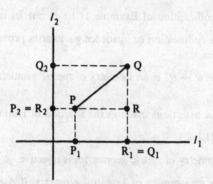

Figure 11-10

since $\square PRR_1P_1$ is a rectangle. Likewise

$$d(Q, R) = |f_2(Q_2) - f_2(R_2)|.$$

Since $R_1 = Q_1$ and $R_2 = P_2$ (Why?) we have

$$d(P, R) = |f_1(P_1) - f_1(Q_1)| \quad \text{and} \quad d(Q, R) = |f_2(P_2) - f_2(Q_2)|$$

so that by the Pythagorean Theorem

$$\begin{aligned}
(d(P, Q))^2 &= (d(P, R))^2 + (d(Q, R))^2 \\
&= (f_1(P_1) - f_1(Q_1))^2 + (f_2(P_2) - f_2(Q_2))^2 \\
&= (d_E(\varphi P, \varphi Q))^2.
\end{aligned}$$

Hence

$$d(P, Q) = d_E(\varphi P, \varphi Q). \qquad \square$$

There is a similar theorem that says that every hyperbolic geometry is isometric to the Poincaré Plane provided that the distance scale (Section 8.3) has been normalized so that $\Pi(\ln(\sqrt{2} + 1)) = 45$. The proof starts out in somewhat the same fashion as above by choosing a coordinate system. However, the Pythagorean Theorem is not available and other results must be used. These results are essentially the trigonometry theory of hyperbolic geometry. You can find the proof in Chapter 33 of Martin [1975] or Chapter 10 of Greenberg [1980].

PROBLEM SET 11.1

Part A.

1. Show that $\varphi:\mathbb{R}^2 \to \mathbb{R}^2$ by $\varphi(x, y) = (2x + y, 1 - y)$ is a collineation of $\{\mathbb{R}^2, \mathscr{L}_E\}$.

2. Show that $\varphi:\mathbb{R}^2 \to \mathbb{R}^2$ by $\varphi(x, y) = (2x, 2y)$ is a collineation of $\{\mathbb{R}^2, \mathscr{L}_E, d_E\}$ but not an isometry.

3. Prove that the collineation of Example 11.1.1 is not an isometry.

*4. If $\varphi:\mathscr{S} \to \mathscr{S}'$ is a collineation of incidence geometries prove that $\varphi^{-1}:\mathscr{S}' \to \mathscr{S}$ is also a collineation.

*5. If the bijection $\varphi:\mathscr{S} \to \mathscr{S}'$ is an isometry of metric geometries prove that φ^{-1} is also an isometry.

6. Verify the various assertions made in the solution of Example 11.1.3.

7. Prove Lemma 11.1.4.

8. Prove that an isometry of metric geometries is injective.

*9. Let $\varphi:\mathscr{S} \to \mathscr{S}$ be an isometry of a neutral geometry. If A and B are points in $\mathscr{S}$ with $\varphi A = B$ and $\varphi B = A$, then prove that $\varphi M = M$ where M is the midpoint of $\overline{AB}$.

10. Let $m > 0$ and define $\varphi:\mathbb{H} \to \mathbb{H}$ by $\varphi(x, y) = (mx + 1, my)$. Prove that φ is a collineation of $\{\mathbb{H}, \mathscr{L}_H\}$.

11. Let $\varphi:\mathbb{R}^2 \to \mathbb{R}^2$ by $\varphi P = P$. Then φ may be thought of as a function from the Euclidean Plane $\mathscr{E}$ to the Taxicab Plane $\mathscr{T}$. Show that φ is a collineation which preserves angle measure but is not an isometry.

12. Let $\theta \in \mathbb{R}$ and let $\varphi_\theta:\mathbb{R}^2 \to \mathbb{R}^2$ be defined by

$$\varphi_\theta(x, y) = (x \cos \theta - y \sin \theta, x \sin \theta + y \cos \theta).$$

Show that φ_θ is an isometry of $\mathscr{E}$. φ_θ is called the **special orthogonal transformation by θ**. In matrix terms φ_θ can be defined by

$$\varphi_\theta \begin{pmatrix} x \\ y \end{pmatrix} = \begin{pmatrix} \cos \theta & -\sin \theta \\ \sin \theta & \cos \theta \end{pmatrix} \begin{pmatrix} x \\ y \end{pmatrix}.$$

13. If we view $\mathbb{R}^2$ as the set $\mathbb{C}$ of complex numbers then

$$\mathbb{H} = \{z = x + iy \in \mathbb{C} | y > 0\}.$$

Let a, b, c, d be real numbers with $ad - bc > 0$ and set

$$S = \begin{pmatrix} a & b \\ c & d \end{pmatrix} \quad \text{and} \quad \varphi_S(z) = \left(\frac{az + b}{cz + d} \right).$$

a. Prove that if $z \in \mathbb{H}$ then $\varphi_S(z) \in \mathbb{H}$.
b. Find S^{-1}.

c. If det $S > 0$ show that det $S^{-1} > 0$.

d. Show that φ_S is a bijection by showing that $\varphi_{S^{-1}}$ is its inverse.

e. Show that if $a > 0$ and $S = \begin{pmatrix} a & b \\ 0 & 1 \end{pmatrix}$ then φ_S is an isometry of $\mathbb{H}$. (We will see later that φ_S is an isometry for any S with $\det(S) > 0$.)

14. Prove Lemma 11.1.8.

15. Let $\varphi: \mathscr{S} \to \mathscr{S}'$ be an isometry of neutral geometries. If A, B, C are noncollinear points of $\mathscr{S}$, prove that $\varphi(\angle ABC) = \angle \varphi A \varphi B \varphi C$.

16. Prove Lemma 11.1.10.

17. Let $\varphi: \mathscr{S} \to \mathscr{S}$ be a collineation of a hyperbolic geometry that preserves angle measurement. Prove that φ is an isometry. Show that the corresponding statement for a Euclidean geometry is false.

18. Let $\varphi: \mathscr{S} \to \mathscr{S}'$ be an isometry of neutral geometries. Prove that two lines of $\mathscr{S}$ intersect (resp. are divergently parallel, or are asymptotically parallel) if and only if their images under φ intersect (resp. are divergently parallel, or are asymptotically parallel).

*19. If $\varphi: \mathscr{S}_1 \to \mathscr{S}_2$ and $\psi: \mathscr{S}_2 \to \mathscr{S}_3$ are collineations, prove that $\psi \circ \varphi: \mathscr{S}_1 \to \mathscr{S}_3$ is a collineation.

*20. If $\varphi: \mathscr{S}_1 \to \mathscr{S}_2$ and $\psi: \mathscr{S}_2 \to \mathscr{S}_3$ are isometries prove that $\psi \circ \varphi: \mathscr{S}_1 \to \mathscr{S}_3$ is an isometry.

21. Complete the proof of Theorem 11.1.20 for the cases where $\overrightarrow{PQ}$ is perpendicular to either l_1 or l_2.

22. If $\varphi: \mathscr{S} \to \mathscr{S}'$ is an isometry of neutral geometries and $\mathscr{S}$ satisfies HPP prove that $\mathscr{S}'$ satisfies HPP.

11.2 The Klein and Poincaré Disk Models

In this optional section we shall present two other important models of a hyperbolic geometry. We saw in the last section that if $\{\mathscr{S}, \mathscr{L}\}$ is an incidence geometry and if $\varphi: \mathscr{S} \to \mathscr{S}'$ is a bijection, then there is an induced incidence geometry $\{\mathscr{S}', \mathscr{L}'\}$ where $\mathscr{L}' = \{\varphi(l) \,|\, l \in \mathscr{L}\}$. In this section we will see that a bijection can also induce distance functions and angle measures. This idea will be used to develop the new models and verify that they satisfy the axioms of a hyperbolic geometry.

Definition. Let $\{\mathscr{S}, \mathscr{L}\}$ be an incidence geometry and let $\varphi: \mathscr{S} \to \mathscr{S}'$ be a bijection. If d is a distance function on $\mathscr{S}$ then the **distance function** d' on $\mathscr{S}'$ **induced by** φ is given by

$$d'(A', B') = d(\varphi^{-1}A', \varphi^{-1}B').$$

If m is an angle measure for the Pasch geometry $\{\mathscr{S}, \mathscr{L}, d\}$ then the **angle measure** m' on $\mathscr{S}'$ **induced** by φ is given by

$$m'(\angle A'B'C') = m(\angle \varphi^{-1}A'\varphi^{-1}B'\varphi^{-1}C').$$

Example 11.2.1. Let $\varphi: \mathbb{R}^2 \to \mathbb{R}^2$ be given by $\varphi(x, y) = (x, y^3)$ as in Example 11.1.5. Find the line in $\varphi(\mathscr{E})$ determined by $A' = (3, 8)$ and $B' = (2, 27)$ and find the distance between these points in $\varphi(\mathscr{E})$.

SOULTION. $\varphi^{-1}A' = (3, 2)$ and $\varphi^{-1}B' = (2, 3)$. The Euclidean line joining $(2, 3)$ to $(3, 2)$ is

$$l = \{(x, y) \mid y = -x + 5\}$$

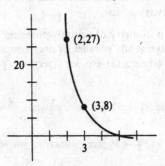

Figure 11-11

so that $P = (s, t)$ is on the line l' through A' and B' if and only if $(s, \sqrt[3]{t}) \in l$, i.e., $\sqrt[3]{t} = -s + 5$. Thus

$$l' = \varphi(l) = \{(s, t) \in \mathbb{R}^2 \mid t = (-s + 5)^3\}.$$

See Figure 11-11.

$$d'(A', B') = d_E((2, 3), (3, 2)) = \sqrt{2}. \qquad \square$$

Theorem 11.2.2. *If $\{\mathscr{S}, \mathscr{L}, d\}$ is a metric geometry and if $\{\mathscr{S}', \mathscr{L}', d'\}$ is the geometry induced by the bijection $\varphi: \mathscr{S} \to \mathscr{S}'$, then $\{\mathscr{S}', \mathscr{L}', d'\}$ is a metric geometry.*

PROOF. By Lemma 11.1.4 we know that $\{\mathscr{S}', \mathscr{L}'\}$ is an incidence geometry. The proof that d' is a distance function is left to Problem A2.

We may obtain rulers for d' by carrying over the rulers from $\mathscr{S}$. Let $l' \in \mathscr{L}'$ so that $l' = \varphi(l)$ and choose a ruler f for l. Then $f' = f \circ \varphi^{-1}$ is a ruler for l' (Problem A2). Note that φ is an isometry. $\qquad \square$

The proof of the next result is left to Problems A3 and A4.

Theorem 11.2.3. *If $\{\mathcal{S}, \mathcal{L}, d, m\}$ is a protractor geometry and $\varphi: \mathcal{S} \to \mathcal{S}'$ is a bijection then the geometry induced by φ, $\{\mathcal{S}', \mathcal{L}', d', m'\}$, is also a protractor geometry.*

Theorem 11.2.4. *If $\{\mathcal{S}, \mathcal{L}, d, m\}$ is a neutral geometry and $\varphi: \mathcal{S} \to \mathcal{S}'$ is a bijection then the geometry $\{\mathcal{S}', \mathcal{L}', d', m'\}$ induced by φ is also a neutral geometry.*

PROOF. We need to show that SAS is satisfied. Suppose that A', B', C', D', E', F' are points in $\mathcal{S}'$ with $\angle A'B'C' \simeq \angle D'E'F'$, $\overline{A'B'} \simeq \overline{D'E'}$, and $\overline{B'C'} \simeq \overline{E'F'}$. Let A, B, C, D, E, F be the corresponding points of $\mathcal{S}$ (i.e., $A = \varphi^{-1}(A')$, etc.). Then $\angle ABC \simeq \angle DEF$, $\overline{AB} \simeq \overline{DE}$, and $\overline{BC} \simeq \overline{EF}$ because of the definitions of d' and m'. Since $\mathcal{S}$ is a neutral geometry

$$\triangle ABC \simeq \triangle DEF.$$

Hence $\angle BCA \simeq \angle EFD$, $\angle CAB \simeq \angle FDE$, and $\overline{AC} \simeq \overline{DF}$. This implies that $\angle B'C'A' \simeq \angle E'F'D'$, $\angle C'A'B' \simeq \angle F'D'E'$, and $\overline{A'C'} \simeq \overline{D'F'}$, again by the definition of d' and m'. Hence $\triangle A'B'C' \simeq \triangle D'E'F'$ and the geometry induced by φ satisfies SAS. □

In the two applications of Theorem 11.2.4 that follow we will actually start with $\{\mathcal{S}', \mathcal{L}'\}$ and find a bijection $\psi: \mathbb{H} \to \mathcal{S}'$ such that $\mathcal{L}' = \psi(\mathcal{L}_H)$; i.e., such that ψ is a collineation. We will then know that the incidence geometry induced by ψ is a neutral geometry. (Actually, we have not proved that $\mathcal{H}$ satisfies SAS yet but will in Section 11.8. Once that has been proved we will know our new models are also neutral geometries.)

Our two new models will have the same underlying set $\mathcal{S}'$ (the unit disk $\mathbb{D}$) but will have quite different sets of lines. The first model will be the Klein Plane. It is due to three 19th century mathematicians, Felix Klein (1849–1925), Arthur Cayley (1821–1895), and Eugenio Beltrami (1835–1900). The German mathematician Klein is well known for the introduction of transformation geometry and the application of group theory to geometry. In his famous *Erlangen Program* he proposed that geometry should be viewed as the study of the invariants of a group acting on a set. See Millman [1977]. The British mathematician Cayley was instrumental in uniting projective and metrical geometry. It is commonly felt (E. T. Bell [1937]) that this paved the way for Klein's disk model. He also did fundamental work in matrix theory as well as introducing n-dimensional space and, with J. J. Sylvester, discovering and thoroughly investigating an algebraic phenomenon called invariance theory. Besides his work in proving the relative consistency of hyperbolic geometry in 1868, the Italian Beltrami is known for his work on physical problems, abstract algebra, invariance theory, and differential equations.

Definition. Let $\mathbb{D} = \{(u,v) \in \mathbb{R}^2 \,|\, u^2 + v^2 < 1\}$ be the unit disk. A **K-line** in $\mathbb{D}$ is the intersection of $\mathbb{D}$ with a Euclidean line $l \subset \mathbb{R}^2$. The **Cayley-Klein-Beltrami Plane** (or more simply the **Klein Plane**) is the incidence geometry $\{\mathbb{D}, \mathscr{L}_K\}$ where $\mathscr{L}_K$ is the set of all K-lines in $\mathbb{D}$.

We should note that $\{\mathbb{D}, \mathscr{L}_K\}$ really is an incidence geometry because of Problem A3 of Section 2.1. Some K-lines are illustrated in Figure 11-12. Note that through the point P there are several lines parallel to l. Hence $\{\mathbb{D}, \mathscr{L}_K\}$ cannot possibly satisfy EPP.

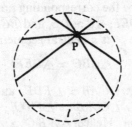

Figure 11-12

In order to apply Theorem 11.2.4 we must find a collineation between $\{\mathbb{H}, \mathscr{L}_H\}$ and $\{\mathbb{D}, \mathscr{L}_K\}$. Actually we will find a collineation $\varphi : \mathbb{D} \to \mathbb{H}$ and then apply Theorem 11.2.4 to $\psi = \varphi^{-1}$. The choice of φ given below will be motivated after the proof.

Proposition 11.2.5. *The function $\varphi : \mathbb{D} \to \mathbb{H}$ by*

$$\varphi(u,v) = \left(\frac{u}{1-v}, \frac{\sqrt{1-u^2-v^2}}{1-v} \right)$$

is a collineation from $\{\mathbb{D}, \mathscr{L}_K\}$ to $\{\mathbb{H}, \mathscr{L}_H\}$.

PROOF. First we note that if $(u,v) \in \mathbb{D}$ then $v < 1$ so that $1 - v > 0$. Since $u^2 + v^2 < 1$, $\sqrt{1-u^2-v^2} > 0$ and $\varphi(u,v) \in \mathbb{H}$.

By Problem A5, φ is a bijection—its inverse is

$$\varphi^{-1}(x,y) = \left(\frac{2x}{1+x^2+y^2}, \frac{x^2+y^2-1}{1+x^2+y^2} \right). \tag{2-1}$$

We must show that the image of a K-line is a line in $\mathscr{L}_H$. If l is a K-line then there are real numbers a, b, c with $c^2 < a^2 + b^2$ (see Problem A6) and

$$l = \{(u,v) \in \mathbb{D} \,|\, au + bv = c\}.$$

If $(u,v) \in l$ and $(x,y) = \varphi(u,v)$ then by Equation (2-1)

$$u = \frac{2x}{1+x^2+y^2} \quad \text{and} \quad v = \frac{x^2+y^2-1}{1+x^2+y^2} \text{ with } y > 0.$$

Hence $(u, v) \in l$ if and only if

$$a\left(\frac{2x}{1 + x^2 + y^2}\right) + b\left(\frac{x^2 + y^2 - 1}{1 + x^2 + y^2}\right) = c$$

or if and only if

$$2ax + b(x^2 + y^2) - b = c + c(x^2 + y^2)$$

or if and only if

$$(b - c)(x^2 + y^2) + 2ax - b - c = 0. \qquad (2\text{-}2)$$

If $b \neq c$ then Equation (2-2) describes a Euclidean circle in $\mathbb{R}^2$ with its center at $(-a/(b - c), 0)$ and radius $\sqrt{(a^2 + b^2 - c^2)/(c - b)^2}$. If $b = c$ then Equation (2-2) describes a vertical line $x = b/a$. (Note if $b = c$ then since $c^2 < a^2 + b^2$, a is not zero.) Hence, since $\varphi(l) \subset \mathbb{H}$, $\varphi(l)$ is part of a type II line or a type I line.

On the other hand $\varphi^{-1}(_aL)$ is contained in the K-line $\{(u, v) \in \mathbb{D} \,|\, u + av = a\}$ while $\varphi^{-1}(_cL_r)$ is contained in the K-line $\{(u, v) \in \mathbb{D} \,|\, \alpha u + \beta v = \gamma\}$ where $\alpha = -c$, $\beta = (r^2 - c^2 + 1)/2$ and $\gamma = (r^2 - c^2 - 1)/2$. By Lemma 11.1.2, φ is thus a collineation.   □

The function φ may seem very artificial. Actually it can be described geometrically as the composition of two geometric functions φ_1 and φ_2: $\varphi = \varphi_2 \circ \varphi_1$. φ_1 takes $\mathbb{D}$ to the right half of the sphere of radius 1 by linear projection: $\varphi_1(u, v) = (u, \sqrt{1 - u^2 - v^2}, v)$. φ_2 takes this right hemisphere to

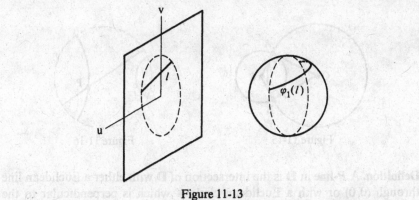

Figure 11-13

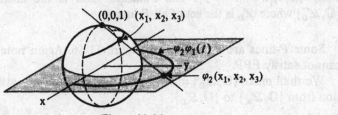

Figure 11-14

H by what is called *stereographic projection* from the North Pole $(0,0,1)$: a point $(x_1, x_2, x_3) \neq (0,0,1)$ on the unit sphere S^2 is sent to the point where the Euclidean line determined by (x_1, x_2, x_3) and $(0,0,1)$ intersects the plane $x_3 = 0$. See Figures 11-13 and 11-14.

By Problem A4 of Section 11.1, φ^{-1} is also a collineation and the set of lines it induces on $\mathbb{D}$ is precisely $\mathscr{L}_K$. φ^{-1} then induces a neutral geometry $\mathscr{K} = \{\mathbb{D}, \mathscr{L}_K, d_K, m_K\}$ where

$$d_K(A, B) = d_H(\varphi A, \varphi B)$$
$$m_K(\angle ABC) = m_H(\angle \varphi A \varphi B \varphi C).$$

Since $\mathscr{K}$ satisfies HPP it is a model of hyperbolic geometry. While it is easy to find the line through two points in $\mathscr{K}$, the computation of distance and angle measure is more involved. It can be shown that if the vertex of an angle in $\mathscr{K}$ is at $(0, 0)$ then the angle measure is given by the Euclidean measure. (See Problem B13 for a computation.) However, the angle measure of an angle whose vertex is not $(0, 0)$ is hard to compute.

Our second new model is due to Henri Poincaré, who is also responsible for $\{H, \mathscr{L}_H\}$. This example makes use of the idea of two circles being perpendicular.

Definition. Two circles $\mathscr{C}_r(C)$ and $\mathscr{C}_s(O)$ in $\mathbb{R}^2$ are **perpendicular** if they intersect in two points A and B and if both $\angle OAC$ and $\angle OBC$ are right angles. (See Figure 11-15.)

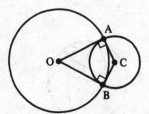

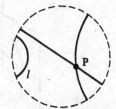

Figure 11-15   Figure 11-16

Definition. A *P-line* in $\mathbb{D}$ is the intersection of $\mathbb{D}$ with either a Euclidean line through $(0,0)$ or with a Euclidean circle $\mathscr{C}$ which is perpendicular to the circle $\{(u,v) \mid u^2 + v^2 = 1\}$. The **Poincaré Disk** is the abstract geometry $\{\mathbb{D}, \mathscr{L}_D\}$ where $\mathscr{L}_D$ is the set of all *P-lines*.

Some *P-lines* are illustrated in Figure 11-16. Again note that $\{\mathbb{D}, \mathscr{L}_D\}$ cannot satisfy EPP.

We shall make $\{\mathbb{D}, \mathscr{L}_D\}$ into a hyperbolic geometry by giving a collineation from $\{\mathbb{D}, \mathscr{L}_K\}$ to $\{\mathbb{D}, \mathscr{L}_D\}$.

Proposition 11.2.6. *The function* $\psi : \mathbb{D} \to \mathbb{D}$ *given by*

$$\psi(u,v) = \left(\frac{u}{1 + \sqrt{1 - u^2 - v^2}}, \frac{v}{1 + \sqrt{1 - u^2 - v^2}} \right) \qquad (2\text{-}3)$$

is a collineation from $\{\mathbb{D}, \mathscr{L}_K\}$ *to* $\{\mathbb{D}, \mathscr{L}_D\}$.

PROOF. By Problem A10, ψ is a bijection whose inverse is given by

$$\psi^{-1}(x,y) = \left(\frac{2x}{1 + x^2 + y^2}, \frac{2y}{1 + x^2 + y^2} \right). \qquad (2\text{-}4)$$

To show that ψ is a collineation we must show that the image of a K-line $l = \{(u,v) \in \mathbb{D} \,|\, au + bv = c\}$ is a P-line. If $\psi(u,v) = (x,y)$ then by Equation (2-4)

$$u = \frac{2x}{1 + x^2 + y^2} \quad \text{and} \quad v = \frac{2y}{1 + x^2 + y^2}.$$

Thus $au + bv = c$ if and only if

$$\frac{2ax}{1 + x^2 + y^2} + \frac{2by}{1 + x^2 + y^2} = c$$

or

$$2ax + 2by = c + c(x^2 + y^2)$$

or

$$c(x^2 + y^2) - 2ax - 2by + c = 0. \qquad (2\text{-}5)$$

There are two cases to consider. If $c = 0$ then the original K-line l went through the origin, and Equation (2-5) reduces to $ax + by = 0, (x,y) \in \mathbb{D}$. This is a P-line. (Note that a K-line through the origin gets sent to a P-line through the origin!)

If $c \neq 0$, then Equation (2-5) describes a Euclidean circle $\mathscr{C}$ with center $C = (a/c, b/c)$ and radius $\sqrt{(a^2/c^2) + (b^2/c^2) - 1}$. We must show that $\mathscr{C}$ is perpendicular to $\mathscr{C}' = \{(x,y) \,|\, x^2 + y^2 = 1\}$.

$\mathscr{C}'$ has center $O = (0,0)$ and radius 1. Since $d_E(O, C) = \sqrt{(a^2/c^2) + (b^2/c^2)}$ and $1 + \sqrt{(a^2/c^2) + (b^2/c^2) - 1} > \sqrt{(a^2/c^2) + (b^2/c^2)}$, the Two Circle Theorem for the Euclidean Plane (Theorem 6.6.5) states that $\mathscr{C} \cap \mathscr{C}'$ consists of exactly two points. If $A \in \mathscr{C} \cap \mathscr{C}'$ then

$$d_E(O, A) = 1, \qquad d_E(A, C) = \sqrt{\frac{a^2 + b^2}{c^2} - 1}, \qquad d_E(O, C) = \sqrt{\frac{a^2 + b^2}{c^2}}$$

so that

$$(d_E(O, A))^2 + (d_E(A, C))^2 = (d_E(O, C))^2.$$

By the Pythagorean Theorem in $\mathscr{E}$, $\angle OAC$ is a right angle. Thus $\mathscr{C}$ is perpendicular to $\mathscr{C}'$ and the image of l is a P-line in this case also. Hence ψ is a collineation. $\qquad \square$

The function ψ can be described geometrically in a manner similar to φ in Proposition 11.2.5. $\psi = \psi_2 \circ \varphi_1$ where φ_1 is the projection of $\mathbb{D}$ onto the right hemisphere as before. ψ_2 is a stereographic projection also but this time from the point $(0, -1, 0)$ and to the plane $x_2 = 0$. See Figures 11-17 and 11-18.

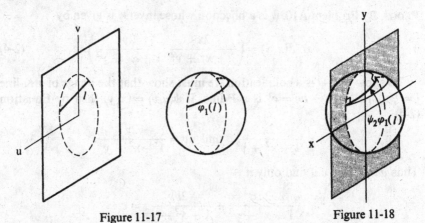

Figure 11-17   Figure 11-18

By Theorem 11.2.4, $\mathscr{D} = \{\mathbb{D}, \mathscr{L}_D, d_D, m_D\}$ is a neutral geometry (and by Problem A7 it is a hyperbolic geometry) where

$$d_D(A, B) = d_K(\psi^{-1}A, \psi^{-1}B) = d_H(\varphi\psi^{-1}A, \varphi\psi^{-1}B)$$
$$m_D(\angle ABC) = m_K(\angle \psi^{-1}A\psi^{-1}B\psi^{-1}C) = d_H(\angle \varphi\psi^{-1}A\varphi\psi^{-1}B\varphi\psi^{-1}C).$$

It can be shown that the angle measure in $\mathscr{D}$ is essentially Euclidean; that is, the Poincaré measure of an angle is given by the Euclidean measure of the angle formed by the Euclidean tangent rays just as in the case of $\mathscr{H}$. This makes it easier to do computations in $\mathscr{D}$ than in $\mathscr{H}$.

In Problem A12 you will show that $\varphi \circ \psi^{-1} : \mathbb{D} \to \mathbb{H}$ is given by

$$\varphi \circ \psi^{-1}(z) = \frac{z + i}{1 + iz}$$

where we have identified the point $(x, y) \in \mathbb{R}^2$ with the complex number $z = x + iy$. Note that since φ and ψ^{-1} are isometries $\varphi \circ \psi^{-1}$ is also an isometry.

PROBLEM SET 11.2A

1. Let $\varphi : \mathbb{R}^2 \to \mathbb{R}^2$ be given by $\varphi(x, y) = (x + y, x - y)$. In the induced geometry $\varphi(\mathscr{E}) = \{\mathbb{R}^2, \mathscr{L}', d', m'\}$
   i. Let $A' = (2, 1)$, $B' = (4, 6)$ and find the line in $\varphi(\mathscr{E})$ determined by A' and B'.
   ii. Find $d'(A', B')$.
   iii. Prove for any P', Q' in $\mathbb{R}^2$ that $d'(P', Q') = (1/\sqrt{2})d_E(P', Q')$.

2. In the proof of Theorem 11.2.2, prove that d' is a distance and $f' = f \circ \varphi^{-1}$ is a ruler.

3. If $\{\mathscr{S}, \mathscr{L}, d\}$ is a metric geometry which satisfies PSA and $\varphi: \mathscr{S} \to \mathscr{S}'$ is a bijection prove that the geometry $\{\mathscr{S}', \mathscr{L}', d'\}$ induced by φ also satisfies PSA.

4. If $\{\mathscr{S}, \mathscr{L}, d, m\}$ is a protractor geometry and $\varphi: \mathscr{S} \to \mathscr{S}'$ is a bijection, prove that the geometry $\{\mathscr{S}', \mathscr{L}', d', m'\}$ induced by φ is also a protractor geometry.

5. In Proposition 11.2.5, prove that Equation (2-1) does indeed give the inverse of φ.

6. Show that if $ax + by = c$ describes a K-line then $c^2 < a^2 + b^2$.

7. Show that the Poincaré Disk satisfies HPP.

8. Describe geometrically all the lines in $\mathscr{K}$ that are sent to type I lines of $\mathscr{H}$ by φ.

9. In the Klein Plane give an example of two asymptotically parallel lines and two divergently parallel lines.

10. In Proposition 11.2.6 prove that Equation (2-4) does give the inverse of ψ.

11. In the Poincaré Disk give an example of two asymptotically parallel lines and two divergently parallel lines.

12. Let $\varphi: \mathbb{D} \to \mathbb{H}$ as given in Proposition 11.2.5 and let $\psi: \mathbb{D} \to \mathbb{D}$ as given in Proposition 11.2.6. View $\mathbb{R}^2$ as the set of complex numbers via the identification $(x, y) \leftrightarrow z = x + iy$. Show that the collineation $\varphi \circ \psi^{-1}: \mathbb{D} \to \mathbb{H}$ is given by $\varphi \circ \psi^{-1}(z) = (z + i)/(1 + iz)$.

Part B.

13. In the Klein Plane $\mathscr{K}$ let $A = (\frac{1}{2}, 0)$, $B = (0, 0)$, and $C = (\frac{1}{2}, \frac{1}{2})$. Find $d_K(A, B)$, $d_K(C, B)$, and $m_K(\angle ABC)$.

14. In the Poincaré Disk $\mathscr{D}$ let $A = (\frac{1}{2}, 0)$, $B = (0, 0)$, and $C = (\frac{1}{2}, \frac{1}{2})$. Find $d_D(A, B)$, $d_D(A, C)$, and $m_D(\angle ABC)$. (Hint: Use Problem A12.)

11.3 Reflections and the Mirror Axiom

From now on we shall only be interested in isometries from a geometry to itself. A primary goal is a classification theorem which partitions the set of all isometries of a neutral geometry according to their fixed point properties. This process begins with the study of a special type of isometry called a reflection.

We start this section with proofs that if an isometry fixes two points then it fixes the line they determine, whereas if it fixes three noncollinear points then it must be the identity function. Next we show that any isometry is the composition of three or fewer reflections. Finally we show that for a protractor geometry the SAS axiom is equivalent to the existence of "many" reflections.

Definition. A function $\varphi: \mathscr{S} \to \mathscr{S}$ **fixes** the point $A \in \mathscr{S}$ if $\varphi A = A$.

Lemma 11.3.1. *Let $\varphi: \mathscr{S} \to \mathscr{S}$ be an isometry of a neutral geometry. If φ fixes the points A and B then φ fixes each point on $\overleftrightarrow{AB}$.*

PROOF. Let f be a coordinate system for $\overleftrightarrow{AB}$ with A as origin and B positive. Suppose that $C \in \overleftrightarrow{AB}$ and $C \neq A$, $C \neq B$. If $C' = \varphi C$, then we need to show that $C = C'$. Now $d(A, C') = d(A, C)$ since $\varphi A = A$ and φ is an isometry. Hence $|f(C')| = |f(C)|$ and $f(C') = \pm f(C)$. Since φ preserves betweenness (Lemma 11.1.6) either

$$A-B-C \quad \text{so that} \quad A-B-C'$$

or

$$A-C-B \quad \text{so that} \quad A-C'-B$$

or

$$C-A-B \quad \text{so that} \quad C'-A-B$$

None of these cases permit $f(C') = -f(C)$. Hence $f(C') = +f(C)$ and so $\varphi C = C' = C$. $\qquad\qquad\square$

Lemma 11.3.2. *Let $\varphi: \mathscr{S} \to \mathscr{S}$ be an isometry of a neutral geometry. If φ fixes three noncollinear points A, B, C then φ is the identity.*

PROOF. φ fixes each point of the lines $\overleftrightarrow{AB}$, $\overleftrightarrow{BC}$ and $\overleftrightarrow{AC}$ by Lemma 11.3.1 and hence each of the points of $\triangle ABC$. Let D be any point in $\mathscr{S}$ and let $E \neq D$ be a point in $\text{int}(\overline{AB})$. By Pasch's Theorem $\overleftrightarrow{DE}$ intersects $\triangle ABC$ at some point $F \neq E$. Since both E and F belong to $\triangle ABC$ they are both fixed. Hence every point of $\overleftrightarrow{EF}$, in a particular D, is fixed by φ. Thus $\varphi D = D$ for any point D and φ is the identity isometry. $\qquad\qquad\square$

We are now ready to define a reflection across a line l. The basic properties of a reflection are that it is an isometry, leaves l fixed, and interchanges the half planes determined by l.

Definition. Let l be a line in a neutral geometry. For each $P \in \mathscr{S}$ let P_l be the foot of the perpendicular from P to l. The **reflection across** l is the function $\rho_l: \mathscr{S} \to \mathscr{S}$ given by

$$\begin{cases} \rho_l P = P', \text{ where } P-P_l-P' \text{ and } \overline{PP_l} \simeq \overline{P'P_l}, \text{ if } P \notin l \\ \rho_l P = P, \text{ if } P \in l \end{cases} \qquad (3\text{-}1)$$

Note that we are defining $\rho_l P$ to be the point P' such that P_l is the midpoint of $\overline{PP'}$ if $P \notin l$. Examples in $\mathscr{E}$ and $\mathscr{H}$ are illustrated in Figure 11-19.

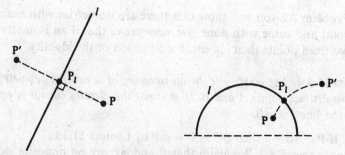

Figure 11-19

Theorem 11.3.3. *A reflection in a neutral geometry $\mathscr{S}$ is an isometry.*

PROOF. Let $A, B \in \mathscr{S}$, let l be a line in $\mathscr{S}$, and, for convenience, write ρ_l as ρ. We must show that $d(A, B) = d(\rho A, \rho B)$. There are several cases to consider: (i) A and B on the same side of l, (ii) A and B on opposite sides of l, (iii) $A \in l$, $B \notin l$, (iv) $A, B \in l$. We shall complete the proof only for the first case and leave the others to Problem A3.

Assume that A and B are on the same side of l. If $\overleftrightarrow{AB} \perp l$ then $A_l = B_l = Q$ for some Q. Let f be a ruler for $\overleftrightarrow{AB}$ with origin Q and A positive. Then for $P \in \overleftrightarrow{AB}$, $f(\rho P) = -f(P)$. See Figure 11-20. Hence

$$d(\rho A, \rho B) = |f(\rho A) - f(\rho B)|$$
$$= |-f(A) + f(B)|$$
$$= d(A, B)$$

Now suppose that $\overleftrightarrow{AB}$ is not perpendicular to l. Then $A_l \neq B_l$. Let $A_l = P$ and $B_l = Q$. See Figure 11-21. $\triangle PQB \simeq \triangle PQ\rho B$ by SAS so that $\overline{PB} \simeq \overline{P\rho B}$ and $\angle BPQ \simeq \angle \rho BPQ$. Because $\overleftrightarrow{AP} \| \overleftrightarrow{BQ}$ (Why?), B and Q lie on the same side of $\overleftrightarrow{AP}$ and $B \in \text{int}(\angle APQ)$.

Since ρA and ρB lie on the same side of l (namely the opposite side from A) a similar argument shows that $\rho B \in \text{int}(\angle \rho APQ)$. By Angle Subtraction, $\angle APB \simeq \angle \rho AP\rho B$. Then $\triangle APB \simeq \triangle \rho AP\rho B$ by SAS and $\overline{AB} \simeq \overline{\rho A\rho B}$ so that $d(A, B) = d(\rho A, \rho B)$. $\square$

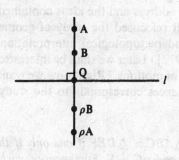

Figure 11-20

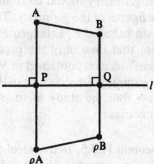

Figure 11-21

In Problem A2 you will show that there are isometries with exactly one fixed point and some with none. We now prove that if an isometry has at least two fixed points then l is either a reflection or the identity.

Theorem 11.3.4. *Let $\varphi: \mathscr{S} \to \mathscr{S}$ be an isometry of a neutral geometry which fixes two distinct points A and B. If φ is not the identity then it is reflection across the line $l = \overleftrightarrow{AB}$.*

PROOF. If $P \in \overleftrightarrow{AB} = l$ then $\varphi P = P = \rho_l P$ by Lemma 11.3.1.

Suppose that $P \notin l$. We claim that P and φP are on opposite sides of l. Since φ is an isometry $\triangle ABP \simeq \triangle AB\varphi P$ by SSS (see Figure 11-22). If P and φP were on the same side of l then since $\angle AB\varphi P \simeq \angle ABP$, the Angle Construction Axiom would imply that $\overrightarrow{BP} \simeq \overrightarrow{B\varphi P}$. Because $\overrightarrow{B\varphi P} \simeq \overrightarrow{BP}$ this would imply that $\varphi P = P$ and so φ fixes three noncollinear points. By Lemma 11.3.2, φ must be the identity, which is contrary to the hypothesis. Hence P and φP are on opposite sides of l so that $\overline{P\varphi P}$ intersects l at a unique point Q.

We must show that $\overline{PQ} \perp l$ and $\overline{PQ} \simeq \overline{\varphi PQ}$. Let $R \neq Q$ be any other point of l. Then $\varphi R = R$ and $\triangle PQR \simeq \triangle \varphi PQR$ by SSS (see Figure 11-23). Hence $\angle PQR$ is a right angle. Since $\varphi P\!-\!Q\!-\!P$ and $\overline{\varphi PQ} \simeq \overline{PQ}$ we have $\varphi P = \rho_l P$. Thus $\varphi = \rho_l$.   □

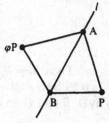

Figure 11-22

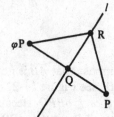

Figure 11-23

In 1872, during an address at Erlangen, Germany, Felix Klein proposed that geometry should be studied by the "group of motions" which preserve the figures of the geometry. This famous address and the ideas contained in it are called the "Erlangen Program." It refocused the study of geometry from that time until the present. (A modern topological interpretation of Klein's ideas is contained in Millman [1977].) Later we shall be interested in determining the structure of this "group of motions." Right now we want to show that the study of triangle congruences corresponds to the study of isometries.

Theorem 11.3.5. *In a neutral geometry $\triangle ABC \simeq \triangle DEF$ if and only if there is an isometry φ with $\varphi A = D$, $\varphi B = E$, and $\varphi C = F$. Furthermore, such an isometry is uniquely determined.*

PROOF. If there is such an isometry then by SSS, $\triangle ABC \simeq \triangle \varphi A \varphi B \varphi C = \triangle DEF$. Hence we will assume that $\triangle ABC \simeq \triangle DEF$ and construct the desired isometry φ. φ will turn out to be the composition of three isometries, $\varphi = \rho \tau \sigma$, each of which is either a reflection or the identity. See Figure 11-24.

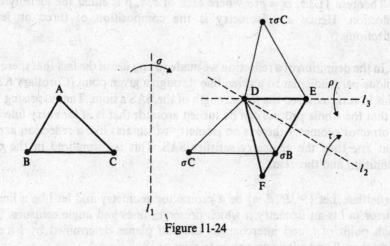

Figure 11-24

If $A = D$ let σ be the identity. If $A \neq D$ let l_1 be the perpendicular bisector of $\overline{AD}$ and let σ be the reflection across l_1. In either case $\triangle \sigma A \sigma B \sigma C = \triangle D \sigma B \sigma C$ is congruent to $\triangle DEF$ since both are congruent to $\triangle ABC$.

We now proceed in a similar fashion with the congruent triangles $\triangle D \sigma B \sigma C$ and $\triangle DEF$. If $\sigma B = E$ then let τ be the identity. If $\sigma B \neq E$ then let l_2 be the perpendicular bisector of $\overline{\sigma B E}$ and let τ be the reflection across l_2. In either case $\tau \sigma B = E$. Note that $\overline{D \sigma B} = \overline{\sigma A \sigma B} \simeq \overline{DE}$ so that D is an element of the perpendicular bisector of $\overline{\sigma B E}$. Hence $\tau D = D$. Also note that

$$\triangle DE \tau \sigma C = \triangle \tau \sigma A \tau \sigma B \tau \sigma C \simeq \triangle ABC \simeq \triangle DEF.$$

We repeat the process one more time. If $\tau \sigma C = F$ let ρ be the identity. Otherwise let ρ be the reflection across the line $l_3 = \overline{DE} = \overline{\tau \sigma A \tau \sigma B}$. Note that l_3 is the perpendicular bisector of $\overline{F \tau \sigma C}$ in this case. Hence.

$$\rho \tau \sigma A = \rho \tau D = \rho D = D$$
$$\rho \tau \sigma B = \rho E = E$$
$$\rho \tau \sigma C = F.$$

Thus $\varphi = \rho \tau \sigma$ gives the desired isometry. All that remains is to show that φ is unique.

Now suppose that ψ is also an isometry such that $\psi A = D, \psi B = E, \psi C = F$. Then $\psi^{-1} \varphi$ fixes A, B, and C so that by Lemma 11.3.2, $\psi^{-1} \varphi$ is the identity and $\varphi = \psi$. Hence there is a unique isometry sending A, B, C to D, E, F. $\square$

Corollary 11.3.6. *In a neutral geometry every isometry is the composition of three or fewer reflections.*

PROOF. Let φ be an isometry and let $\triangle ABC$ be any triangle. Define D, E, F by $\varphi A = D$, $\varphi B = E$, $\varphi C = F$. Then $\triangle ABC \simeq \triangle DEF$ by SSS. By the proof of Theorem 11.3.5, $\varphi = \rho\tau\sigma$ where each of σ, τ, ρ is either the identity or a reflection. Hence any isometry is the composition of three or fewer reflections. $\square$

In the definition of a reflection we made strong use of the fact that there is a unique perpendicular to a given line through a given point (Corollary 6.3.4). This in turn required the full strength of the SAS axiom. The surprising fact is that the whole process can be turned around; that is, if for every line in a protractor geometry there is an isometry which acts like a reflection across that line then the geometry satisfies SAS. This is formalized in the next definition and theorem.

Definition. Let $\{\mathscr{S}, \mathscr{L}, d, m\}$ be a protractor geometry and let l be a line. A **mirror in** l is an isometry μ which preserves lines and angle measure, fixes each point of l, and interchanges the half planes determined by l (i.e., if $P \notin l$ then P and μP lie on opposite sides of l).

A protractor geometry satisfies the **Mirror Axiom** if for each line l there is a mirror in l.

In Section 11.1 we proved that an isometry of a *neutral* geometry was a collineation and preserved angle measure. Since we want to discuss mirrors in the context of a more general protractor geometry, we need to assume that a mirror is an isometry which also preserves lines and angle measures. Note that because mirrors preserve length and angle measure, for any mirror μ, $\triangle ABC \simeq \triangle \mu A \mu B \mu C$.

Theorem 11.3.7. *A protractor geometry is a neutral geometry if and only if it satisfies the Mirror Axiom.*

PROOF. Suppose that $\mathscr{G} = \{\mathscr{S}, \mathscr{L}, d, m\}$ is a protractor geometry. If $\mathscr{G}$ is a neutral geometry and l is a line then the reflection ρ_l is a mirror in l so that the Mirror Axiom is satisfied.

Suppose now that $\mathscr{G}$ satisfies the Mirror Axiom. We must show that SAS is satisfied. Suppose that $\overline{AB} \simeq \overline{DE}$, $\angle A \simeq \angle D$, and $\overline{AC} \simeq \overline{DF}$. We must prove that $\angle B \simeq \angle E$, $\angle C \simeq \angle F$, and $\overline{BC} \simeq \overline{EF}$. We will accomplish this with a slight variation of the proof of Theorem 11.3.5. We will find at most three mirrors σ, τ, ρ such that $\rho\tau\sigma(\triangle ABC) = \triangle DEF$. This will "move" or "superimpose" $\triangle ABC$ onto $\triangle DEF$.

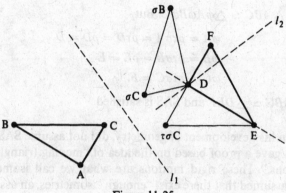

Figure 11-25

If $A = D$ let σ be the identity collineation. If $A \neq D$ let σ be a mirror in the perpendicular bisector l_1 of $\overline{AD}$. Such a bisector exists in a protractor geometry by Corollary 5.3.7. By Problem A6, $\sigma A = D$. See Figure 11-25.

If $\sigma B = E$ let τ be the identity. If σB—D—E let τ be the mirror in the perpendicular l_2 to $\overleftrightarrow{DE}$ at D. Otherwise let τ be the mirror in the angle bisector l_2 of $\angle ED\sigma B$. In any case $D \in l_2$ so that $\tau D = D$. (Note we could not let l_2 be the perpendicular bisector of $\overline{E\sigma B}$ because in a protractor geometry it need not be the case that $D \in l_2$).

We claim $\tau\sigma B = E$. If σB—D—E, then since

$$\overline{D\sigma B} = \overline{\sigma A\sigma B} \simeq \overline{AB} \simeq \overline{DE}$$

l_2 is the perpendicular bisector of $\overline{\sigma BE}$ and $\tau\sigma B = E$ by Problem A6. If σB, D, E are not collinear then l_2 bisects $\angle ED\sigma B$. σB and $\tau\sigma B$ lie on opposite sides of l_2 so that $l_2 \cap \overline{\sigma B\tau\sigma B} = \{Q\}$ for some Q. $Q \in \operatorname{int}(\angle ED\sigma B)$ (Why?). Now $\angle QD\sigma B \simeq \angle QDE$ since l_2 is an angle bisector. $\angle QD\sigma B \simeq \angle QD\tau\sigma B$ since τ is a mirror. $\tau\sigma B$ and E lie on the opposite side of l_2 from σB. By the Angle Construction Theorem $\angle QD\tau\sigma B = \angle QDE$ so that $\overrightarrow{D\tau\sigma B} = \overrightarrow{DE}$. Since

$$\overline{D\tau\sigma B} \simeq \overline{D\sigma B} \simeq \overline{AB} \simeq \overline{DE}$$

$E = \tau\sigma B$. Thus $E = \tau\sigma B$ in all cases.

Finally, if $\tau\sigma C$ is on the same side of $\overleftrightarrow{DE}$ as F let ρ be the identity. Otherwise let ρ be the mirror in $\overleftrightarrow{DE}$. By using the Angle Construction Theorem again, we can show $\rho\tau\sigma C = F$ just as we showed $\tau\sigma B = E$.

$\varphi = \rho\tau\sigma$ is an isometry which preserves angle measure since ρ, τ, and σ

do. Hence $\triangle ABC \simeq \triangle \varphi A \varphi B \varphi C$. But

$$\varphi A = \rho \tau \sigma A = \rho \tau D = \rho D = D$$
$$\varphi B = \rho \tau \sigma B = \rho E = E$$
$$\varphi C = \rho \tau \sigma C = F.$$

Hence $\triangle ABC \simeq \triangle DEF$ and SAS is satisfied. $\qquad\square$

Euclid, in his development of geometry, did not assume SAS as an axiom but instead gave a proof based on the idea of "moving triangles around by rigid motions." These rigid motions are what we call isometries. Euclid essentially assumed that there exist "enough" isometries, an assumption that Theorem 11.3.7 reduces to the existence of mirrors. Euclid's assumption, which was called the "principle of superposition," was that a geometric figure could be picked up and moved to another position without any distortion.

PROBLEM SET 11.3

Part A.

1. Let $\varphi : \mathscr{S} \to \mathscr{S}$ be an isometry of a metric geometry. If A is a fixed point of φ, prove that φ preserves circles centered at A. More precisely, prove that $\varphi(\mathscr{C}_r(A)) \subset \mathscr{C}_r(A)$. Is it true that for all $B \in \mathscr{C}_r(A)$, $\varphi B = B$?

2. Prove that in $\mathscr{E}$ there are isometries with no fixed points and others with exactly one fixed point.

3. Complete the proof of Theorem 11.3.3.

4. Let $l = {}_aL$ be a type I line in the Poincaré Plane. Find $\rho_l : \mathbb{H} \to \mathbb{H}$.

*5. If $\rho_l : \mathscr{S} \to \mathscr{S}$ is the reflection across l in a neutral geometry, show that
   a. $\rho_l = \rho_l^{-1}$;
   b. $(\rho_l)^2 = Id_{\mathscr{S}}$.

6. Let μ be a mirror in the perpendicular bisector of $\overline{AB}$ in a protractor geometry. Prove that $\mu A = B$.

7. Let l be a line in a neutral geometry. If μ is a mirror in l prove that $\mu = \rho_l$, where ρ_l is the reflection defined by Equation (3-1).

Part B.

8. Because of Problem A5 we know that every reflection ρ has the property $\rho^2 = Id$. Find an example in the Euclidean plane of an isometry φ which is neither a reflection nor the identity but which satisfies $\varphi^2 = Id$.

11.4 Pencils and Cycles

In this section we shall introduce two new concepts in a neutral geometry—pencils and cycles. A cycle will be a generalization of a circle while a pencil will be a special collection of lines. These ideas will be useful later as we classify isometries by their fixed points. In this section we will see how pencils can be used to extend certain Euclidean results to arbitrary neutral geometries. For example, in a Euclidean geometry the perpendicular bisectors of the sides of a triangle are concurrent. In a neutral geometry, this need not be true, but they will all belong to the same pencil.

We now define three different kinds of pencils—pointed, parallel, and asymptotic. Each consists of a family of lines with a certain incidence property. Each will have associated with it an object called its center.

Definition. Let $\{\mathcal{S}, \mathcal{L}, d, m\}$ be a neutral geometry. The **pointed pencil** $\mathcal{P}_C$ with **center** C is the set of all lines through the point C.

The **parallel pencil** $\mathcal{P}_l$ **perpendicular to the line** l is the set of all lines perpendicular to l. The **center** of $\mathcal{P}_l$ is l if the geometry is hyperbolic and is the set of all lines parallel to l if the geometry is Euclidean.

If the geometry is hyperbolic, the **asymptotic pencil** $\mathcal{P}_{\overrightarrow{AB}}$ **along the ray** $\overrightarrow{AB}$ is the set of all lines which contain a ray asymptotic to $\overrightarrow{AB}$. The **center** of $\mathcal{P}_{\overrightarrow{AB}}$ is the pencil $\mathcal{P}_{\overrightarrow{AB}}$ itself.

A **pencil** is any set which is either a pointed pencil, a parallel pencil, or an asymptotic pencil.

Note that each pencil has a unique center which may be either a point, a line, or a pencil. In Figure 11-26 parts (a) and (b) illustrate pointed pencils in $\mathcal{E}$ and $\mathcal{H}$, (c) is a parallel pencil in $\mathcal{E}$ while (d) and (e) are parallel pencils in $\mathcal{H}$, and (f) and (g) are asymptotic pencils in $\mathcal{H}$.

The center of an asymptotic pencil may be thought of as an "ideal" point in the following way. If two rays are asymptotic then intuitively they meet "at infinity". Any other ray asymptotic to these two rays also meets them "at infinity". Classically, this place "at infinity" is referred to as an "ideal" point. In the Poincaré Plane the "ideal" points are represented by the points along the x-axis together with one other point. (See Problem B13 of Section 8.1.) In the Poincaré Disk and Klein Plane the "ideal" points are represented by the points on the boundary of the disk: $x^2 + y^2 = 1$. An asymptotic pencil consists of all lines through an "ideal" point and thus in some sense is similar to a pointed pencil.

In Theorem 9.1.5 we saw that the perpendicular bisectors of the sides of a triangle are concurrent in a Euclidean geometry. This means that the three lines belong to the same pointed pencil. Our first result generalizes this result to a neutral geometry.

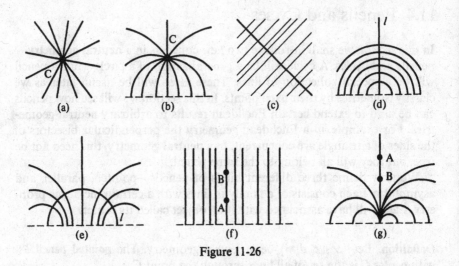

Figure 11-26

Theorem 11.4.1. *In a neutral geometry the perpendicular bisectors of the sides of a triangle $\triangle ABC$ all belong to the same pencil.*

PROOF. Let the perpendicular bisectors of $\overline{AB}$, $\overline{BC}$, and $\overline{AC}$ be l, m, n respectively. There are three possible incidence relations between l and m which lead to three cases: $l \cap m \neq \varnothing$, l and m have a common perpendicular t, or l and m are asymptotic.

Case 1. $l \cap m \neq \varnothing$. Let $P \in l \cap m$. By Problem A10 of Section 6.4, $P \in n$ also so that l, m, n all belong to the pointed pencil $\mathscr{P}_P$.

Case 2. l and m have a common perpendicular t. We will show that n is also perpendicular to t so that $l, m, n \in \mathscr{P}_t$. Let P be the midpoint of $\overline{AB}$ and let Q be the midpoint of $\overline{BC}$. By Problem A2 none of A, B, C, P, and Q belong to t. Let A', B', C', P', Q' be the feet of the perpendiculars from A, B, C, P, Q to t. In Problem A2 you will also show that $P' \neq A'$, $P' \neq B'$, $Q' \neq B'$, $Q' \neq C'$ and $A' \neq C'$. This means that the various angles and segments in the next paragraph all exist. We will show that $\square A'ACC'$ is a Saccheri quadrilateral.

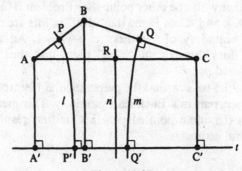

Figure 11-27

Now $\overleftrightarrow{AA'}$, $\overleftrightarrow{BB'}$, $\overleftrightarrow{CC'}$, $\overleftrightarrow{PP'}$, and $\overleftrightarrow{QQ'}$ all belong to the pencil $\mathscr{P}_t$ and are pairwise parallel to each other. Hence, since $A{-}P{-}B$ and $B{-}Q{-}C$, we have $A'{-}P'{-}B'$ and $B'{-}Q'{-}C'$. $\triangle APP' \simeq \triangle BPP'$ and $\triangle CQQ' \simeq \triangle BQQ'$ by SAS. Hence $\angle AP'P \simeq \angle BP'P$ and $\angle CQ'Q \simeq \angle BQ'Q$. Now $A \in$ int($\angle A'P'P$), $C \in$ int($\angle C'Q'Q$), $B \in$ int($\angle B'P'P$), and $B \in$ int($\angle B'Q'Q$)(Why?). Hence we may use Angle Subtraction to obtain $\angle AP'A' \simeq \angle BP'B'$ and $\angle C'Q'C \simeq \angle B'Q'B$. Thus $\triangle AP'A' \simeq \triangle BP'B'$ and $\triangle BQ'B' \simeq \triangle CQ'C'$ by HA. Hence $\overline{AA'} \simeq \overline{BB'} \simeq \overline{CC'}$ and $\square A'ACC'$ is a Saccheri quadrilateral. Therefore the perpendicular bisector n of $\overline{AC}$ must also be the perpendicular bisector of $\overline{A'C'}$. Hence $n \perp t$ and $n \in \mathscr{P}_t$.

Case 3. $l \| m$. Note first that $l \| m$ implies that $n \cap l = n \cap m = \varnothing$, for if this were not true then Case 1 would show that l, m, and n are concurrent which is impossible since $l \| m$ implies $l \| m$. Notice also that Case 2 prohibits n from having a common perpendicular with either l or m. Thus $m | n$ and $l | n$. To show that l, m, n belong to the same asymptotic pencil means we must show that l, m, and n are "asymptotic at the same end." We do this by first finding a common transversal for l, m, n.

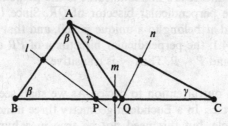

Figure 11-28

Suppose that $\overline{BC}$ is a longest side of $\triangle ABC$ so that $\angle A \geq \angle B$ and $\angle A \geq \angle C$. Then there are points P and Q on $\overline{BC}$ with $\angle BAP \simeq \angle B$ and $\angle CAQ \simeq \angle C$. See Figure 11-28. Thus $\overline{PA} \simeq \overline{PB}$ so that P is on l, the perpendicular bisector of $\overline{AB}$. Likewise $Q \in n$. Thus the line $\overleftrightarrow{BC}$ intersects l at P, $\overleftrightarrow{BC}$ intersects n at Q, and $\overleftrightarrow{BC}$ intersects m at the midpoint of $\overline{BC}$. Thus l, m, and n have a common transversal, namely $\overleftrightarrow{BC}$.

Finally, by Problems A3 and A4, the lines l, m, n cannot form a trebly asymptotic triangle but instead must all belong to the same asymptotic pencil. $\qquad\qquad \square$

Definition. Let $\mathscr{P}$ be a pencil. Two points P and Q are **equivalent** with respect to $\mathscr{P}$, written $P \sim_{\mathscr{P}} Q$, if there is a line $l \in \mathscr{P}$ such that Q is the image of P under the reflection across l; i.e., $Q = \rho_l P$.

We shall normally omit the subscript $\mathscr{P}$ in $P \sim_{\mathscr{P}} Q$ if there is no danger of confusion.

Theorem 11.4.2. *If $\mathscr{P}$ is a pencil in a neutral geometry then $\sim_{\mathscr{P}}$ is an equivalence relation.*

PROOF. Let P be any point. Then there is a line $l \in \mathscr{P}$ with $P \in l$ (Why?). Since $\rho_l P = P$, $P \sim P$ and $\sim$ is reflexive.

Suppose $P \sim Q$ so that there is an $l \in \mathscr{P}$ with $Q = \rho_l P$. Then $\rho_l Q = \rho_l \rho_l P = P$ and $Q \sim P$. Thus $\sim$ is symmetric.

To show that $\sim$ is transitive we assume that $P \sim Q$ and $Q \sim R$. If P, Q, R are not distinct points then either $P = Q$ so that $Q \sim R$ implies $P \sim R$, or $P = R$ so that $P \sim R$ by the first part, or $Q = R$ so that $P \sim Q$ implies that $P \sim R$. Hence we assume that P, Q and R are distinct. There will be two cases depending on whether P, Q and R are collinear or not. Note that $Q = \rho_l P$ and $R = \rho_m Q$ for some lines $l, m \in \mathscr{P}$.

Case 1. P, Q, R are collinear. Since P, Q, and R are distinct, $l \neq m$ and both l and m are perpendicular to $\overleftrightarrow{PQ}$. Hence $l, m \in \mathscr{P}_{\overleftrightarrow{PQ}}$ so that $\mathscr{P} = \mathscr{P}_t$ where $t = \overleftrightarrow{PQ}$ by Problem A1. Let n be the perpendicular bisector of $\overline{PR}$. Then $n \in \mathscr{P}_t = \mathscr{P}$ and $R = \rho_n P$. Hence $P \sim R$.

Case 2. P, Q, and R are noncollinear. l is the perpendicular bisector of $\overline{PQ}$ and m is the perpendicular bisector of $\overline{QR}$. Since $l \neq m$ Problem A1 implies that l and m belong to a unique pencil, and that pencil must be $\mathscr{P}$. By Theorem 11.4.1, the perpendicular bisector n of $\overline{PR}$ also belongs to $\mathscr{P}$. Hence $R = \rho_n P$ and $P \sim R$. Thus $\sim$ is transitive.   □

We now turn our attention to cycles. As we shall see, cycles generalize the notion of circles. In a Euclidean geometry three noncollinear points lie on a unique circle, but this need not be true in a hyperbolic geometry. However, three noncollinear points will always lie on a unique cycle. There will also be results on tangents to cycles which are quite similar to those with circles. Most of these will be left as exercises.

Definition. Let $\mathscr{P}$ be a pencil in a neutral geometry. A **cycle** $\mathscr{C}$ of $\mathscr{P}$ is an equivalence class with respect to $\sim_{\mathscr{P}}$. A cycle is **degenerate** if it is a single point (the center of $\mathscr{P} = \mathscr{P}_C$) or a line (the line l of $\mathscr{P} = \mathscr{P}_l$). All other cycles are called **nondegenerate**. The **center** of a cycle is the center of the associated pencil $\mathscr{P}$.

Theorem 11.4.3. *In a neutral geometry a nondegenerate cycle with respect to a pointed pencil $\mathscr{P}_C$ is a circle with center C. If $\mathscr{C}$ is a nondegenerate cycle of a parallel pencil $\mathscr{P}_l$ then every point of $\mathscr{C}$ is the same distance from l and $\mathscr{C}$ lies on one side of l.*

The proof of the above result is left as Problem A5. The second part of the theorem is quite interesting because it says that a cycle $\mathscr{C}$ of $\mathscr{P}_l$ is an "equidistant curve" of l. Although this curve is a line in the Euclidean case, it is not in a hyperbolic geometry. Can you find an example in $\mathcal{H}$?

By Problem A6 the set of all points a distance $r > 0$ from a line l consists of two cycles. See Figure 11-29.

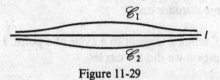

Figure 11-29

Theorem 11.4.4. *In a neutral geometry any three distinct points lie on a unique cycle.*

PROOF. Let the three points be P, Q and R. Let l be the perpendicular bisector of $\overline{PQ}$ and m be the perpendicular bisector of $\overline{QR}$. Then $Q = \rho_l P$ and $R = \rho_m Q$. Since P, Q and R are distinct, $l \neq m$ so that l and m belong to a unique pencil $\mathcal{P}$. With respect to this pencil we have $P \sim Q$ and $Q \sim R$. Hence P, Q, R belong to the same equivalence class and thus the same cycle $\mathcal{C}$. Since $\mathcal{P}$ is unique and P can belong to only one cycle of $\mathcal{P}$, there is only one cycle that contains P, Q, and R. $\square$

The next idea is a generalization of the idea of a fixed point. It is extremely important in many areas of mathematics and will be very useful as we prove the classification theorem in Sections 11.5 and 11.6.

Definition. A set $\mathcal{A}$ is an **invariant set** of a collineation φ if $\varphi(\mathcal{A}) \subset \mathcal{A}$.

Note carefully that the definition says that $\mathcal{A}$ is invariant if for each $a \in \mathcal{A}$, $\varphi(a) \in \mathcal{A}$. It is quite possible that $\varphi(a) \neq a$ for $a \in \mathcal{A}$. For example, if $\varphi: \mathbb{R}^2 \to \mathbb{R}^2$ by $\varphi(x, y) = (x + 1, y)$ then $\mathcal{A} = \{(x, y) \in \mathbb{R}^2 | -7 < y < 3\}$ is an invariant set, but no point of $\mathcal{A}$ of fixed. The set $\mathcal{B} = \{(x, y) \in \mathbb{R}^2 | x > 0\}$ is also invariant, but $\varphi(\mathcal{B}) \neq \mathcal{B}$. We shall usually say "$\mathcal{A}$ is invariant" rather than "$\mathcal{A}$ is an invariant set for φ" if it is clear which collineation we are referring to.

It makes sense to ask if a pencil $\mathcal{P}$ is invariant. $\mathcal{P}$ is a set of lines so that $\varphi(\mathcal{P}) = \{\varphi(l) | l \in \mathcal{P}\}$. Thus $\mathcal{P}$ will be invariant under φ if $\varphi(l) \in \mathcal{P}$ for every $l \in \mathcal{P}$.

The remaining results are left as homework.

Theorem 11.4.5. *Let φ be an isometry of a neutral geometry and let $\mathcal{P}$ be a pencil. If φ can be written as a composition of reflections across lines of $\mathcal{P}$ then the center of $\mathcal{P}$, each cycle of $\mathcal{P}$, and $\mathcal{P}$ itself are invariant under φ.*

Theorem 11.4.6. *Let $\mathscr{C}_1$ and $\mathscr{C}_2$ be two nondegenerate cycles of the pencil $\mathscr{P}$ in a neutral geometry. Let $l, m \in \mathscr{P}$. Suppose that l intersects $\mathscr{C}_1$ and $\mathscr{C}_2$ at A_1, A_2 and m intersects $\mathscr{C}_1$ and $\mathscr{C}_2$ at B_1, B_2. If $\mathscr{P} = \mathscr{P}_C$ so that $\mathscr{C}_1$ and $\mathscr{C}_2$ are actually circles with center C suppose further that C is not between A_1 and A_2 and C is not between B_1 and B_2. Then $\overline{A_1 A_2} \simeq \overline{B_1 B_2}$. Hence two "concentric cycles" are the same distance apart.*

It is possible to define a tangent to a cycle at a point. However, we cannot use the same definition we did for circles.

Definition. Let $\mathscr{C}$ be a nondegenerate cycle of the pencil $\mathscr{P}$ in a neutral geometry. If $P \in \mathscr{C}$ then the **tangent line to $\mathscr{C}$ at P** is the line through P which is perpendicular to the unique line of $\mathscr{P}$ through P.

Theorem 11.4.7. *In a neutral geometry, if the line l is tangent to the nondegenerate cycle $\mathscr{C}$ at P then $l \cap \mathscr{C}$ contains just the point P.*

The converse of Theorem 11.4.7 is false (Problem A11) which is why we had to define tangents differently for cycles than we did for circles.

Theorem 11.4.8. *In a neutral geometry, if l is tangent to the nondegenerate cycle $\mathscr{C}$ at P then the set $\mathscr{C} - \{P\}$ lies on one side of l.*

Because of this theorem we can define the interior and exterior of a cycle.

Definition. Let $\mathscr{C}$ be a nondegenerate cycle in a neutral geometry $\{\mathscr{S}, \mathscr{L}, d, m\}$. For each $P \in \mathscr{C}$ let H_P be the half plane determined by the tangent line to $\mathscr{C}$ at P which contains $\mathscr{C} - \{P\}$. The **interior** of $\mathscr{C}$ is

$$\text{int}(\mathscr{C}) = \bigcap_{P \in \mathscr{C}} H_P.$$

*The **exterior** of $\mathscr{C}$ is*

$$\text{ext}(\mathscr{C}) = \mathscr{S} - \mathscr{C} - \text{int}(\mathscr{C}).$$

The next theorem is proved almost the same way as Theorem 6.5.10. The hard part is finding a replacement for Theorem 6.5.9 to use.

Theorem 11.4.9. *If $\mathscr{C}$ is a nondegenerate cycle in a neutral geometry and $P \in \text{ext}(\mathscr{C})$ then there are exactly two lines through P tangent to $\mathscr{C}$.*

PROBLEM SET 11.4

Part A.

1. If l and m are distinct lines of a neutral geometry, prove that there is a unique pencil $\mathscr{P}$ with $l, m \in \mathscr{P}$.

2. In the proof of Theorem 11.4.1 show that
   a. none of $A, B, C, P,$ and Q belong to t;
   b. $P' \neq A', P' \neq B', Q' \neq B', Q' \neq C', A' \neq C'$.

3. Let l, m, n be three distinct lines in a neutral geometry with $l|m, l|n$ and $m|n$. Prove that either there is a ray $\overrightarrow{PQ}$ asymptotic to all three lines (so that all three belong to $\mathscr{P}_{\overrightarrow{PQ}}$) or else $l = \overleftrightarrow{AB}, m = \overleftrightarrow{CD}, n = \overleftrightarrow{EF}$ with $\overrightarrow{AB}|\overrightarrow{DC}, \overrightarrow{CD}|\overrightarrow{FE}, \overrightarrow{EF}|\overrightarrow{BA}$ (so that we have a **trebly asymptotic triangle**). See Figure 11-30.

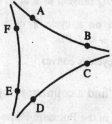

Figure 11-30

4. Let l, m, n be three distinct lines in a neutral geometry with $l|m, m|n,$ and $l|n$. Prove that if l, m, n have a common transversal t then l, m, n all belong to the same pencil.

5. Prove Theorem 11.4.3.

6. Let $r > 0$ and let l be a line in a neutral geometry. Prove that $\mathscr{B} = \{P \in \mathscr{S} \mid d(P, l) = r\}$ is the union of two cycles.

7. Let φ_θ be the isometry of Problem A12 of Section 11.1. What are the invariant sets of φ_θ?

8. Prove Theorem 11.4.5.

9. Prove Theorem 11.4.6.

10. Prove Theorem 11.4.7.

11. Find a nondegenerate cycle $\mathscr{C}$, a point $P \in \mathscr{C}$, and a line l which intersects $\mathscr{C}$ *only* at P but is not tangent to $\mathscr{C}$.

12. Let $\mathscr{C}$ be a nondegenerate cycle in a neutral geometry. If $P \notin \mathscr{C}$ and there is a line through P tangent to $\mathscr{C}$, prove there are exactly two lines through P tangent to $\mathscr{C}$.

13. Describe the cycles of the different possible pencils of $\mathscr{E}$.

14. Let $\mathscr{C}$ be a nondegenerate cycle and let $P \in \mathscr{C}$. How many lines intersect $\mathscr{C}$ just at P? (Your answer will depend on the type of pencil associated with $\mathscr{C}$.)

15. Let $l = \overleftrightarrow{AB}$ and $\overrightarrow{AP}, \overrightarrow{BQ} \in \mathscr{P}_l = \mathscr{P}$. Prove $P \sim_{\mathscr{P}} Q$ if and only if $\boxed{S}APQB$.

16. If $m, n \in \mathscr{P}_l = \mathscr{P}$ and $P \in m$ prove there is a unique $Q \in n$ with $P \sim_{\mathscr{P}} Q$.

17. Repeat Problem A16 for the pencil $\mathscr{P}_{\overrightarrow{AB}}$.

18. If l is divergently parallel to m let $\mathcal{T} = \{\text{lines } t | t \text{ is transversal to } l, m \text{ with}$ alternate interior angles congruent$\}$. Prove that $\mathcal{T}$ is contained in a pointed pencil if the geometry is hyperbolic.

19. In a neutral geometry let l and m be distinct lines in a pencil $\mathcal{P}$ with $P \in l, Q \in m$. Prove that $P \sim Q$ if and only if $P = Q$ or $\overleftrightarrow{PQ}$ is transversal to l, m with alternate interior angles supplementary.

20. In a neutral geometry let l, m be two lines of a pencil $\mathcal{P}$ that intersect a cycle $\mathcal{C}$ of $\mathcal{P}$ at points $A \neq B$. Choose P on the same side of l as B and Q on the same side of m as A and with $\overrightarrow{AP}$ and $\overrightarrow{BQ}$ tangent to $\mathcal{C}$ at A, B. Prove that $\angle PAB \simeq \angle QBA$.

21. If $\square ABCD$ has its vertices on a cycle $\mathcal{C}$ in a neutral geometry prove that $m(\angle A) + m(\angle C) = m(\angle B) + m(\angle D)$.

22. Prove that the interior of a cycle is convex.

Part B. "Prove" may mean "find a counterexample".

23. Describe the possible cycles in the Poincaré Plane $\mathcal{H}$.

24. Prove Theorem 11.4.8.

25. Prove Theorem 11.4.9.

26. Repeat Problem A16 for the pencil $\mathcal{P}_A$.

27. For the situation in the proof of Theorem 11.4.1 find an example where $A' = Q'$. Thus it need not be true that A', B', C', P', Q' are distinct.

11.5 Double Reflections and Their Invariant Sets

In the next section we shall classify isometries according to their geometric properties. That is, we will partition the set of all isometries of a neutral geometry into classes with two isometries in the same class if and only if they act in a similar fashion. The primary geometric property that will be used in this partitioning will be invariant sets. We shall start the process in this section by studying reflections and isometries which are the composition of two reflections (double reflections).

> **Convention.** Throughout Sections 11.5 and 11.6 all results refer to a neutral geometry $\{\mathcal{S}, \mathcal{L}, m, d\}$.

Theorem 11.5.1. *If ρ_l is the reflection across the line l then*

(i) *A point A is fixed by ρ_l if and only if $A \in l$.*
(ii) *A line $m \neq l$ is invariant under ρ_l if and only if $m \perp l$.*
(iii) *A pencil $\mathcal{P}$ is invariant under ρ_l if and only if either $l \in \mathcal{P}$ or $\mathcal{P} = \mathcal{P}_l$.*
(iv) *ρ_l interchanges the half planes determined by l.*
(v) *$\rho_l^{-1} = \rho_l$.*

This theorem, whose proof is left to Problem A1, fairly well sums up all the important properties of a reflection. We may thus turn our attention to isometries which can be written as the composition of two reflections.

Definition. A **double reflection** φ is an isometry which can be written as the composition of two distinct reflections: $\varphi = \rho_l \rho_m$ with $l \neq m$. φ is a **rotation with center** A if $l \cap m = \{A\}$. φ is **translation along** n if n is a common perpendicular of l and m. φ is a **parallel displacement** if l and m are asymptotically parallel.

First we would like to see that rotations and translations behave essentially as we would expect from our Euclidean experience. This is done in the next two theorems. The proof of the first is left to Problem A2.

Theorem 11.5.2. *Let l and m be distinct lines with $l \cap m = \{C\}$ so that $\varphi = \rho_l \rho_m$ is a rotation about C. Let θ be the smaller of the measures of the angles formed by l and m. If l is not perpendicular to m then $m(\angle PC\varphi P) = 2\theta$ for any $P \neq C$. (See Figure 11-31.) If $l \perp m$ and $P \neq C$ then P—C—φP and $\overline{PC} \simeq \overline{C\varphi P}$.*

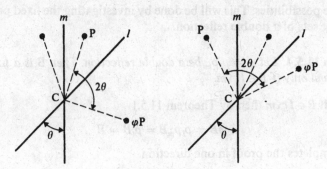

Figure 11-31

Theorem 11.5.3. *Let l and m be distinct lines with common perpendicular t so that $\varphi = \rho_l \rho_m$ is a translation along t. If l and m intersect t at A and B respectively then $d(P, \varphi P) = 2d(A, B)$ for all $P \in t$. If $Q \notin t$ then Q and φQ lie on the same side of t and $d(Q, \varphi Q) \geq 2d(A, B)$ with equality if and only if the geometry is Euclidean.*

PROOF. Let f be a coordinate system for t with $f(B) = 0$ and $f(A) = r > 0$ so that $d(A, B) = r$. Suppose $P \in t$ and $f(P) = s$. Then $\rho_m P \in t$ and has coordinate $-s$. Hence $\varphi P = \rho_l \rho_m P \in t$ has coordinate $r - (-s - r) = 2r + s$. Thus $d(P, \varphi P) = |2r + s - s| = 2r = 2d(A, B)$.

If $Q \notin t$ then $\overline{Q\rho_m Q}$ is perpendicular to m. Since $m \perp t$, we have $\overline{Q\rho_m Q} \| t$. Hence Q and $\rho_m Q$ lie on the same side of t. Similarly $\rho_m Q$ and $\rho_l \rho_m Q = \varphi Q$

lie on the same side of t. Hence Q and φQ lie on the same side of t.

Let P be the foot of the perpendicular from Q to t. Since φ preserves perpendicularity, the foot of the perpendicular from φQ to t must be φP. Now $d(P, Q) = d(\varphi P, \varphi Q)$ so that $\square PQ\varphi Q\varphi P$ is a Saccheri quadrilateral. Hence

$$d(Q, \varphi Q) \geq d(P, \varphi P) = 2d(A, B).$$

Since the upper base of a Saccheri quadrilateral is congruent to the lower base only when the geometry is Euclidean, we are done. $\square$

There are no parallel displacements in a Euclidean geometry because parallel lines always have a common perpendicular there. In Problem A3 you will describe the parallel displacement in $\mathcal{H}$ determined by the asymptotically parallel lines ${}_aL$ and ${}_bL$.

Clearly if φ is a double reflection then φ is either a rotation, a translation, or a parallel displacement. However, it is conceivable that $\varphi = \rho_l\rho_m = \rho_n\rho_p$ where l and m have a common perpendicular while n and p don't. This would mean that the ideas of reflection, translation, and parallel displacement are not disjoint. Our first task is to show a double reflection can be only one of the three possibilities. This will be done by investigating the fixed points and invariant sets of a double reflection.

Theorem 11.5.4. *Let $\varphi = \rho_l\rho_m$ be a double reflection. Then B is a fixed point of φ if and only if $B \in l \cap m$.*

PROOF. If $B \in l \cap m$ then by Theorem 11.5.1

$$\varphi B = \rho_l\rho_m B = \rho_l B = B.$$

This completes the proof in one direction.

Suppose now that φ fixes the point B so that $\rho_l\rho_m B = B$. Since $(\rho_l)^2 = $ identity,

$$\rho_l B = \rho_l(\rho_l\rho_m B) = (\rho_l\rho_l)\rho_m B = \rho_m B.$$

Let $C = \rho_l B = \rho_m B$. We will show that $C = B$. This will imply that $B = \rho_l B = \rho_m B$ and $B \in l \cap m$ by Theorem 11.5.1.

If $B \neq C$ then l and m are both perpendicular to $\overline{BC}$ at its midpoint by the definition of a reflection. But this implies $l = m$, which is impossible. Thus $C = B$ and ρ_l, ρ_m each fix B. Hence B belongs to both l and m so that $B \in l \cap m$. $\square$

Corollary 11.5.5. *An isometry φ has exactly one fixed point if and only if φ is a rotation.*

PROOF. Thanks to the previous theorem a rotation $\varphi = \rho_l\rho_m$ fixes exactly one point, namely the unique intersection of l and m. On the other hand if φ fixes exactly one point then in order to prove that φ is a rotation it is sufficient to prove that φ is a double reflection. We do this by showing that $\rho_l\varphi$ is a reflection for some line l.

Suppose the isometry φ fixes the point D. Let $E \neq D$ be any other point. By hypothesis $E \neq \varphi E$ so that we may let l be the perpendicular bisector of $\overline{E\varphi E}$. Then $d(D, E) = d(\varphi D, \varphi E) = d(D, \varphi E)$ so that D belongs to the perpendicular bisector of $\overline{E\varphi E}$, i.e., $D \in l$. Since $\rho_l(\varphi E) = E$, the isometry $\rho_l\varphi$ fixes both D and E and so it fixes every point on $m = \overline{DE}$ by Lemma 11.3.1. If $\rho_l\varphi$ is the identity then $\varphi = \rho_l^{-1} = \rho_l$ and φ has more than one fixed point, contrary to the hypothesis. Hence by Theorem 11.3.4, $\rho_l\varphi = \rho_m$ and

$$\varphi = \rho_l^{-1}\rho_m = \rho_l\rho_m.$$

Since $D \in l$ and $D \in m$, φ is a rotation with center D.   □

Corollary 11.5.6. If φ is a translation or a parallel displacement then φ does not fix any points and so φ is not a rotation.

Now that we know that a rotation is neither a translation nor a parallel displacement we turn our attention to showing that a translation is not a parallel displacement. Whereas fixed points were the key to distinguishing rotations, invariant lines will be the deciding factor in the next step.

Theorem 11.5.7. Let $\varphi = \rho_l\rho_m$ be a double reflection which is not a rotation. If the geometry is Euclidean then φ is a translation. If the geometry is hyperbolic then φ is a translation along t if and only if φ leaves the line t invariant. Furthermore, in this case φ is not a parallel displacement.

PROOF. In the Euclidean case every double reflection is a rotation or a translation so there is nothing to prove. Furthermore if φ is a translation along t then t is left invariant.

In the hyperbolic case we first assume that t is left invariant by φ and show that φ is a translation along t. Let $\mathscr{P}$ be the unique pencil containing both l and m. We shall show that $\mathscr{P}$ is the pencil $\mathscr{P}_t$ consisting of all lines perpendicular to t. This will mean that $l \perp t$ and $m \perp t$ so that φ is a translation along t. The proof will be by contradiction. We assume that t is not the center of $\mathscr{P}$. Note that in a hyperbolic geometry t is the center of $\mathscr{P}$ if and only if $\mathscr{P} = \mathscr{P}_t$.

Since t is not the center of $\mathscr{P}$ there is a point $A \in t$ which is not in the center of $\mathscr{P}$. (Note that if $\mathscr{P}$ is an asymptotic pencil there are no points at all in the center of $\mathscr{P}$.) Let $\mathscr{C}$ be the unique cycle of $\mathscr{P}$ through A. Since A is not in the center of $\mathscr{P}$ and the geometry is hyperbolic, $\mathscr{C}$ is a nondegenerate cycle. Let $B = \varphi A$. $B \neq A$ because φ is not a rotation. $\mathscr{C}$ is invariant under φ so that $B \in \mathscr{C}$. Since t is invariant, $B \in t$ also. Thus $B \in \mathscr{C} \cap t$. We claim $\varphi B = A$.

Now $\varphi B \in \mathscr{C} \cap t$ since $\mathscr{C}$ and t are invariant and $B \in \mathscr{C} \cap t$. If $\varphi B \neq A$ then A, B, φB are three distinct points which belong to two distinct cycles $\mathscr{C}$ and t. (Note t may not be a cycle of $\mathscr{P}$ but it certainly is a cycle of some pencil.) This contradicts Theorem 11.4.4 which says that three distinct points belong to a unique cycle. Hence we must have $\varphi B = A$.

Thus the isometry φ interchanges A and B. By Problem A9 of Section 11.1, φ fixes the midpoint of $\overline{AB}$. This contradicts the assumption that φ is not a rotation. Hence it must be that t is the center of $\mathscr{P}$ after all and l and m are both perpendicular to t. Thus φ is a translation along t and is not a parallel displacement. $\square$

Theorem 11.5.8. *In a Euclidean geometry a translation along t leaves invariant only those lines parallel to t. In a hyperbolic geometry a translation along t leaves only the line t invariant. A parallel displacement has no invariant lines.*

PROOF. The Euclidean case is left to Problem A7. In the hyperbolic case if $\varphi = \rho_l \rho_m$ is a translation along t then t is the center of the (parallel) pencil determined by l and m. By Theorem 11.4.5 this center is invariant under φ.

On the other hand, if $\varphi = \rho_l \rho_m$ is not a rotation and if φ leaves a line invariant, then by Theorem 11.5.7 that line must be the center of the unique pencil $\mathscr{P}$ determined by l and m. Hence φ is not a parallel displacement and t is the only line invariant under φ. $\square$

Corollary 11.5.9. *If $\varphi = \rho_l \rho_m$ is a double reflection, then φ is exactly one of a rotation, a translation, or a parallel displacement. φ is not a reflection.*

Theorem 11.5.10. *Two distinct asymptotic pencils $\mathscr{P}_{\overline{AB}}$ and $\mathscr{P}_{\overline{CD}}$ have a unique line l in common.*

PROOF. Let P be any point not in $\overline{AB} \cup \overline{CD}$. Choose points R, S so that $\overline{PR} | \overline{AB}$ and $\overline{PS} | \overline{CD}$. If P, R, S are collinear then we may let $l = \overline{PR}$. If P, R, S are not collinear let l be the line of enclosure of $\angle RPS$ (Problem A6 of Section 8.2). Either way, $l \in \mathscr{P}_{\overline{AB}} \cap \mathscr{P}_{\overline{CD}}$.

Suppose $l' \in \mathscr{P}_{\overline{AB}} \cap \mathscr{P}_{\overline{CD}}$ with $l' \neq l$. Since l, $l' \in \mathscr{P}_{\overline{AB}}$, l and l' are asymptotic at one end. Since l, $l' \in \mathscr{P}_{\overline{CD}}$, l and l' are also asymptotic at the other end. This contradicts Problem A5 of Section 8.3. Hence $l = l'$. $\square$

We have investigated invariant points and invariant lines of a double reflection. We now turn to invariant pencils.

Theorem 11.5.11. *In a hyperbolic geometry a parallel displacement $\varphi = \rho_l \rho_m$ leaves invariant exactly one pencil, namely the pencil determined by l and m.*

PROOF. Let $\varphi = \rho_l\rho_m$ be a parallel displacement and let $\mathscr{P}$ be the (asymptotic) pencil determined by l and m. By Theorem 11.4.5, φ leaves $\mathscr{P}$ invariant.

If φ leaves a pointed pencil $\mathscr{P}_C$ invariant then the center C must be fixed by φ, which is impossible. If the φ leaves the parallel pencil $\mathscr{P}_t$ invariant, then the center t is invariant also, which is impossible since φ is a parallel displacement and so has no invariant lines. Finally if φ leaves an asymptotic pencil $\mathscr{P}' \neq \mathscr{P}$ invariant, then φ must leave the unique line $l \in \mathscr{P}' \cap \mathscr{P}$ invariant. But this is impossible since parallel displacements have no invariant lines. Thus $\mathscr{P}$ is the only pencil invariant under φ. $\square$

The invariant pencils of a translation are fairly simple to determine. The proof of the next result is left to Problem A11. We will leave the determination of the invariant pencils of a rotation until after we discuss half-turns, which are a special type of rotation. This will be done in the next section.

Theorem 11.5.12. *A translation never leaves a pointed pencil invariant. In a Euclidean geometry a translation leaves every parallel pencil invariant. In a hyperbolic geometry, a translation along a line t leaves invariant only the parallel pencil $\mathscr{P}_t$ and the two asymptotic pencils that contain t.*

It is possible to write a double reflection in more than one way as a composition of two reflections. For example, if l, m, n, p all belong to the pointed pencil $\mathscr{P}_C$ and $l \perp m$ while $n \perp p$, then $\rho_l\rho_m = \rho_n\rho_p$ (Problem A12). However, our next result says that even if an isometry can be written as a double reflection in more than one way, all the lines must belong to the same pencil.

Theorem 11.5.13. *If the double reflection φ can be written both as $\varphi = \rho_l\rho_m$ and as $\varphi = \rho_n\rho_p$ then l, m, n, and p all belong to the same pencil.*

PROOF. Let $\mathscr{P}$ be the unique pencil that contains l and m while $\mathscr{P}'$ is the unique pencil that contains n and p. By Theorem 11.4.5, φ leaves the center of $\mathscr{P}$ invariant and leaves the center of $\mathscr{P}'$ invariant. If φ is a rotation then the pencils are pointed and their centers must consist of the same single point. Hence $\mathscr{P} = \mathscr{P}'$.

If φ is a translation then the pencils must be parallel pencils $\mathscr{P} = \mathscr{P}_a$ and $\mathscr{P}' = \mathscr{P}_b$, where a is a common perpendicular of l and m while b is a common perpendicular of n and p. By Theorem 11.5.8, φ leaves a and b invariant. If the geometry is Euclidean then $a \| b$ by the same theorem and $\mathscr{P} = \mathscr{P}_a = \mathscr{P}_b = \mathscr{P}'$. If the geometry is hyperbolic then $a = b$ and $\mathscr{P} = \mathscr{P}'$.

Finally if φ is a parallel displacement then $\mathscr{P}$ and $\mathscr{P}'$ are their own centers. Since a parallel displacement leaves only one pencil invariant by Theorem 11.5.11, $\mathscr{P} = \mathscr{P}'$. Hence in all cases $\mathscr{P} = \mathscr{P}'$ and l, m, n and p all belong to the same pencil. $\square$

Our last result in this section is that the composition of two double reflections is a double reflection. This will require two preliminary results. The first gives a condition for when a composition of three reflections is really a reflection. It illustrates how efficient it is to use the language of pencils.

Theorem 11.5.14. *If l, m, n belong to the same pencil $\mathcal{P}$ and if $\varphi = \rho_l\rho_m\rho_n$ then $\varphi = \rho_p$ for some $p \in \mathcal{P}$.*

PROOF. Let A and B be distinct points on n that are not in the center of $\mathcal{P}$. Let $\mathcal{A}$ and $\mathcal{B}$ be the cycles of $\mathcal{P}$ through A and B. Assume that A is not equivalent to B so that $\mathcal{A} \neq \mathcal{B}$. If $\varphi A = A$ we define p to be n so that $p \in \mathcal{P}$. If $\varphi A \neq A$ let p be the perpendicular bisector of $\overline{A\varphi A}$. In this case, since $\varphi(\mathcal{A}) = \mathcal{A}$, we have $\varphi A \in \mathcal{A}$ and so $A \sim \varphi A$. Since $\rho_p A = \varphi A$ we must have that $p \in \mathcal{P}$. We claim that $\varphi = \rho_p$ in either case.

Now in either case $\rho_p\varphi$ fixes A. Since $p \in \mathcal{P}$, $\mathcal{P}$ is invariant under $\rho_p\varphi$ by Theorem 11.4.5. Hence $\rho_p\varphi$ must leave invariant the line n which is the (unique) line of $\mathcal{P}$ that goes through A. If $\mathcal{P}$ is not a pointed pencil then n intersects $\mathcal{B}$ in exactly one point B (Why?). Since $\rho_p\varphi$ leaves both n and $\mathcal{B}$ invariant, it must fix B. On the other hand, if $\mathcal{P}$ is a pointed pencil $\mathcal{P}_C$ then $\rho_p\varphi$ must fix $C \in n$. Either way $\rho_p\varphi$ fixes two points of n (A, B or A, C) and thus fixes each point of n. Hence $\rho_p\varphi$ is either a reflection (which must be ρ_n) or the identity.

If $\rho_p\varphi = \rho_n$ then $\rho_l\rho_m\rho_n = \varphi = \rho_p\rho_n$ so that $\rho_l\rho_m = \rho_p$, which contradicts Corollary 11.5.9. Hence $\rho_p\varphi = \text{id}$ and $\varphi = \rho_p^{-1}\text{id} = \rho_p$ where $p \in \mathcal{P}$. $\qquad\square$

Note that the above proof is constructive—it tells us how to actually find p if we are given l, m, and n.

Corollary 11.5.15. *If l, m, n belong to the pencil $\mathcal{P}$ then there exists $p, q \in \mathcal{P}$ such that $\rho_l\rho_m = \rho_n\rho_p$ and $\rho_l\rho_m = \rho_q\rho_n$.*

PROOF. By Theorem 11.5.14, $\rho_n\rho_l\rho_m = \rho_p$ for some $p \in \mathcal{P}$. Hence $\rho_l\rho_m = \rho_n^{-1}\rho_p = \rho_n\rho_p$. Likewise $\rho_l\rho_m\rho_n = \rho_q$ for some $q \in \mathcal{P}$ so that $\rho_l\rho_m = \rho_q\rho_n^{-1} = \rho_q\rho_n$. $\qquad\square$

Theorem 11.5.16. *The composition of two double reflections is a double reflection or the identity.*

PROOF. This insidious proof is based on producing two pencils $\mathcal{P}'$ and $\mathcal{P}_A$ and applying Theorem 11.5.14 twice. $\mathcal{P}_A$ is simply the pointed pencil at an arbitrary point A of the line d. $\mathcal{P}'$ will be constructed below.

Let $\varphi = \rho_a\rho_b\rho_c\rho_d$ be the composition of two double reflections. If $b = c$ then $\varphi = \rho_a\rho_d$ which is a double reflection (or the identity if $a = d$). Hence we assume that $b \neq c$. Let $\mathcal{P}$ be the unique pencil that contains b and c. Let A be any point on d and let l be the line of $\mathcal{P}$ through A. By Theorem

11.5.14, $\rho_b\rho_c\rho_l = \rho_m$ for some $m \in \mathscr{P}$. Let $\mathscr{P}'$ be any pencil containing a and m ($\mathscr{P}'$ is unique if $a \neq m$). Let n be a line of $\mathscr{P}'$ through A. (n is unique if $\mathscr{P}' \neq \mathscr{P}_A$.) Then

$$\varphi = \rho_a\rho_b\rho_c\rho_d = \rho_a(\rho_b\rho_c\rho_l)\rho_l\rho_d = \rho_a\rho_m\rho_l\rho_d = (\rho_a\rho_m\rho_n)(\rho_n\rho_l\rho_d).$$

Since $a, m, n \in \mathscr{P}'$, $\rho_a\rho_m\rho_n$ is a reflection across a line of $\mathscr{P}'$. Since n, l, d are in the pointed pencil $\mathscr{P}_A$, $\rho_n\rho_l\rho_d$ is a reflection across a line of $\mathscr{P}_A$. Hence φ is a double reflection (or the identity if these two reflections are the same). □

Corollary 11.5.17. *The composition of three reflections is not a double reflection.*

PROOF. If $\rho_a\rho_b\rho_c = \rho_l\rho_m$ then $\rho_a\rho_b\rho_c\rho_m = \rho_l$. By the previous theorem $\rho_a\rho_b\rho_c\rho_m$ is a double reflection. But a double reflection is not a reflection (Corollary 11.5.9) so that $\rho_a\rho_b\rho_c$ cannot be a double reflection. □

PROBLEM SET 11.5

Part A.

Throughout this set all geometries are assumed to be neutral.

1. Prove Theorem 11.5.1.

2. Prove Theorem 11.5.2.

3. Let $l = {}_aL$ and $m = {}_bL$ be two distinct type I lines in $\mathscr{H}$ so that $\varphi = \rho_l\rho_m$ is a parallel displacement. Find a (simple) formula for $\varphi(x, y)$ where $(x, y) \in \mathbb{H}$.

4. Let $\varphi = \rho_l\rho_m$ be a double reflection and A a point with $\varphi A \neq A$. If n is the perpendicular bisector of $\overline{A\varphi A}$ prove that l, m, n all belong to the same pencil.

5. Let φ be a rotation about A and let ψ be a rotation about B. If $A \neq B$ prove that $\varphi\psi \neq \psi\varphi$.

6. Prove Corollary 11.5.6.

7. Prove the Euclidean part of Theorem 11.5.8.

8. Let φ be a translation along l and let ψ be a translation along m with $l \neq m$. Prove that $\varphi\psi = \psi\varphi$ if and only if the geometry is Euclidean.

9. Prove Corollary 11.5.9.

10. Let $\mathscr{P}$ and $\mathscr{P}'$ be two pencils. By considering cases, describe $\mathscr{P} \cap \mathscr{P}'$.

11. Prove Theorem 11.5.12.

12. If $l, m, n, p \in \mathscr{P}_C$ with $l \perp m$ and $n \perp p$ prove that $\rho_l\rho_m = \rho_n\rho_p$.

13. Let l, m, n be the perpendicular bisectors of $\overline{AB}$, $\overline{BC}$ and $\overline{AC}$ respectively. Prove that $\rho_l \rho_m \rho_n$ is a reflection in a line through A.

14. If l, m, n do not belong to the same pencil then prove that $\rho_l \rho_m \rho_n$ is not a reflection.

15. Let φ and ψ be two double reflections associated with the pencil $\mathscr{P}$ (i.e., $\varphi = \rho_l \rho_m$ with l, $m \in \mathscr{P}$). If $\varphi A = \psi A$ and $\mathscr{P} \neq \mathscr{P}_A$ then prove that $\varphi = \psi$.

16. Let φ and ψ be two double reflections and suppose $A \neq B$. If $\varphi A = \psi A$ and $\varphi B = \psi B$ prove that $\varphi = \psi$. (Thus double reflections which agree on two distinct points are equal.)

17. Given $\triangle ABC$ let φ be the translation along $\overrightarrow{AB}$ such that $\varphi A = B$, let ψ be the translation along $\overrightarrow{BC}$ with $\psi B = C$, and let τ be the translation along $\overrightarrow{AC}$ with $\tau C = A$. Prove that $\sigma = \tau \psi \varphi$ is a rotation about A and that if $P \neq A$ then $m \angle PA\sigma P = \delta(\triangle ABC)$. (This result is a special case of an important result in differential geometry called the Gauss-Bonnet Formula. See Millman-Parker [1977].)

11.6 The Classification of Isometries

As mentioned before, we want to classify isometries according to their geometric properties. A classification theorem will partition the set of isometries of a neutral geometry into a collection of disjoint geometrically meaningful subsets. A simple but rather useless classification is given by {identity} $\cup$ {all other isometries}. The classification which we will eventually prove will show that every isometry is either the identity, a reflection, a rotation, a translation, a parallel displacement, or a glide. (Glides will be defined later.) Our first approach to the classification theorem will be a parity check involving the number of reflections needed to write an isometry. By Corollary 11.3.6 every isometry can be written as a composition of (three or fewer) reflections.

Definition. An isometry φ is an **even isometry** if it can be written as the composition of an even number of reflections. φ is an **odd isometry** if it can be written as a composition of an odd number of reflections.

We wish to prove that an isometry is either even or odd but not both.

Theorem 11.6.1. *Every even isometry is either the identity or a double reflection. Every odd isometry is either a reflection or a product of three reflections. An isometry cannot be both even and odd.*

PROOF. Suppose that φ can be written as a composition of k reflections. If $k \geq 4$ then we can use Theorem 11.5.16 to rewrite φ as a composition of $k - 2$ or $k - 4$ reflections. This may be repeated until we have φ written as a composition of three or fewer reflections. This will not affect the even- or oddness of φ since k, $k - 2$, and $k - 4$ are either all odd or all even. Thus every even isometry is either the identity or a double reflection while every odd isometry is either a reflection or the composition of three reflections. Since the latter two are neither the identity nor double reflections (Corollary 11.5.9 and Theorem 11.5.17) the theorem is proved.   □

The next result is left as Problem A1.

Theorem 11.6.2. *If* φ *is an isometry then* $\varphi \rho_l \varphi^{-1} = \rho_{\varphi(l)}$.

Definition. An **involution** is an isometry $\varphi \neq$ identity such that $\varphi = \varphi^{-1}$. A **half-turn** about the point A is a rotation about A which is also an involution.

Every reflection is an involution. Intuitively a half-turn is a "rotation through 180 degrees." We will show in Theorem 11.6.5 that every involution is either a half-turn or a reflection. This will use the following fact which says that distinct reflections commute exactly when they are across perpendicular lines.

Theorem 11.6.3. *If* $l \neq m$ *then* $\rho_l \rho_m = \rho_m \rho_l$ *if and only if* $l \perp m$.

PROOF. Let $\varphi = \rho_m$. If $l \perp m$ then $\varphi(l) = l$ so that by Theorem 11.6.2, $\rho_l = \rho_{\varphi(l)} = \varphi \rho_l \varphi^{-1} = \rho_m \rho_l \rho_m^{-1}$. Hence $\rho_l \rho_m = \rho_m \rho_l$. On the other hand, if $\rho_l \rho_m = \rho_m \rho_l$ then $\rho_l = \rho_m \rho_l \rho_m^{-1} = \rho_{\varphi(l)}$ so that $\varphi(l) = l$. Since $l \neq m$, we must have $l \perp m$ by Theorem 11.5.1.   □

Theorem 11.6.4. *A double reflection* $\varphi = \rho_l \rho_m$ *is an involution if and only if* $l \perp m$. *In this case* φ *is a half-turn. For each point* A *there is a unique half-turn* η_A *about* A.

PROOF. $\rho_l \rho_m$ is an involution if and only if $l \neq m$ and $\rho_l \rho_m = (\rho_l \rho_m)^{-1}$. Since

$$(\rho_l \rho_m)^{-1} = \rho_m^{-1} \rho_l^{-1} = \rho_m \rho_l$$

the first assertion follows from Theorem 11.6.3. If $l \perp m$ then $\rho_l \rho_m$ is a half-turn about the point where l and m intersect.

For each A we can find lines l and m which are perpendicular at A. Hence there is at least one half-turn $\rho_l \rho_m$ about A. Suppose that $\rho_n \rho_p$ is also a half-turn about A. By Corollary 11.5.15 there is a reflection ρ_s with $\rho_n \rho_p = \rho_l \rho_s$. Since $\rho_n \rho_p = \rho_l \rho_s$ is a half-turn about A, $l \perp s$ and $l \cap s = \{A\}$. Thus $s = m$ and $\rho_l \rho_m = \rho_l \rho_s = \rho_n \rho_p$. Hence there is only one half-turn about A.   □

Theorem 11.6.5. *Every involution* φ *is either a half-turn or a reflection.*

PROOF. Let A be any point with $A \neq \varphi A$. Since $\varphi \varphi A = A$, φ interchanges A and φA. Hence φ fixes the midpoint M of $\overline{A \varphi A}$.

If M is the only fixed point of φ then φ is a rotation about M by Corollary 11.5.5 and is thus a half-turn since it is an involution. If φ also fixes the point N then φ is the reflection in the line $\overline{MN}$ by Theorem 11.3.4. $\square$

We are now able to determine the invariant pencils of a rotation. The proof of the following result is left to Problem A8.

Theorem 11.6.6. *A rotation which is not a half-turn does not have any invariant lines. The invariant lines of a half-turn about A are the lines through A. The only invariant pencil of a rotation which is not a half-turn is the pointed pencil associated with the fixed point of the rotation. The invariant pencils of the half-turn η_A are $\mathscr{P}_A$ and all the parallel pencils $\mathscr{P}_l$ for lines l with $A \in l$.*

We already know that every even isometry is either a rotation, a translation, a parallel displacement, or is the identity. The only isometries we have not really studied are the triple reflections $\rho_l \rho_m \rho_n$ when l, m, n do not belong to the same pencil. The first step will be to identify a special type of triple reflection called a glide. A glide along a line l will consist of a translation along l followed by a reflection across l. In Figure 11-32 we see the result of applying a glide several times to a geometric figure. The result reminds us of the gliding strokes of an ice skater.

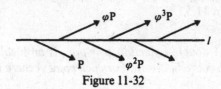

Figure 11-32

Definition. A **glide along** l is an isometry $\varphi = \rho_l \rho_m \rho_n$ where $\rho_m \rho_n$ is a translation along l.

Theorem 11.6.7. *If $\varphi = \rho_l \rho_m \rho_n$ is a glide along l then*

(i) $\rho_l \rho_m \rho_n = \rho_m \rho_n \rho_l$
(ii) *φ has no fixed points.*
(iii) *l is the only invariant line of φ.*
(iv) *φ interchanges the half planes determined by l.*

PROOF.
(i) Since $\rho_m \rho_n$ is a translation along l both m and n are perpendicular to l. Hence by Theorem 11.6.3, $\rho_m \rho_l = \rho_l \rho_m$ and $\rho_n \rho_l = \rho_l \rho_n$ so that

$$\rho_l \rho_m \rho_n = \rho_m \rho_l \rho_n = \rho_m \rho_n \rho_l.$$

(ii) If φ fixes exactly one point then it is a rotation by Corollary 11.5.5. But then φ is both even and odd which is impossible. If φ fixes two points A and B then it fixes every point on the line $\overline{AB}$. φ is not the identity so that by Theorem 11.3.4, $\rho_l\rho_m\rho_n = \varphi = \rho_t$ for some t. Then $\rho_l\rho_m = \rho_t\rho_n^{-1} = \rho_t\rho_n$ so that by Theorem 11.5.13, l, m, n and t belong to the same pencil. Since $m \neq n$, $m \perp l$, and $n \perp l$, this pencil must be the parallel pencil $\mathscr{P}_l$. This is impossible because $l \notin \mathscr{P}_l$, and φ cannot have any fixed points.

(iii, iv) Clearly φ leaves l invariant and interchanges the half planes of l. If $t \neq l$ is left invariant and $A \in t$ then $\varphi A \neq A$ and $t = \overline{A\varphi A}$. $A \notin l$ so that A and φA are on opposite sides of l. Thus t intersects l at a point B. Since both l and t are invariant, $\varphi B \in l \cap t$ so that $\varphi B = B$. This contradicts the second part of the theorem so that l must be the only invariant line of φ.   □

We leave to Problem A9 the determination of which pencils are invariant under a glide.

Theorem 11.6.8. *If r is a line and $B \notin r$ then $\rho_r\eta_B$ is a glide.*

PROOF. Let m be the perpendicular to r through B and let n be the perpendicular to m through B as in Figure 11-33. Since $B \notin r$, $n \neq r$ and $\rho_r\rho_n$ is a translation along m while the half-turn η_B is $\eta_B = \rho_n\rho_m$. Hence

$$\rho_m(\rho_r\rho_n) = (\rho_r\rho_n)\rho_m = \rho_r(\rho_n\rho_m) = \rho_r\eta_B$$

and $\rho_r\eta_B$ is a glide along m.   □

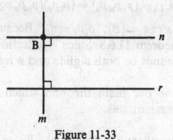

Figure 11-33

The next result classifies all odd isometries.

Theorem 11.6.9. *A triple reflection $\varphi = \rho_l\rho_m\rho_n$ is either a reflection or a glide but not both.*

PROOF. If l, m, n belong to the same pencil then φ is a reflection by Theorem 11.5.14. Hence we will assume that l, m, n do not belong to the same pencil and will show that φ is a glide. We do this by rewriting φ in a different manner in Equation (6-3).

Let $A \in l$ and let $p \neq l$ be the line through A that is in the pencil $\mathscr{P}$ determined by m and n. Then there is a $t \in \mathscr{P}$ with

$$\rho_p \rho_m \rho_n = \rho_t. \tag{6-1}$$

See Figure 11-34. Let B be the foot of the perpendicular s from A to t. Since l, p, and s all intersect at A, they all belong to the same pencil $\mathscr{P}_A$. Thus there is an $r \in \mathscr{P}_A$ with

$$\rho_l \rho_p \rho_s = \rho_r. \tag{6-2}$$

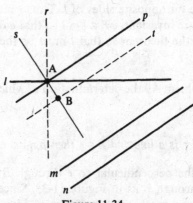

Figure 11-34

Now $B \notin r$ (or else $r = \overrightarrow{AB} = s$, and Equation (6-2) implies $\rho_l = \rho_p$ so that $l = p$). Thus, using Equations (6-1) and (6-2) we have

$$\rho_r \rho_s \rho_t = (\rho_l \rho_p \rho_s) \rho_s (\rho_p \rho_m \rho_n) = (\rho_l \rho_p)(\rho_p \rho_m \rho_n) = \rho_l \rho_m \rho_n = \varphi. \tag{6-3}$$

Since $s \perp t$ and $s \cap t = \{B\}$, $\rho_s \rho_t = \eta_B$. Because $B \notin r$, $\varphi = \rho_r \rho_s \rho_t = \rho_r \eta_B$ is a glide by Theorem 11.6.8. Since a reflection has a fixed point and a glide does not, φ cannot be both a glide and a reflection. □

We can now prove our main theorem which classifies all isometries based upon their invariant sets.

Theorem 11.6.10 (Classification Theorem). *Every isometry of a neutral geometry is exactly one of the following*

(i) *identity*   (ii) *reflection*
(iii) *rotation*   (iv) *translation*
(v) *parallel displacement*   (vi) *glide*

PROOF. If φ is an isometry then it is either even or odd but not both. If φ is even it is the identity or a double reflection. By Corollary 11.5.9 every double reflection is either a rotation, a translation, or a parallel displacement but not any two of these. If φ is odd then by Theorem 11.6.9, φ is either a reflection or a glide but not both. □

Our classification theorem was proved by considering isometries as compositions of reflections. Interesting theorems regarding even isometries can be found by considering double reflections as the basic building blocks.

Theorem 11.6.11. *An isometry φ is a translation if and only if it is the composition of two distinct half-turns.*

PROOF. Let $\varphi = \eta_A \eta_B$ be the composition of two half-turns with $A \neq B$. Let $l = \overleftrightarrow{AB}$, let m be perpendicular to l at A, and let n be perpendicular to l at B. Then

$$\varphi = \eta_A \eta_B = (\rho_m \rho_l)(\rho_l \rho_n) = \rho_m \rho_n$$

so that φ is a translation along l.

On the other hand, if $\varphi = \rho_m \rho_n$ is a translation along l then there are points A and B with m perpendicular to l at A and with n perpendicular to l at B. Since $m \neq n$ we have $A \neq B$ and $\varphi = \rho_m \rho_n = \rho_m \rho_l \rho_l \rho_n = \eta_A \eta_B$. □

Theorem 11.6.12. *In a Euclidean geometry the composition of three half-turns $\eta_A \eta_B \eta_C$ is a half-turn η_D. In a hyperbolic geometry $\eta_A \eta_B \eta_C$ is a half-turn if and only if A, B, C are collinear.*

PROOF. If A, B, C all lie on the line p and if l, m, n are perpendicular to p at A, B, C then

$$\eta_A \eta_B \eta_C = (\rho_l \rho_p)(\rho_p \rho_m)(\rho_n \rho_p) = \rho_l \rho_m \rho_n \rho_p = \rho_s \rho_p$$

for some $s \perp p$ by Theorem 11.5.14. Let $s \cap p = \{D\}$. Then $\eta_A \eta_B \eta_C = \eta_D$.

Now assume that A, B, C are not collinear. Let $l = \overleftrightarrow{AB}$ and let E be the foot of the perpendicular m from C to l. The isometry $\varphi = \eta_A \eta_B$ may be written, thanks to Theorem 11.6.11, as $\rho_n \rho_m$ for some $n \perp l$. Let F be the point where n intersects l. See Figure 11-35. Let p be the line through C perpendicular to m so that $\eta_C = \rho_m \rho_p$. Then

$$\eta_A \eta_B \eta_C = \varphi \eta_C = (\rho_n \rho_m)(\rho_m \rho_p) = \rho_n \rho_p.$$

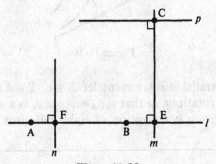

Figure 11-35

Thus for A, B, C noncollinear $\eta_A \eta_B \eta_C$ will be a half-turn if and only if $n \perp p$.

Now $n \perp l$, $l \perp m$, and $m \perp p$. If $n \cap p = \{D\}$ then $\square CEFD$ is a Lambert quadrilateral. If the geometry is Euclidean $n \perp p$ and $\eta_A \eta_B \eta_C = \rho_n \rho_p = \eta_D$. If the geometry is hyperbolic the rotation $\rho_n \rho_p$ is not a half-turn since $\angle FDC$ is not a right angle. If n is divergently parallel to p then $\rho_n \rho_p$ is a translation and if n is asymptotically parallel to p then $\rho_n \rho_p$ is a parallel displacement. Hence in the Euclidean case $\eta_A \eta_B \eta_C$ is always a half-turn and in the hyperbolic case it is a half-turn if and only if A, B, and C are collinear. $\square$

Note that, courtesy of Theorem 11.6.11 every translation is the composition of half-turns. By Theorems 11.6.11 and 11.6.12 the only rotations in a Euclidean geometry which are the compositions of half-turns are half-turns. However, the situation is different in the hyperbolic case.

Theorem 11.6.13. *In a hyperbolic geometry, every double reflection is a composition of half-turns.*

PROOF. We know the result is true for translations (Theorem 11.6.11). Suppose that $\rho_a \rho_b$ is a rotation about C (so $a \neq b$). Let $a = \overleftrightarrow{AC}$, $b = \overleftrightarrow{BC}$, and l be the line of enclosure of $\angle ACB$. Set D to be the foot of the perpendicular from C to l and choose E with $C{-}D{-}E$. Let m be perpendicular to $\overleftrightarrow{CD}$ at E. $l|a$ and $l|b$ while m is divergently parallel to both a and b. See Figure 11-36. Hence $\rho_a \rho_m$ and $\rho_m \rho_b$ are translations since divergently parallel lines have a common perpendicular. Then $\rho_a \rho_b = \rho_a \rho_m \rho_m \rho_b$ is a composition of two translations and hence a composition of four half-turns.

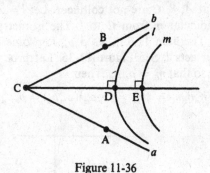

Figure 11-36

If $\rho_c \rho_d$ is a parallel displacement let $X \in c$, $Y \in d$ and $l = \overleftrightarrow{XY}$. Then $\rho_c \rho_l$ and $\rho_l \rho_d$ are rotations so that $\rho_c \rho_d = \rho_c \rho_l \rho_l \rho_d$ is a composition of two rotations and hence a composition of eight half-turns. $\square$

PROBLEM SET 11.6

Part A.

Throughout this set all geometries are assumed to be neutral.

1. Prove Theorem 11.6.2.

2. If l is a line and $A \notin l$ then prove that $\eta_A(l)$ is divergently parallel to l.

3. Prove that if φ is an isometry then $\varphi \eta_A \varphi^{-1} = \eta_{\varphi A}$.

4. If φ is an isometry prove that $\varphi \rho_l = \rho_l \varphi$ if and only if $\varphi(l) = l$.

5. If φ is an isometry prove that $\varphi \eta_A = \eta_A \varphi$ if and only if $\varphi A = A$.

6. Let φ be a rotation about A. If the line l is invariant under φ prove that $\varphi = \eta_A$ and $A \in l$.

7. If $\rho_l \rho_m \rho_n = \rho_n \rho_m \rho_l$ prove that l, m, n belong to the same pencil.

8. Prove Theorem 11.6.6.

9. If φ is a glide along l, what are the invariant pencils of φ?

10. If $P \neq Q$ how many rotations send P to Q? Half-turns? Translations? Parallel displacements? Reflections? Glides? (Be careful to distinguish between the Euclidean and hyperbolic cases.)

11. If φ and ψ are glides along l, prove that $\varphi \psi$ is either a translation along l or the identity.

12. If $\varphi = \rho_l \rho_m \rho_n$ prove that φ^2 is either a translation or the identity.

13. If φ is a glide along l and $B \in l$ prove there is a line r with $\varphi = \rho_r \eta_B$.

14. If φ is a glide along l and ψ is a glide along m with $l \perp m$ prove that $\varphi \psi$ is a half-turn if and only if the geometry is Euclidean.

15. Prove Hjelmslev's Lemma: Let φ be a glide along l and let m be a line which is not invariant under φ. If $P \in m$ then the midpoint of $\overline{P \varphi P}$ lies on l. If Q is any other point of m, the midpoint of $\overline{Q \varphi Q}$ is the same as the midpoint of $\overline{P \varphi P}$ if and only if $l \perp m$.

16. Given a glide φ how would you find the invariant line l along which φ is a glide?

17. Prove Hjelmslev's Theorem: If φ is an isometry and m is a line which is not invariant then there is a line n such that for any $P \in m$, the midpoint of $\overline{P \varphi P}$ is on n. Furthermore, the midpoints of all such segments $\overline{P \varphi P}$ with $P \in m$ are either distinct or are all the same.

18. Is Hjelmslev's Theorem still true if m is an invariant line of φ?

Part B. "Prove" may mean "find a counterexample".

19. Prove that any translation is the composition of two rotations.

20. Prove that any translation is the composition of two glides.

21. In a hyperbolic geometry, prove that any translation is the composition of two parallel displacements.

22. In a hyperbolic geometry prove that any rotation is the composition of two parallel displacements.

23. In a hyperbolic geometry prove that any rotation is the composition of two translations.

24. In a hyperbolic geometry prove that any parallel displacement is the composition of two rotations.

25. In a hyperbolic geometry prove that any parallel displacement is the composition of two translations.

26. In a hyperbolic geometry prove that any even isometry is the composition of three half-turns.

27. In a Euclidean geometry show that a rotation is never the composition of two translations.

11.7 The Isometry Group

We shall see in this section that the collection of all isometries of a metric geometry forms a special algebraic structure called a group. Groups are a major object of study in a course in abstract algebra. (For example, see Herstein [1990] or McCoy-Janusz [1987].) We shall assume in this section that the reader has some familiarity with the subject and will give only a few introductory words about the theory of groups. This language will then be applied to the group of isometries of a neutral geometry.

Definition. A **group** $\{G, \cdot\}$ is a set G together with an operation, $\cdot$, for combining elements of G such that

(i) if $g, h \in G$ then $g \cdot h \in G$;

(ii) if $g, h, k \in G$ then $(g \cdot h) \cdot k = g \cdot (h \cdot k)$;

(iii) there is an element $e \in G$, called the **identity**, such that $e \cdot g = g \cdot e = g$ for every $g \in G$;

(iv) for each $g \in G$ there is an element $g^{-1} \in G$, called an **inverse** of g, such that $g \cdot g^{-1} = g^{-1} \cdot g = e$.

Among the first results in group theory are the facts that each group has exactly one identity element and each element has a *unique* inverse. Furthermore, in order to show that h is the inverse of g it is only necessary to show that $gh = e$ (and not $hg = e$ also). Simple examples of groups are given by the integers under addition ($e = 0$ and inverse means negative) and the positive real numbers under multiplication ($e = 1$ and inverse means reciprocal). Both of these examples have the property that they are *commutative*: the order two elements are combined does not matter. This is not typical of groups in general.

Example 11.7.1. Show that the set $SL(2, \mathbb{R})$ of 2×2 matrices with real entries and determinant $+1$ form a group under matrix multiplication.

SOLUTION. Recall that the product of two matrices $A = \begin{pmatrix} a & b \\ c & d \end{pmatrix}$ and $B = \begin{pmatrix} e & f \\ g & h \end{pmatrix}$ is given by

$$AB = \begin{pmatrix} ae + bg & af + bh \\ ce + dg & cf + dh \end{pmatrix}.$$

If $\det(A) = (ad - bc) = 1$ and $\det(B) = (eh - fg) = 1$ then a straightforward calculation shows that

$$\det(AB) = (ae + bg)(cf + dh) - (af + bh)(ce + dg)$$
$$= (ad - bc)(eh - fg) = 1 \cdot 1 = 1.$$

Thus the product of two elements of $SL(2, \mathbb{R})$ is in $SL(2, \mathbb{R})$ and axiom (i) is satisfied.

A straightforward but tedious calculation shows that $(AB)C = A(BC)$ so that axiom (ii) is satisfied. The identity element is $I = \begin{pmatrix} 1 & 0 \\ 0 & 1 \end{pmatrix}$. Finally the inverse of $\begin{pmatrix} a & b \\ c & d \end{pmatrix}$ is $\begin{pmatrix} d & -b \\ -c & a \end{pmatrix}$ because

$$\begin{pmatrix} a & b \\ c & d \end{pmatrix}\begin{pmatrix} d & -b \\ -c & a \end{pmatrix} = \begin{pmatrix} ad - bc & -ab + ab \\ cd - cd & -bc + ad \end{pmatrix} = \begin{pmatrix} 1 & 0 \\ 0 & 1 \end{pmatrix} = I.$$

Note that in this example, AB need not equal BA; e.g., let $A = \begin{pmatrix} 1 & 1 \\ 0 & 1 \end{pmatrix}$ and $B = \begin{pmatrix} 2 & 0 \\ 0 & \frac{1}{2} \end{pmatrix}$.   □

The group $SL(2, \mathbb{R})$ will be very important when we study the collection of isometries of the Poincaré Plane $\mathcal{H}$. We will see that it is essentially the set of even isometries of $\mathcal{H}$.

Theorem 11.7.2. *The collection of all collineations of an incidence geometry with itself forms a group under composition of functions.*

PROOF. Let $\varphi : \mathcal{S} \to \mathcal{S}$ and $\psi : \mathcal{S} \to \mathcal{S}$ be collineations. Then $\varphi \circ \psi$ is a collineation by Problem A19 of Section 11.1. Thus axiom (i) is satisfied.

The composition of functions is associative so that axiom (ii) is satisfied. The identity element is given by the identity function

$$\text{id} : \mathcal{S} \to \mathcal{S} \quad \text{by} \quad \text{id}(P) = P \quad \text{for all } P \in \mathcal{S},$$

which is clearly a collineation.

If φ is a collineation let φ^{-1} denote the inverse of the bijection φ. By Problem A4 of Section 11.1, φ^{-1} is also a collineation. Hence the set of collineations of $\mathcal{S}$ forms a group.   □

We leave the proof of the next result as Problem A1.

Theorem 11.7.3. *If $\mathcal{G} = \{\mathcal{S}, \mathcal{L}, d\}$ is a metric geometry then the set $\mathcal{I}(\mathcal{G})$ of all bijections of $\mathcal{S}$ which preserve distance forms a group under composition of functions. In particular, the set of isometries of a neutral geometry forms a group under composition.*

Definition. The **isometry group** of a neutral geometry $\mathcal{N} = \{\mathcal{S}, \mathcal{L}, d, m\}$ is the group $\mathcal{I}(\mathcal{N})$ of all isometries of $\mathcal{N}$.

The isometry group of a neutral geometry is a group in its own right and also a subset of the group of all collineations. This arrangement has a formal name.

Definition. If $\{G, \cdot\}$ is a group and if H is a subset of G, then H is a **subgroup** of G if $\{H, \cdot\}$ is a group.

Theorem 11.7.4. *The set of all even isometries of a neutral geometry $\mathcal{N}$ is a subgroup of the isometry group $\mathcal{I}(\mathcal{N})$.*

Theorem 11.7.5. *Let $\mathcal{P}$ be a pencil in a neutral geometry and set*

$$\mathcal{D} = \{\rho_l \rho_m \,|\, l \text{ and } m \text{ are in } \mathcal{P}\}.$$

Then $\mathcal{D}$ is a subgroup of the isometry group and is commutative.

PROOF. Suppose $\varphi = \rho_l \rho_m$ and $\psi = \rho_n \rho_p$ belong to $\mathcal{D}$. Then by Corollary 11.5.15 there is a $t \in \mathcal{P}$ with $\psi = \rho_n \rho_p = \rho_m \rho_t$. Hence $\varphi \psi = \rho_l \rho_m \rho_n \rho_p = \rho_l \rho_m \rho_m \rho_t = \rho_l \rho_t \in \mathcal{D}$. Hence axiom (i) is satisfied. Axiom (ii) is trivial and id $= \rho_l \rho_l \in \mathcal{D}$ if $l \in \mathcal{P}$ so axiom (iii) is satisfied.

If $\varphi = \rho_l \rho_m \in \mathcal{D}$ then $\varphi^{-1} = \rho_m^{-1} \rho_l^{-1} = \rho_m \rho_l \in \mathcal{D}$ so that φ has an inverse in $\mathcal{D}$. Hence $\mathcal{D}$ is a subgroup of the isometry group.

Finally we must show that if $\varphi, \psi \in \mathcal{D}$ then $\varphi \psi = \psi \varphi$. Let $\varphi = \rho_l \rho_m$ and $\psi = \rho_n \rho_p$. By Corollary 11.5.15 there are lines $s, t \in \mathcal{P}$ with $\psi = \rho_s \rho_l = \rho_m \rho_t$, $\psi \varphi = \rho_s \rho_l \rho_l \rho_m = \rho_s \rho_m$. Thus since $\psi^{-1} = (\rho_m \rho_t)^{-1} = \rho_t \rho_m$,

$$\psi \varphi = \rho_s \rho_m = \rho_s (\psi \psi^{-1}) \rho_m = \rho_s (\rho_s \rho_l \rho_t \rho_m) \rho_m$$
$$= \rho_l \rho_t = \rho_l \rho_m \rho_m \rho_t = \varphi \psi.$$

Hence l is commutative. $\square$

By taking different kinds of pencils in Theorem 11.7.5 we get subgroups of rotations, translations or parallel displacements.

Theorem 11.7.6. *In a Euclidean geometry the set of all translations together with the identity form a commutative subgroup of the isometry group.*

The statement of Theorem 11.7.6 needs the words "together with the identity" because by definition the identity is not a translation. However, it is the product of two reflections: Id $= \rho_l \rho_l$.

Definition. Let S be a subset of a group G and let H be a subgroup of G which contains S. Then H is **generated** by S if every subgroup of G which contains S also contains H. (This is the same as saying that every element of H can be written as a product of elements of S or their inverses.)

Theorem 11.7.7. *In a hyperbolic geometry the subgroup H of even isometries is generated by the set of half-turns.*

PROOF. By Theorem 11.6.13 every double reflection is a product of half-turns, as is the identity. Hence H is generated by the set of half-turns.   □

There are several other special sets that generate the even isometries. See Problems A6, A7, and A8.

The isometry group of a neutral geometry is very large. The classification theorem partitions it into subsets (*not* subgroups) according to certain geometric properties. These geometric properties are reflected as certain algebraic properties involving the order of a group element.

Definition. If $\{G, \cdot\}$ is a group and $g \in G$ then g has **finite order** if $g^n = e$ for some positive integer n. In this case the **order** of g is the smallest positive integer r such that $g^r = e$.

Clearly the identity has order 1 and every involution has order 2. If n is an integer greater than 2 then it is possible to find a rotation with order n (Problem A9). However, not every rotation has finite order (Problem A10). The next result, whose proof is left to Problem A11, says that if an isometry φ has order $n > 2$ then φ must be a rotation.

Theorem 11.7.8. *In a neutral geometry if φ is either a translation, a glide, or a parallel displacement then φ does not have finite order.*

Just as two geometries are equivalent if they are isometric, there is a notion of equivalence of groups. This is given by saying that there is a bijection that preserves the group operation.

Definition. Two groups $\{G, \cdot\}$ and $\{G', *\}$ are **isomorphic** if there is a bijection $f: G \to G'$ such that $f(g_1 \cdot g_2) = f(g_1) * f(g_2)$ for every two elements g_1, g_2 of G.

In the last section we will determine the isometry groups of $\mathscr{E}$ and $\mathscr{H}$. Because of the next result we will then know the (abstract) structure of the isometry group of any neutral geometry.

Theorem 11.7.9. *Let $\mathscr{N}_1, \mathscr{N}_2$ be two neutral geometries and let φ be an isometry from $\mathscr{N}_1$ to $\mathscr{N}_2$. Then the isometry groups of $\mathscr{N}_1$ and $\mathscr{N}_2$ are isomorphic.*

PROOF. If $\varphi: \mathcal{S}_1 \to \mathcal{S}_2$ is an isometry we define a function $f: \mathcal{I}(\mathcal{N}_1) \to \mathcal{I}(\mathcal{N}_2)$ by

$$f(\sigma) = \varphi\sigma\varphi^{-1} \quad \text{if } \sigma \in \mathcal{I}(\mathcal{N}_1).$$

Note first that $f(\sigma) \in \mathcal{I}(\mathcal{N}_2)$ because it is the composition of isometries and $f(\sigma): \mathcal{N}_2 \to \mathcal{N}_2$. We must show that f is a bijection and that $f(\sigma_1\sigma_2) = f(\sigma_1)f(\sigma_2)$ for all $\sigma_1, \sigma_2 \in \Phi(\mathcal{N}_1)$.

If $f(\sigma_1) = f(\sigma_2)$ then $\varphi\sigma_1\varphi^{-1} = \varphi\sigma_2\varphi^{-1}$ so that

$$\varphi^{-1}(\varphi\sigma_1\varphi^{-1})\varphi = \varphi^{-1}(\varphi\sigma_2\varphi^{-1})\varphi \quad \text{or} \quad \sigma_1 = \sigma_2.$$

Hence f is injective.

If $\tau \in \mathcal{I}(\mathcal{N}_2)$ let $\sigma = \varphi^{-1}\tau\varphi$. $\sigma \in \mathcal{I}(\mathcal{N}_1)$ because it is the composition of isometries and $\sigma: \mathcal{N}_1 \to \mathcal{N}_1$. Clearly $f(\sigma) = \tau$ so that f is surjective and hence a bijection.

Finally, if $\sigma_1, \sigma_2 \in \mathcal{I}(\mathcal{N}_1)$ then

$$f(\sigma_1\sigma_2) = \varphi\sigma_1\sigma_2\varphi^{-1} = \varphi\sigma_1\varphi^{-1}\varphi\sigma_2\varphi^{-1}$$
$$= f(\sigma_1)f(\sigma_2).$$

Hence $\mathcal{I}(\mathcal{N}_1)$ is isomorphic to $\mathcal{I}(\mathcal{N}_2)$.   □

PROBLEM SET 11.7

Part A.

1. Prove Theorem 11.7.3.

2. Prove Theorem 11.7.4.

3. Prove Theorem 11.7.6.

4. Prove that the isometry group of a neutral geometry is generated by the set of reflections.

5. Prove that the isometry group of a neutral geometry is generated by the set of glides.

6. Prove that the subgroup of even isometries of a neutral geometry is generated by the set of rotations.

7. Prove that the subgroup of even isometries of a hyperbolic geometry is generated by the set of translations.

8. Prove that the subgroup of even isometries of a hyperbolic geometry is generated by the set of parallel displacements.

9. Let A be a point in a neutral geometry and let n be an integer greater than 2. Find a rotation about A whose order is n.

10. Let A be a point in a neutral geometry. Find a rotation about A which does not have finite order.

11. Prove Theorem 11.7.8.

12. If $\mathscr{E}_1$ and $\mathscr{E}_2$ are Euclidean geometries prove that $\mathscr{I}(\mathscr{E}_1)$ is isomorphic to $\mathscr{I}(\mathscr{E}_2)$.

Part B. "Prove" may mean "find a counterexample".

13. If N is a subgroup of G then N is a **normal** subgroup if $gng^{-1} \in N$ for every $g \in G$ and $n \in N$. Prove that the subgroup of even isometries of a neutral geometry is a normal subgroup.

14. Prove that the subgroup of even isometries of a neutral geometry is generated by the set of translations.

15. A **transformation** of a neutral geometry is a collineation which preserves betweenness, segment congruence, and angle congruence. Prove that the set of transformations of a neutral geometry forms a group. Is it the same as the isometry group?

The remaining problems completely determine the finite subgroups of the isometry group. A group is **finite** if it has only a finite number of elements. A group is **cyclic** if it is generated by a single element g. (In this case the group consists just of powers, positive and negative, of g.) A finite group is a **dihedral group** if it is generated by two elements g and h such that
 (i) g has finite order n for some n
 (ii) h has order 2
 (iii) $gh = hg^{n-1}$.

16. Prove that a finite group of isometries does not contain translations, parallel displacements, or glides.

17. If a finite group of isometries contains more than one rotation, they all have the same center. (Hint: Suppose that φ is a rotation about A and ψ is a rotation about B with $A \neq B$. Let $l = \overrightarrow{AB}$, show that $\psi^{-1}\varphi^{-1}\psi\varphi$ can be written as $(\rho_n\rho_l\rho_m)^2$ and use Problem A12 of Section 11.6.)

18. If G is a finite group of isometries with more than 2 elements, then prove there is a unique point fixed by each of the isometries in G.

19. If a finite group G of isometries does not contain any reflections then prove G is a cyclic group.

20. If a finite group G of isometries contains a reflection then prove that G is a dihedral group.

11.8 The SAS Axiom in $\mathscr{H}$

In this section we shall finally verify that the Poincaré Plane $\mathscr{H} = \{\mathbb{H}, \mathscr{L}_H, d_H, m_H\}$ actually satisfies the Side-Angle-Side Axiom. Because of Theorem 11.3.7, we may prove that $\mathscr{H}$ is a neutral geometry by showing that there is a mirror for every line $l \in \mathscr{L}_H$.

The first step in this program is to introduce an alternative description of $\mathbb{H}$ which comes from viewing $\mathbb{R}^2$ as the set $\mathbb{C}$ of complex numbers. The point $(x, y) \in \mathbb{R}^2$ may be identified with $z = x + iy \in \mathbb{C}$, so that

$$\mathbb{H} = \{z = x + iy \in \mathbb{C} \,|\, y > 0\}.$$

That is, we view $\mathbb{H}$ as the set of complex numbers with positive imaginary part. This observation is the basis for a great deal of advanced mathematics (e.g., geometric function theory, Riemann surface theory, eigenvalue problems for the Laplace operator among others) because it blends complex variables and geometry. We shall describe below the notion of a fractional linear transformation from the theory of complex functions. (The only prerequisite for the material that follows is how to add, subtract, multiply and divide complex numbers.)

Definition. A **real positive fractional linear transformation** (which we abbreviate as FLT) is a function $\varphi : \mathbb{H} \to \mathbb{H}$ of the form

$$\varphi(z) = \frac{az + b}{cz + d}$$

where $a, b, c, d \in \mathbb{R}$ and $ad - bc > 0$.

Note that the defining equation gives $\varphi(z)$ as a complex number. In Problem A1 you will show that the imaginary part of $\varphi(z)$ is positive if $z \in \mathbb{H}$ so that $\varphi(\mathbb{H}) \subset \mathbb{H}$. One of our goals is to show that an FLT is an isometry of $\mathbb{H}$. In fact it will turn out that the set of FLT's is precisely the set of even isometries. To this end we set up a correspondence between matrices and FLT's.

Definition. If $\Phi = \begin{pmatrix} a & b \\ c & d \end{pmatrix}$ is a matrix with a positive determinant, then the FLT **associated** with Φ is

$$\varphi(z) = \frac{az + b}{cz + d}.$$

If we just say that φ is the FLT associated with the matrix Φ then it is assumed that $\det(\Phi) > 0$. Note that the FLT associated with the identity matrix I is the identity FLT, $\varphi(z) = z$. Every 2×2 matrix of positive determinant determines an FLT. On the other hand, it is possible that two different matrices give the same FLT. We will see when this happens in Proposition 11.8.3. First we need a result whose proof is left to Problem A2.

Proposition 11.8.1. *If φ and ψ are the FLT's associated with the matrices Φ and Ψ then $\varphi \circ \psi$ is the FLT associated with the product matrix $\Phi\Psi$.*

Proposition 11.8.2. *If $\varphi : \mathbb{H} \to \mathbb{H}$ is an FLT then φ is a bijection.*

PROOF. Let φ be associated with the matrix Φ. Since det $\Phi > 0$, Φ has an inverse Φ^{-1}, whose determinant is also positive. Let ψ be the FLT determined by Φ^{-1}. Since $\Phi^{-1}\Phi = I = \Phi\Phi^{-1}$, an application of Proposition 11.8.1 shows that both $\psi\varphi$ and $\varphi\psi$ are the identity FLT. Hence ψ is the inverse of φ, and φ is a bijection. $\square$

Proposition 11.8.3. *Let* $\lambda \neq 0$ *and let* φ *and* ψ *be the* FLT*'s associated with* Φ *and* Ψ. *If* $\Psi = \lambda\Phi$ *then* $\varphi = \psi$. *On the other hand, if* Φ *and* Ψ *determine the same* FLT *then* $\Psi = \lambda\Phi$ *for some nonzero real number* λ.

PROOF. First, if $\Phi = \begin{pmatrix} a & b \\ c & d \end{pmatrix}$ then $\Psi = \begin{pmatrix} \lambda a & \lambda b \\ \lambda c & \lambda d \end{pmatrix}$ so that

$$\psi(z) = \frac{\lambda az + \lambda b}{\lambda cz + \lambda d} = \frac{az + b}{cz + d} = \varphi(z)$$

and $\psi = \varphi$.

Next, suppose that the identity FLT $\varphi(z) = z$ is associated with the matrix $\Phi = \begin{pmatrix} a & b \\ c & d \end{pmatrix}$ so that

$$\frac{az + b}{cz + d} = z \quad \text{for all } z \in \mathbb{H}.$$

Thus $az + b = cz^2 + dz$ or

$$cz^2 + (d - a)z - b = 0 \quad \text{for all } z \in \mathbb{H}. \tag{8-1}$$

Since the quadratic polynomial equation in Equation (8-1) has more than two solutions (in fact, all of $\mathbb{H}$), its coefficients must all be zero: $c = 0$, $d - a = 0$, $-b = 0$. Hence $a = d = \lambda$ for some λ and $b = c = 0$. Thus

$$\Phi = \begin{pmatrix} \lambda & 0 \\ 0 & \lambda \end{pmatrix} = \lambda\begin{pmatrix} 1 & 0 \\ 0 & 1 \end{pmatrix} = \lambda I$$

and $\lambda \neq 0$ since $\lambda^2 = \det \Phi \neq 0$.

Finally suppose that both Φ and Ψ determine the same FLT φ. Then by Proposition 11.8.1, $\Psi\Phi^{-1}$ determines the identity FLT, $\varphi\varphi^{-1}$. Hence by the second part of the proof,

$$\Psi\Phi^{-1} = \lambda I \quad \text{or} \quad \Psi = \lambda\Phi \quad \text{for some } \lambda \neq 0. \qquad \square$$

The value of this result is that it allows us to "normalize" the matrix Φ *when desired* either by assuming that $\det(\Phi) = 1$ (i.e., let $\lambda = \sqrt{(\det \Phi)^{-1}}$), that a particular entry of Φ is ≥ 0 ($\lambda = \pm 1$), or that a particular nonzero entry of Φ is 1 ($\lambda =$ reciprocal of that entry).

Definition. A **special translation** is an FLT τ whose associated matrix may be written as $\begin{pmatrix} m & b \\ 0 & 1 \end{pmatrix}$ where $m > 0$. In this case $\tau(z) = mz + b$.

The **special inversion** is the FLT σ whose associated matrix is $\begin{pmatrix} 0 & -1 \\ 1 & 0 \end{pmatrix}$ so that $\sigma(z) = -1/z$.

A **special rotation** is an FLT ζ_θ whose associated matrix is $\begin{pmatrix} \cos\theta & -\sin\theta \\ \sin\theta & \cos\theta \end{pmatrix}$ for some $\theta \in \mathbb{R}$.

We will eventually see that a special translation is a translation (in the sense of Section 11.5) if $m \neq 1$ and is a parallel displacement if $m = 1$ and $b \neq 0$. σ will turn out to be the half-turn about $i \leftrightarrow (0,1)$ and ζ_θ will be a rotation about i. For convenience we may omit the word "special" in this section.

Proposition 11.8.4. *Every FLT is either a (special) translation or can be written as a composition $\tau_1 \sigma \tau_2$, where τ_1 and τ_2 are (special) translations and σ is the (special) inversion.*

PROOF. If the FLT φ corresponds to the matrix $\begin{pmatrix} a & b \\ c & d \end{pmatrix}$ we may assume that $c \geq 0$ by Proposition 11.8.3. If $c = 0$ we can assume that $d = 1$ by the same result and so φ is a translation (possibly the identity).

If $c > 0$ then by matrix multiplication we have

$$\begin{pmatrix} \dfrac{ad - bc}{c} & \dfrac{a}{c} \\ 0 & 1 \end{pmatrix} \begin{pmatrix} 0 & -1 \\ 1 & 0 \end{pmatrix} \begin{pmatrix} c & d \\ 0 & 1 \end{pmatrix} = \begin{pmatrix} a & b \\ c & d \end{pmatrix}.$$

Thus we may let τ_1 be the FLT determined by

$$\begin{pmatrix} \dfrac{ad - bc}{c} & \dfrac{a}{c} \\ 0 & 1 \end{pmatrix}$$

and τ_2 the FLT determined by $\begin{pmatrix} c & d \\ 0 & 1 \end{pmatrix}$. $\qquad\square$

This proposition is useful because it allows us to show that an FLT is a collineation by considering only translations and the special inversion. The first part of this program is a routine calculation and is left to Problem A4.

Proposition 11.8.5.

(i) *If τ is the (special) translation $\tau(z) = mz + b$ then*

$$\tau(_aL) = {}_{ma+b}L$$
$$\tau(_cL_r) = {}_{mc+b}L_{mr}$$

(ii) *If σ is the (special) inversion then $\sigma(_0L) = {_0}L$; if $a \neq 0$, $\sigma(_aL) = {_d}L_s$ with $d = -1/2a$, $s = |d|$; if $c \neq \pm r$, $\sigma(_cL_r) = {_d}L_s$ with $d = c/(r^2 - c^2)$, $s = |r/(r^2 - c^2)|$; $\sigma(_{\pm r}L_r) = {_{\pm a}}L$ with $a = -1/2r$.*

(iii) *A special translation or special inversion is a collineation.*

Proposition 11.8.6. *An FLT φ is a collineation.*

PROOF. If φ is a translation then the result follows from Proposition 11.8.5. If φ is not a translation then $\varphi = \tau_1 \sigma \tau_2$. By Proposition 11.8.5, τ_1, τ_2 and σ are collineations, so their composition is also a collineation.   □

Our next step is to show an FLT preserves distance. This will use the following lemma whose proof is left to Problem A5.

Lemma 11.8.7. *If $\varphi: \mathcal{S} \to \mathcal{S}$ is a collineation of a metric geometry then φ preserves distance if and only if for each line l there is a ruler $f : \varphi(l) \to \mathbb{R}$ such that $f \circ \varphi : l \to \mathbb{R}$ is a ruler for l.*

Proposition 11.8.8. *If φ is an FLT then φ preserves distance.*

PROOF.

Case 1. φ is a special translation. Suppose $\varphi(z) = mz + b$ and $l = {_a}L$ so that $\varphi(l) = {_{ma+b}}L$. The standard ruler for $\varphi(l)$ is $f(x, y) = \ln y$. Then

$$(f \circ \varphi)(x, y) = f(mx + b, my) = \ln(my) = \ln y + \ln m.$$

$f \circ \varphi$ is certainly a ruler for l.

Likewise if $l = {_c}L_r$ so that $\varphi(l) = {_{mc+b}}L_{mr}$ and if f is the standard ruler for $\varphi(l)$ then

$$(f \circ \varphi)(x, y) = f(mx + b, my) = \ln\left(\frac{mx + b - (mc + b) + mr}{my}\right) = \ln\left(\frac{x - c + r}{y}\right)$$

and $f \circ \varphi$ is a ruler for l.

Case 2. φ is the special inversion σ. The proof in this case proceeds along similar lines and considers four subcases (as in Proposition 11.8.5(ii)). In each case choose f to be the standard ruler for $\varphi(l)$. The calculations are messy but not difficult. We will consider only the case where $l = {_a}L$ with $a \neq 0$ and leave the rest to Problem B8.

$\varphi(_aL) = {_d}L_s$ where $d = -1/2a$ and $s = |d|$. The standard ruler for $_dL_s$ is $f(x, y) = \ln((x - d + s)/y)$. Hence, since

$$\varphi(x, y) = \left(\frac{-x}{x^2 + y^2}, \frac{y}{x^2 + y^2}\right),$$

we have

$$(f \circ \varphi)(x, y) = f\left(\frac{-x}{x^2 + y^2}, \frac{y}{x^2 + y^2}\right) = \ln\left(\frac{\dfrac{-x}{x^2 + y^2} - d + s}{\dfrac{y}{x^2 + y^2}}\right)$$

$$= \ln\left(\frac{\dfrac{-a}{a^2 + y^2} + \dfrac{1}{2a} + \left|\dfrac{1}{2a}\right|}{\dfrac{y}{a^2 + y^2}}\right) = \ln\left(\frac{-a + (a^2 + y^2)\left(\dfrac{1}{2a} + \left|\dfrac{1}{2a}\right|\right)}{y}\right)$$

$$= \begin{cases} \ln\left(\dfrac{y}{a}\right) = \ln y - \ln a & \text{if } a > 0 \\ \ln\left(\dfrac{-a}{y}\right) = \ln(-a) - \ln y & \text{if } a < 0. \end{cases}$$

In either case ($a > 0$, $a < 0$) $f \circ \varphi$ is a ruler for $_aL$.

Case 3. $\varphi = \tau_1 \sigma \tau_2$. By the first two cases and Lemma 11.8.7, translations and the inversion preserve distance. Hence so do compositions of them and in particular $\varphi = \tau_1 \sigma \tau_2$ preserves distance. $\qquad \square$

Now we want to show that an FLT preserves angle measure. In this case we will not use the factorization $\varphi = \tau_1 \sigma \tau_2$ of an arbitrary FLT but will exploit the homogeneity of $\mathbb{H}$. By this we mean that we will prove that FLT's preserve the measure of angles whose vertex is at $B = (0, 1) \leftrightarrow i$ and then translate, via a special translation, the general case to this case.

Recall that if $\angle ABC$ is an angle in $\mathscr{H}$ then

$$m_H(\angle ABC) = \cos^{-1}\left(\frac{\langle T_{BA}, T_{BC}\rangle}{\|T_{BA}\|\|T_{BC}\|}\right)$$

where the tangent vector T_{BA} of the ray $\overline{BA}$ is defined by

$$T_{BA} = \begin{cases} (0, y_A - y_B) & \text{if } \overline{AB} = _aL \\ (y_B, c - x_B) & \text{if } \overline{AB} = _cL, \text{ and } x_B < x_A \\ -(y_B, c - x_B) & \text{if } \overline{AB} = _cL, \text{ and } x_B > x_A. \end{cases}$$

Proposition 11.8.9. *If τ is a (special) translation then τ preserves angle measure.*

PROOF. First note that if $\tilde{T}_{BA} = T_{BA}/\|T_{BA}\|$ then

$$m_H(\angle ABC) = \cos^{-1}(\langle \tilde{T}_{BA}, \tilde{T}_{BC}\rangle).$$

We claim that if $\tau(z) = mz + b$ then $\tilde{T}_{BA} = \tilde{T}_{\tau B \tau A}$.

Case 1. $\overrightarrow{AB} = {}_aL$ with $A = (a, y_A)$ and $B = (a, y_B)$. Then

$$\tau A = (ma + b, my_A) \quad \text{and} \quad \tau B = (ma + b, my_B)$$

so that

$$T_{\tau B \tau A} = (0, my_A - my_B) \quad \text{and} \quad \tilde{T}_{\tau B \tau A} = \left(0, \frac{y_A - y_B}{|y_A - y_B|}\right) = \tilde{T}_{BA}$$

since $m > 0$.

Case 2. $\overrightarrow{AB} = {}_cL_r$. Here $\tau({}_cL_r) = {}_{mc+b}L_{mr}$ so that by a routine calculation

$$T_{\tau B \tau A} = \pm(my_B, mc + b - (mx_B + b)) = \pm(my_B, mc - mx_B)$$

where the sign is the same as the sign of

$$(mx_A + c) - (mx_B + c) = m(x_A - x_B).$$

Since $T_{BA} = \pm(y_B, c - x_B)$ where the sign is that of $x_A - x_B$, and since $m > 0$, we see that $\tilde{T}_{\tau B \tau A} = \tilde{T}_{BA}$. Thus

$$m_H(\angle \tau A \tau B \tau C) = \cos^{-1}(\langle \tilde{T}_{\tau B \tau A}, \tilde{T}_{\tau B \tau C} \rangle$$
$$= \cos^{-1}(\langle \tilde{T}_{BA}, \tilde{T}_{BC} \rangle = m_H(\angle ABC)$$

and τ preserves angle measurement. $\square$

Proposition 11.8.10. *If* $\varphi = \zeta_\theta$ *is a (special) rotation and* $B = (0, 1) \in \mathbb{H}$, *then* $m_H(\angle ABC) = m_H(\angle \varphi A \varphi B \varphi C)$.

PROOF. Since $B = (0, 1)$ we have

$$T_{BA} = \begin{cases} (0, y_A - 1) & \text{if } \overrightarrow{AB} = {}_0L \\ (1, c) & \text{if } \overrightarrow{AB} = {}_cL_r, \quad 0 < x_A \\ -(1, c) & \text{if } \overrightarrow{AB} = {}_cL_r, \quad 0 > x_A. \end{cases}$$

Since the vector $\tilde{T}_{BA}$ has length 1, it can be written as $\tilde{T}_{BA} = (\sin \omega, \cos \omega)$ where

$$\omega = 0 \quad \text{if } \overrightarrow{AB} = {}_0L, \quad y_A > 1$$
$$\omega = 180 \quad \text{if } \overrightarrow{AB} = {}_0L, \quad y_A < 1$$
$$0 < \omega < 180 \quad \text{and} \quad \cot \omega = c \quad \text{if } \overrightarrow{AB} = {}_cL_r, \quad 0 < x_A$$
$$180 < \omega < 360 \quad \text{and} \quad \cot \omega = c \quad \text{if } \overrightarrow{AB} = {}_cL_r, \quad 0 > x_A$$

Case 1. $\varphi = \zeta_0 = \text{id}$. The proof in this case is immediate.

Case 2. $\varphi = \zeta_{90} = \sigma$. If $A = (0, y) \in {}_0L = \overrightarrow{AB}$ then $\varphi A = (0, 1/y)$ and $\tilde{T}_{BA} = -\tilde{T}_{\varphi B \varphi A}$. If $A = (x, y) \in {}_cL_r = \overrightarrow{AB}$ then as noted above

$$T_{BA} = \pm(1, c) \quad \text{where the sign is that of } x.$$

Furthermore

$$\varphi A = \left(\frac{-x}{x^2 + y^2}, \frac{y}{x^2 + y^2} \right) \in {}_cL_r$$

so that

$$T_{\varphi B \varphi A} = \pm(1, c) \quad \text{where the sign is that of } x_{\varphi A} = \frac{-x}{x^2 + y^2}.$$

Hence $T_{\varphi B \varphi A} = -T_{BA}$ and $\tilde{T}_{\varphi B \varphi A} = -\tilde{T}_{BA}$.

Since in all cases $\tilde{T}_{\varphi B \varphi A} = -\tilde{T}_{BA}$ is clear that φ preserves the measure of angles with vertex B if $\varphi = \zeta_{90} = \sigma$.

Case 3. $\varphi = \zeta_\theta$ where $\cos \theta \sin \theta \neq 0$. By Problem A6, φ sends $_0L$ to $_cL_r$ where $c = \cot 2\theta$ and $r = |\csc 2\theta|$. The tangent to $_cL_r$ at $B = (0, 1)$ is $T = \pm(1, \cot 2\theta)$. Let $D = (0, 2)$ so that $T_{BD} = (0, 1)$. Now

$$\varphi(0, 2) = \left(\frac{3 \sin \theta \cos \theta}{4 \sin^2 \theta + \cos^2 \theta}, \frac{2}{4 \sin^2 \theta + \cos^2 \theta} \right).$$

When $(3 \sin \theta \cos \theta)/(4 \sin^2 \theta + \cos^2 \theta) > 0$ we need to take the plus sign for T:

$$T_{\varphi B \varphi D} = (1, \cot 2\theta).$$

In this case $\sin \theta \cos \theta > 0$ so that $\sin 2\theta > 0$ and

$$\tilde{T}_{\varphi B \varphi D} = (\sin 2\theta, \cos 2\theta).$$

When $\sin \theta \cos \theta < 0$ we have

$$T_{\varphi B \varphi D} = -(1, \cot 2\theta) \quad \text{and} \quad \tilde{T}_{\varphi B \varphi D} = (\sin 2\theta, \cos 2\theta).$$

Hence either way $\tilde{T}_{\varphi B \varphi D} = (\sin 2\theta, \cos 2\theta)$.

Now suppose that $\overrightarrow{BA}$ is a ray with $\tilde{T}_{BA} = (\sin 2\alpha, \cos 2\alpha)$ so that $\overrightarrow{BA}$ is the image of $\overrightarrow{BD}$ under ζ_α. Then

$$\zeta_\theta(\overrightarrow{BA}) = \zeta_\theta \zeta_\alpha(\overrightarrow{BD}) = \zeta_{\theta + \alpha}(\overrightarrow{BD})$$

which has tangent $(\sin 2(\theta + \alpha), \cos 2(\theta + \alpha))$. Hence the ray with $\tilde{T} = (\sin 2\alpha, \cos 2\alpha)$ is sent to the ray with tangent $(\sin 2(\theta + \alpha), \cos 2(\theta + \alpha))$. If $\tilde{T}_{BC} = (\sin 2\beta, \cos 2\beta)$ then

$$
\begin{aligned}
m(\angle \varphi A \varphi B \varphi C) &= \cos^{-1}(\langle(\sin 2(\theta + \alpha), \cos 2(\theta + \alpha)), \\
&\quad (\sin 2(\theta + \beta), \cos 2(\theta + \beta))\rangle) \\
&= \cos^{-1}(\sin 2(\theta + \alpha) \sin 2(\theta + \beta) + \cos 2(\theta + \alpha) \cos 2(\theta + \beta)) \\
&= \cos^{-1}(\cos 2(\theta + \alpha - \theta - \beta)) \\
&= \cos^{-1}(\cos 2(\alpha - \beta)) \\
&= \cos^{-1}(\langle(\sin 2\alpha, \cos 2\alpha), (\sin 2\beta, \cos 2\beta)\rangle) \\
&= m(\angle ABC).
\end{aligned}
$$

Hence $\varphi = \zeta_\theta$ preserves angle measure of angles with vertex at $B = (0, 1)$. $\square$

Proposition 11.8.11. *If l is a line in $\mathscr{H}$ and $B = (p,q) \in l$, then there is a translation τ that sends B to $(0,1)$ (and l to a line through $(0,1)$).*

PROOF. Let τ be the FLT associated with the matrix

$$\begin{pmatrix} 1/q & -p/q \\ 0 & 1 \end{pmatrix}.$$ □

Proposition 11.8.12. *If an FLT ψ fixes the point $B = (0,1)$ then ψ is a special rotation.*

PROOF. Since $(0,1) \leftrightarrow i$, $\psi(i) = i$. If ψ is associated with the matrix $\Psi = \begin{pmatrix} a & b \\ c & d \end{pmatrix}$ then $\psi(i) = i$ implies that

$$\frac{ai+b}{ci+d} = i \quad \text{or} \quad ai+b = -c + di.$$

Hence $a = d$ and $b = -c$ so that $\Psi = \begin{pmatrix} a & b \\ -b & a \end{pmatrix}$ for some a and b. By Proposition 11.8.3 we may assume that $\det(\Psi) = a^2 + b^2 = 1$ so that $a = \cos\theta$, $b = -\sin\theta$ for some θ and ψ is a special rotation. □

Proposition 11.8.13. *If φ is any FLT then φ preserves angle measure.*

PROOF. Let $\angle ABC$ be given and let τ be a translation sending B to $(0,1)$ while τ' is a translation sending φB to $(0,1)$. (τ and τ' exist by Proposition 11.8.11.) Then $\psi = \tau'\varphi\tau^{-1}$ is an FLT sending $(0,1)$ to $(0,1)$. Hence it is a special rotation by Proposition 11.8.12 and preserves the measure of angles with vertex at $(0,1)$. Therefore

$$m_H(\angle ABC) = m_H(\angle \tau A \tau B \tau C)$$
$$= m_H(\angle \psi\tau A \psi\tau B \psi\tau C)$$
$$= m_H(\angle \varphi A \varphi B \varphi C)$$

since $\varphi = \tau'^{-1}\psi\tau$. □

Recall that a mirror for the line l is a collineation which preserves distance and angle measure, fixes each point of l, and interchanges the half planes determined by l. We will show that $\mathscr{H}$ is a neutral geometry by finding a mirror for $_0L$ and transporting it to any other line by way of FLT's.

Definition. The **special reflection** ρ is the function $\rho: \mathbb{H} \to \mathbb{H}$ by $\rho(x,y) = (-x,y)$ (i.e., $\rho(z) = -\bar{z}$).

Proposition 11.8.14. *ρ is a mirror for $_0L$.*

PROOF.

Step 1. ρ is a bijection since $\rho^{-1} = \rho$. ρ sends ${}_aL$ to ${}_{-a}L$ and ${}_cL_r$ to ${}_{-c}L_r$ so that ρ is a collineation.

Step 2. We must show that ρ preserves distance. Let l be a line and f be the standard ruler for the line $\rho(l)$. An easy calculation shows that $f \circ \rho$ is the standard ruler for l if l is a type I line and is the negative of the standard ruler for l if l is a type II line. Hence by Lemma 11.8.7, ρ preserves distance.

Step 3. We must show that ρ preserves angle measure. If the ray $\overline{AB}$ has tangent $T_{AB} = (u, v)$ then $T_{\rho A \rho B} = (-u, v)$. Since $\langle (u, v), (r, s) \rangle = \langle (-u, v), (-r, s) \rangle$ and $\|(u, v)\| = \|(-u, v)\|$, the result follows from the definition of m_H.

Step 4. ρ certainly interchanges the half planes of ${}_0L$ and leaves each point of ${}_0L$ fixed. Hence ρ is a mirror. $\square$

The next result will be useful in transporting mirrors from one line to another. It is also important in its own right because it is the formal statement that $\mathscr{H}$ is homogeneous.

Proposition 11.8.15. *If l and m are lines in $\mathscr{H}$ then there is an FLT φ with $\varphi(l) = m$.*

PROOF. We will show there is an FLT φ_l sending ${}_0L$ to l. Then $\varphi_m \varphi_l^{-1}$ will be the desired FLT.

If $l = {}_aL$ define φ_l to be the translation $\varphi_l(z) = z + a$. If $l = {}_cL_r$ let $\varphi_l = \tau \zeta_{45}$ where τ is the translation $\tau(z) = rz + c$. The special rotation ζ_{45} sends ${}_0L$ to ${}_0L_1$ while τ sends ${}_0L_1$ to ${}_cL_r$. In both cases φ_l sends ${}_0L$ to l. $\square$

Proposition 11.8.16. $\mathscr{H} = \{\mathbb{H}, \mathscr{L}_H, d_H, m_H\}$ *satisfies the Mirror Axiom and is thus a neutral geometry.*

PROOF. Let l be any line in $\mathscr{H}$, let φ_l be an FLT that sends ${}_0L$ to l, and let ρ be the special reflection across ${}_0L$. Set

$$\rho_l = \varphi_l \rho \varphi_l^{-1}.$$

We claim that ρ_l is the desired mirror. Clearly ρ_l is a collineation, preserves distance, and preserves angle measure since φ_l, ρ, and φ_l^{-1} all do.

Suppose that $A \notin l$. Then $\varphi_l^{-1}A \notin {}_0L$ so that $\varphi_l^{-1}A$ and $\rho \varphi_A^{-1} A$ are on opposite sides of ${}_0L$.

Thus $\overline{\varphi_l^{-1}A \rho \varphi_l^{-1}A}$ intersects ${}_0L$ at a point B and $\overline{A \rho_l A} = \overline{\varphi_l \varphi_l^{-1}A \varphi_l \rho \varphi_l^{-1}A}$ intersects $l = \varphi_l({}_0L)$ at $\varphi_l B$. Hence A and $\rho_l A$ are on opposite sides of l and ρ_l interchanges the half planes of l.

If $A \in l$ then $\varphi_l^{-1}A \in {}_0L$ so that $\rho \varphi_l^{-1}A = \varphi_l^{-1}A$ and $\rho_l A = \varphi_l \rho \varphi_l^{-1}A = \varphi_l \varphi_l^{-1}A = A$. Hence ρ_l fixes each point of l. Thus ρ_l is a mirror. Because the Mirror Axiom is equivalent to SAS (Theorem 11.3.7) $\mathscr{H}$ is a neutral geometry. $\square$

PROBLEM SET 11.8

Part A.

1. Prove $\varphi(\mathsf{H}) \subset \mathsf{H}$ if φ is an FLT.

2. Prove Proposition 11.8.1.

3. Prove that $(0, 1)$ is a fixed point of ζ_θ for each θ.

4. Prove Proposition 11.8.5.

5. Prove Lemma 11.8.7.

6. Prove that $\rho_\theta({}_0L) = {}_cL_r$ where $c = \cot 2\theta$ and $r = |\csc 2\theta|$ if $\theta \neq 90n$ for any integer n.

7. Prove that the special reflection ρ is not an FLT. (Hence the collecton of FLT's is not the entire isometry group of $\mathscr{H}$.)

Part B.

8. Complete the proof of case 2 in Proposition 11.8.8.

9. Let $G = GL^+(2, \mathbb{R})$ be the group of 2×2 matrices with real entries and positive determinants. Let $N = \{A \in G | A = \lambda I \text{ for some } \lambda \neq 0\}$.
   a. Prove that N is a normal subgroup of G.
   b. Prove that the set of all FLT's forms a group F under composition.
   c. Prove that F is isomorphic to the quotient group G/N. (G/N is called the **special projective linear group** and is denoted $PSL(2, \mathbb{R})$.)

10. Show that you can replace G in Problem B9 with $SL(2, \mathbb{R})$ and obtain the same result. That is, let $N' = \{A \in SL(2, \mathbb{R}) | A = \lambda I \text{ for some } \lambda \neq 0\} = \{\pm I\}$ and prove F is isomorphic to $SL(2, \mathbb{R})/N'$.

11.9 The Isometry Groups of $\mathscr{E}$ and $\mathscr{H}$

In this final section we shall explicitly determine the isometry groups of our two basic models—the Euclidean Plane $\mathscr{E} = \{\mathbb{R}^2, \mathscr{L}_E, d_E, m_E\}$ and the Poincaré Plane $\mathscr{H} = \{\mathsf{H}, \mathscr{L}_H, d_H, m_H\}$. In the Euclidean Plane we will show that the isometry group is almost the product of two groups. In particular, every isometry can be written as a rotation about $(0, 0)$ (or possibly the identity) followed by either reflection across the y-axis (or the identity), and then followed by a translation (or the identity). In the Poincaré Plane every isometry will either be an FLT or an FLT preceded by the special reflection.

The keys to determining the two isometry groups will be the classification theorem (Theorem 11.6.10), the fixed point properties of the various types of isometries, and the following observation.

Theorem 11.9.1. *If A, B, C, D are points in a neutral geometry and $\overline{AB} \simeq$ $\overline{CD}$ then there are exactly two isometries which send A to C and B to D. One of these is even and the other odd.*

PROOF. By Problem A1 there is at least one such isometry φ. If $l = \overleftrightarrow{CD}$, then $\rho_l\varphi$ is a second (different) isometry sending A, B to C, D. Note one of φ, $\rho_l\varphi$ is even and the other is odd.

Now suppose ψ is any isometry sending A and B to C and D. Then $\psi\varphi^{-1}$ fixes both C and D. By Theorem 11.3.4, either $\psi\varphi^{-1} = $ identity (so that $\psi = \varphi$) or $\psi\varphi^{-1} = \rho_l$ (so that $\psi = \rho_l\varphi$). Hence φ and $\rho_l\varphi$ are the only isometries sending A and B to C and D. $\square$

We first consider the Euclidean Plane and identify all translations and all rotations about $(0,0)$.

Definition. If $A \in \mathbb{R}^2$, then the **Euclidean translation by A** is the function $\tau_A : \mathbb{R}^2 \to \mathbb{R}^2$ given by
$$\tau_A(P) = P + A.$$

Proposition 11.9.2. *An isometry $\varphi : \mathbb{R}^2 \to \mathbb{R}^2$ of the Euclidean Plane is a translation if and only if it is a Euclidean translation τ_A for some $A \neq (0,0)$.*

PROOF. Since
$$d_E(\tau_A P, \tau_A Q) = \|(P + A) - (Q + A)\| = \|P - Q\| = d_E(P, Q)$$
τ_A is an isometry for each $A \in \mathbb{R}^2$. If $A \neq (0,0)$ then τ_A has no fixed points and is either a glide or a translation. If $O = (0,0)$ then every line parallel to $\overleftrightarrow{OA}$ is invariant under τ_A so that τ_A cannot be a glide. (Glides have only one invariant line by Theorem 11.6.7.) Thus if $A \neq (0,0)$, τ_A is a translation.

Let φ be a translation along a line l. Let l' be the unique line through O parallel to l. l' is invariant so that $A = \varphi O \in l'$. We will show that $\varphi = \tau_A$.

Since $A \in l'$, $\varphi A \in l'$ also. Furthermore
$$d_E(O, A) = d_E(\varphi O, \varphi A) = d_E(A, \varphi A) = d_E(\varphi A, A).$$

Because O, A, and φA are collinear, either $O = \varphi A$ or O—A—φA. If $\varphi A = O$ then the translation φ^2 has a fixed point, which is impossible. Thus O—A—φA. Since $d_E(O, A) = d_E(A, \varphi A)$ this implies that $\varphi A = A + A$.

Thus the two translations φ and τ_A send O to A and A to $A + A$. Since both φ and τ_A are even (they are translations) Theorem 11.9.1 says they are equal: $\varphi = \tau_A$. $\square$

Definition. If θ is a real number, then the **special orthogonal transformation by** θ is the function $\varphi_\theta : \mathbb{R}^2 \to \mathbb{R}^2$ given by
$$\varphi_\theta(x, y) = (x \cos \theta - y \sin \theta, x \sin \theta + y \cos \theta).$$

If we write elements of $\mathbb{R}^2$ as column vectors then φ_θ can be given by matrix multiplication

$$\varphi_\theta \begin{pmatrix} x \\ y \end{pmatrix} = \begin{pmatrix} \cos\theta & -\sin\theta \\ \sin\theta & \cos\theta \end{pmatrix} \begin{pmatrix} x \\ y \end{pmatrix} \tag{9-1}$$

We shall say that the matrix on the right hand side of Equation (9-1) represents the special orthogonal transformation by θ, φ_θ. By Problem A2, φ_θ is an isometry for each θ. Note that $\varphi_\theta = \varphi_{\theta+360}$. (We are assuming the standard extension of $\cos(t)$ and $\sin(t)$ to all values of t, not just those between 0 and 180.) The next result tells us that φ_θ is a rotation.

Proposition 11.9.3. *An isometry* $\varphi: \mathbb{R}^2 \to \mathbb{R}^2$ *is a rotation about O if and only if* $\varphi = \varphi_\theta$ *for some* θ *which is not a multiple of* 360.

PROOF. A point (x, y) is fixed by φ_θ if and only if

$$\begin{cases} x\cos\theta - y\sin\theta = x \\ x\sin\theta + y\cos\theta = y \end{cases}$$

or

$$\begin{cases} x(\cos\theta - 1) - y\sin\theta = 0 \\ x\sin\theta + y(\cos\theta - 1) = 0 \end{cases} \tag{9-2}$$

Equations (9-2) have a unique solution (x, y) if and only if

$$0 \neq (\cos\theta - 1)^2 + \sin^2\theta = \cos^2\theta - 2\cos\theta + 1 + \sin^2\theta$$
$$= 2 - 2\cos\theta.$$

Therefore φ_θ has a unique fixed point if and only if $2\cos\theta \neq 2$, which occurs if and only if θ is not a multiple of 360. Thus φ_θ is a rotation if θ is not a multiple of 360.

On the other hand, let φ be a rotation about O and let $\varphi(1, 0) = (a, b)$. Note $a \neq 1$.

$$1 = d_E((0,0), (1,0)) = d_E(\varphi(0,0), \varphi(1,0)) = d_E((0,0), (a,b)).$$

Hence $a^2 + b^2 = 1$. Let θ be a number such that $a = \cos\theta$ and $b = \sin\theta$. Since $a \neq 1$, θ is not a multiple of 360. The two rotations φ and φ_θ are both even and agree on $(0,0)$ and $(1,0)$. Hence $\varphi = \varphi_\theta$ by Theorem 11.9.1. $\square$

A matrix representing a special orthogonal transformation (Equation (9-1)) has the property that its inverse is its transpose:

$$\begin{pmatrix} \cos\theta & -\sin\theta \\ \sin\theta & \cos\theta \end{pmatrix}^{-1} = \begin{pmatrix} \cos\theta & \sin\theta \\ -\sin\theta & \cos\theta \end{pmatrix}$$

Any matrix with the property that $A^{-1} = A^t$ is called an *orthogonal matrix*. If $A^{-1} = A^t$ then $I = AA^t$ so that

$$1 = \det(I) = \det(AA^t) = \det(A)\det(A^t) = \det(A)\det(A).$$

Thus det $A = \pm 1$. In Problem A3 you will show that the only orthogonal 2×2 matrices with determinant $+1$ are the special orthogonal matrices. We now consider those with determinant -1.

Let R be the matrix $R = \begin{pmatrix} -1 & 0 \\ 0 & 1 \end{pmatrix}$. Clearly $R = R^t = R^{-1}$ so that R is an orthogonal matrix. Let A be any orthogonal matrix. Then

$$(RA)^{-1} = A^{-1}R^{-1} = A^t R^t = (RA)^t$$

so that RA is also an orthogonal matrix. Since

$$\det(RA) = \det(R)\det(A) = -\det(A),$$

if $\det(A) = -1$ then RA is a special orthogonal matrix B. This means that $A = R^{-1}B = RB$. By Problem A4, R represents reflection across the y-axis and is thus an isometry. Hence $A = RB$ also represents an isometry.

Definition. An **orthogonal transformation** of $\mathbb{R}^2$ is any isometry that can be represented by an orthogonal matrix. The set of all 2×2 orthogonal transformations is denoted $O(2)$. The set of all 2×2 special orthogonal transformations is denoted $SO(2)$.

Proposition 11.9.4. *If φ is an isometry of $\mathbb{R}^2$ that fixes the origin O then φ is an orthogonal transformation. If φ is even then it can be uniquely represented by a special orthogonal matrix A. If φ is odd then it can be uniquely represented by RA, where A is a special orthogonal matrix and $R = \begin{pmatrix} -1 & 0 \\ 0 & 1 \end{pmatrix}$.*

PROOF. Suppose φ fixes O. If φ is even then φ is a rotation (or the identity). By Proposition 11.9.3, $\varphi = \varphi_\theta$ for some θ. ($\theta = 0$ gives the identity.) φ_θ is unique even though θ is not.

If φ is odd then $\rho\varphi$ is even where ρ is reflection across the y-axis. Hence $\rho\varphi$ is represented by a unique special orthogonal matrix A and φ is represented by RA. In particular, φ is an orthogonal transformation. $\qquad\square$

Proposition 11.9.5. *Any isometry φ of $\mathbb{R}^2$ can be uniquely written as $\varphi = \tau_A \psi$ when τ_A is a Euclidean translation and $\psi \in O(2)$.*

PROOF. Let $A = \varphi(O)$ so that $\psi = \tau_A^{-1}\varphi$ is an isometry which fixes O. By Proposition 11.9.4, $\psi \in O(2)$. Hence $\varphi = \tau_A \psi$.

Suppose that $\varphi = \tau_B \psi'$ also with $\psi' \in O(2)$. Then $B = \tau_B(O) = \tau_B \psi'(O) = \varphi(O) = A$ so that $B = A$. Hence $\psi' = \tau_B^{-1}\varphi = \tau_A^{-1}\varphi = \psi$. Thus φ can be uniquely written in the form $\varphi = \tau_A \psi$ with $\psi \in O(2)$. $\qquad\square$

Proposition 11.9.5 says that every isometry of $\mathbb{R}^2$ can be written as a product of a Euclidean translation and an orthogonal transformation. The collection of Euclidean translations forms a group isomorphic to $\mathbb{R}^2$: $\tau_A\tau_B = \tau_{A+B}$. By Problem A5 the set of all orthogonal transformations $O(2)$ also forms a group. You might think that $\mathscr{I}(\mathscr{E})$ is thus the direct product of the groups $\mathbb{R}^2$ and $O(2)$. (Recall that the direct product of two groups G and H is $G \times H = \{(g,h)|g \in G, h \in H\}$ with operation $(g_1,h_1)(g_2,h_2) = (g_1g_2, h_1h_2)$.) This is not true as the next result shows. The proof is left to Problem A6.

Proposition 11.9.6. *The group structure of the isometry group of* $\mathscr{E}$ *is given as follows.*

$$\mathscr{I}(\mathscr{E}) = \{\tau_A\psi|A \in \mathbb{R}^2, \psi \in O(2)\}$$

with

$$(\tau_A\psi)(\tau_B\varphi) = (\tau_A\tau_{\psi B})(\psi\varphi) = \tau_{A+\psi B}\psi\varphi \tag{9-3}$$

$$(\tau_A\psi)^{-1} = \tau_{-\psi^{-1}A}\psi^{-1} \tag{9-4}$$

Because of the way ψ affects τ_B and "twists" it to $\tau_{\psi B}$, the group structure on $\mathscr{I}(\mathscr{E})$ is often referred to as a *twisted product* of the groups $\mathbb{R}^2$ and $O(2)$.

We now turn our attention to determining $\mathscr{I}(\mathscr{H})$. The first step is to prove that the set of even isometries is precisely the set of FLT's. This requires an investigation of special translations.

Proposition 11.9.7. *A special translation* φ *of* $\mathscr{H}$ *is either a translation along a type I line or else a parallel displacement whose invariant asymptotic pencil is the pencil of type I lines.*

PROOF. Let φ be the special translation $\varphi(z) = mz + b$ where $m > 0$. We first show that φ has no fixed points. If $m = 1$ and $b \neq 0$ there are clearly no fixed points. Assume $m \neq 1$. If $\varphi(z) = z$ then $mz + b = z$ so that $z = -b/(m-1)$. Since b and m are real numbers, the imaginary part of z is zero and so $z \notin \mathbb{H}$. Thus no point of $\mathbb{H}$ is fixed by φ.

What lines are invariant under φ? By Proposition 11.8.5

$$\varphi(_cL_r) = {}_{mc+b}L_{mr} \quad \text{and} \quad \varphi(_aL) = {}_{ma+b}L.$$

Thus if a type II line is invariant under φ then $mc + b = c$ and $mr = r$. Hence $m = 1$ and $b = 0$ so that φ is the identity. If φ is not the identity, φ has no invariant type II lines. The type I line $_aL$ is invariant if and only if $a = ma + b$, or $a = -b/(m-1)$. Hence φ has a unique invariant type I line if $m \neq 1$. In this case φ must be either a translation or a glide. If $m \neq 1$ and $x > -b/(m-1)$ then

$$mx + b > m\left(\frac{-b}{m-1}\right) + b = \frac{-mb + mb - b}{m-1} = \frac{-b}{m-1}$$

so that $mx + b > -b/(m-1)$ also. This means that φ does not interchange the half planes of $_aL$. Thus φ cannot be a glide and so is a translation along $_aL$ for $a = -b/(m-1)$ if $m \neq 1$.

If $m = 1$ and $b \neq 0$ then φ has no invariant points or lines. Hence φ is a parallel displacement. Since φ leaves the pencil of type I lines invariant, the proof is complete. $\square$

Proposition 11.9.8. *The special inversion σ is a half-turn about $i \leftrightarrow (0, 1)$. In particular σ is a rotation.*

PROOF. $\sigma(z) = z$ implies $-1/z = z$ or $z^2 = -1$. Thus the only fixed point of σ in $\mathbb{H}$ is $i \leftrightarrow (0, 1)$. Therefore σ is a rotation about i. $\sigma^2(z) = \sigma(\sigma(z)) = -1/(-1/z) = z$ so that $\sigma^2 = $ identity. Thus σ is an involution and hence a half-turn. $\square$

Proposition 11.9.9. *φ is an even isometry of $\mathcal{H}$ if and only if φ is an FLT.*

PROOF. If φ is an FLT then φ can be written as a product of special translations and the special inversion by Proposition 11.8.4. Since these isometries are even any FLT is an even isometry.

Suppose φ is an even isometry and that $A \neq B$. Let $C = \varphi A$ and $D = \varphi B$ so that $\overline{AB} \simeq \overline{CD}$. By Problem A7 there is an FLT ψ with $\psi A = C$ and $\psi B = D$. Since both φ and ψ are even they are equal by Theorem 11.9.1. Hence φ is an FLT. $\square$

The next result is left to Problem A10.

Proposition 11.9.10. *The special reflection $\rho : \mathbb{H} \to \mathbb{H}$ by $\rho(z) = \bar{z}$ is a reflection. Any odd isometry ψ of $\mathcal{H}$ can be uniquely written as $\psi = \varphi\rho$ where φ is an FLT.*

If the FLT φ sends z to $(az + b)/(cz + d)$ then $\varphi\rho$ sends z to $(-a\bar{z} + b)/(-c\bar{z} + d)$. The matrix $\begin{pmatrix} a' & b' \\ c' & d' \end{pmatrix} = \begin{pmatrix} -a & b \\ -c & d \end{pmatrix}$ satisfies $a'd' - b'c' < 0$. Hence every odd isometry corresponds to a matrix $\begin{pmatrix} a & b \\ c & d \end{pmatrix}$ with negative determinant.

Definition. A function $\psi : \mathbb{H} \to \mathbb{H}$ by

$$\psi(z) = \frac{a\bar{z} + b}{c\bar{z} + d} \quad \text{with } ad - bc < 0$$

is called a **conjugate fractional linear transformation** (CFLT).

The next two results are left as exercises.

Proposition 11.9.11. *The composition of two CFLT's is an FLT. The composition of an FLT and a CFLT is a CFLT. The matrix associated with the product of two FLT's or CFLT's is the product of the corresponding matrices.*

Proposition 11.9.12. *The set of all 2×2 matrices with nonzero determinant forms a group which is denoted $GL(2, \mathbb{R})$. Every element of $GL(2, \mathbb{R})$ determines an isometry, either an FLT or a CFLT. Every isometry of $\mathscr{H}$ arises in this way. Two matrices Φ, $\Psi \in GL(2, \mathbb{R})$ determine the same isometry if and only if $\Psi = \lambda\Phi$ for some $\lambda \neq 0$.*

The final description of $\mathscr{I}(\mathscr{H})$ involves the ideas of normal subgroups and factor groups. (See Problem B9 of Section 11.8.) To understand the statement of the theorem and its proof requires a knowledge of homomorphism, kernels, and the Fundamental Theorem of Homomorphisms. (See Herstein [1990] or McCoy-Janusz [1987].)

Proposition 11.9.13. *The isometry group of $\mathscr{H}$ is isomorphic to $PGL(2, \mathbb{R}) = GL(2, \mathbb{R})/N$, when N is the normal subgroup of $GL(2, \mathbb{R})$*

$$N = \left\{ \begin{pmatrix} \lambda & 0 \\ 0 & \lambda \end{pmatrix} \middle| \lambda \neq 0 \right\}.$$

PROOF. The function f that takes a matrix $\Phi \in GL(2, \mathbb{R})$ to its associated FLT or CFLT is a homomorphism $(f(\Phi\Psi) = f(\Phi)f(\Psi))$ by Proposition 11.9.11. By Proposition 11.9.12, N is the set of matrices sent to the identity so that N is the kernel of f. Since $f : GL(2, \mathbb{R}) \to \mathscr{I}(\mathscr{H})$ is surjective (Proposition 11.9.12), the Fundamental Theorem of Homomorphisms says that $\mathscr{I}(\mathscr{H})$ is isomorphic to the factor group $GL(2, \mathbb{R})/N$. $\qquad\square$

The group $PGL(2, \mathbb{R}) = GL(2, \mathbb{R})/N$ is called the **real projective linear group.** It arises in the study of projective geometry as the set of transformations of a projective line. By Theorem 11.7.9 and the result (which we did not prove) that any two hyperbolic geometries with the same distance scale are isometric, the isometry group of any hyperbolic geometry (say the Poincaré Disk or the Klein Plane) is isomorphic to $PGL(2, \mathbb{R})$.

PROBLEM SET 11.9

Part A.

1. In a neutral geometry prove there is an isometry φ with $\varphi A = C$ and $\varphi B = D$ if and only if $\overline{AB} \simeq \overline{CD}$.

2. Prove that φ_θ, as given by Equation (9-1), is an isometry of $\mathscr{E}$.

3. Let A be a 2×2 orthogonal matrix. If $\det A = +1$ prove that $A = \begin{pmatrix} \cos\theta & -\sin\theta \\ \sin\theta & \cos\theta \end{pmatrix}$ for some θ.

4. Let R be the orthogonal matrix $\begin{pmatrix} -1 & 0 \\ 0 & 1 \end{pmatrix}$. Show that R corresponds to reflection across that line L_0.

5. Prove that $O(2)$ is a group and that $SO(2)$ is a subgroup of $O(2)$.

6. Prove Proposition 11.9.6.

7. In $\mathscr{H}$ if $\overline{AB} \simeq \overline{CD}$ prove there is an FLT sending A to C and B to D.

8. Let $B = (0, 0)$, $C = (5, 0)$, $D = (0, 5)$, $E = (3, 2)$, $F = (-1, 5)$ and $G = (6, 6)$ in $\mathscr{E}$. Find the unique isometry φ sending B, C, D to E, F, G respectively. Express your answer in the form $\tau_A \psi$ for some $A \in \mathbb{R}^2$ and $\psi \in O(2)$.

9. Let $A = (0, 1)$, $B = (0, 2)$, $C = (3, 6)$ and $D = (3, 3)$ in $\mathscr{H}$. Find the unique FLT which sends A to C and B to D.

10. Prove Proposition 11.9.10.

11. Prove Proposition 11.9.11.

12. Prove Proposition 11.9.12.

Bibliography

Bell, E., *Men of Mathematics*, New York: Simon and Schuster, 1937.

Birkhoff, G., "A Set of Postulates for Plane Geometry, Based on Scale and Protractor," *Annals of Math.*, 33 (1932), 329–345.

Bonola, R., *Non-Euclidean Geometry*, New York: Dover, 1955.

Borsuk, K., and W. Szmielew, *Foundations of Geometry*, Amsterdam: North-Holland, 1960.

Byrkit, D., "Taxicab Geometry—a Non-Euclidean Geometry of Lattice Points," *Mathematics Teacher*, (1971), 418–422.

Coexter, H., *Introduction to Geometry*, New York: Wiley, 1961.

Coolidge, J., *A History of Geometrical Methods*, Oxford, England: Oxford University Press, 1940.

Feldman, E., *Varieties of Visual Experience*, New York: Harry N. Abrams, 1981.

Greenberg, M., *Euclidean and Non-Euclidean Geometry*, 2nd Ed., San Francisco: Freeman, 1980.

Hall, T., *Carl Friedrich Gauss, A Biography*, tr. A. Froderberg, Cambridge, MA.: MIT Press, 1970.

Halstead, G., *Girolamo Saccheri's Euclides Vindicatus*, 2nd Ed., New York: Chelsea, 1986.

Heath, T., *Euclid, The Thirteen Books of the Elements*, 2nd Ed., New York: Dover, 1956.

Herstein, I., *Abstract Algebra*, 2nd Ed., New York: Macmillan, 1990.

Hilbert, D., *The Foundations of Geometry*, 2nd Ed., tr. E. Townsend, Chicago: Open Court, 1921.

————, *Grundlagen der Geometrie*, 8th Ed., Stuttgart: Teubner, 1956.

Hoffer, A., "Geometry is More than Proof," *Mathematics Teacher*, 1981, 11–18.

Kennedy, H., "The Origins of Modern Axiomatics: Pasch to Peano," *Amer. Math. Monthly*, 79 (1972), 133–136.

Kleven, D., "Morley's Theorem and a Converse," *Amer. Math. Monthly*, 85 (1978), 100–105.

Martin, G., *The Foundations of Geometry and the Non-Euclidean Plane*, New York: Intext, 1975; corrected printing, New York: Springer-Verlag, 1982.

McCall Collection of Modern Art, *Modern Americans—Op, Pop and the School of*

Color, New York: McCall Publishing, 1970.

——, *Picasso and the Cubists*, New York: McCall Publishing, 1970.

McCoy, N., and G. Janusz, *Introduction to Modern Algebra*, 4th Ed., Boston: Allyn and Bacon, 1987.

Millman, R., "Kleinian Transformation Geometry," *Amer. Math. Monthly*, **84** (1977), 338–349.

——, "The Upper Half Plane Model for Hyperbolic Geometry," *Amer. Math. Monthly*, **87** (1980), 48–53.

——, "Writing in a Non-Euclidean Geometry Course," in *Using Writing to Teach Mathematics*, A. Sterrett, Ed., Washington: MAA, 1990, 134–137.

——, and G. Parker, *Elements of Differential Geometry*, Englewood Cliffs: Prentice-Hall, 1977.

——, and R. Speranza, "Artist's View of Points and Lines to Motivate Geometric Concepts," *Mathematics Teacher*, **84** (1991), issue #2.

Moise, E., *Elementary Geometry from an Advanced Standpoint*, 3rd Ed., Reading, Mass.: Addison-Wesley, 1990.

Moulton, F. R., "A Simple Non-desarguesian Plane Geometry." Transactions of A.M.S., **3** (1902), 192–195.

Parker, G., "A Metric Model of Plane Euclidean Geometry," *Amer. Math. Monthly*, **87** (1980), 567–572.

Pasch, M., *Vorlesungen über neuere Geometrie*, Leipzig: Teubner, 1882.

Prenwitz, W., and M. Jordan, *Basic Concepts of Geometry*, New York: Blaisdell, 1965; reprinted, New York: Ardsley House, 1989.

Reyes, V., "Sur la théorème relative au carré de l'hypoténuse et le cinquième postulate d'Euclide," *Mathesis* (2) 7 (1897), 96.

Smith, D., *A Source Book in Mathematics*, New York: McGraw-Hill, 1929.

Spivak, M., *A Comprehensive Introduction to Differential Geometry*, vol. II, Boston: Publish or Perish, 1970.

Struik, D., *A Concise History of Mathematics* 3rd Ed., New York: Dover, 1967.

Index

PART II: The Models

The Cartesian Plane

The Euclidean Plane $\mathscr{E} = \{\mathbf{R}^2, \mathscr{L}_E, d_E, m_E\}$

PART III: The Terminology

Undergraduate Texts in Mathematics

(continued from page ii)

Halmos: Naive Set Theory.

Hämmerlin/Hoffmann: Numerical Mathematics.
Readings in Mathematics.

Harris/Hirst/Mossinghoff: Combinatorics and Graph Theory.

Hartshorne: Geometry: Euclid and Beyond.

Hijab: Introduction to Calculus and Classical Analysis.

Hilton/Holton/Pedersen: Mathematical Reflections: In a Room with Many Mirrors.

Iooss/Joseph: Elementary Stability and Bifurcation Theory. Second edition.

Isaac: The Pleasures of Probability.
Readings in Mathematics.

James: Topological and Uniform Spaces.

Jänich: Linear Algebra.

Jänich: Topology.

Jänich: Vector Analysis.

Kemeny/Snell: Finite Markov Chains.

Kinsey: Topology of Surfaces.

Klambauer: Aspects of Calculus.

Lang: A First Course in Calculus. Fifth edition.

Lang: Calculus of Several Variables. Third edition.

Lang: Introduction to Linear Algebra. Second edition.

Lang: Linear Algebra. Third edition.

Lang: Undergraduate Algebra. Second edition.

Lang: Undergraduate Analysis.

Lax/Burstein/Lax: Calculus with Applications and Computing. Volume 1.

LeCuyer: College Mathematics with APL.

Lidl/Pilz: Applied Abstract Algebra. Second edition.

Logan: Applied Partial Differential Equations.

Macki-Strauss: Introduction to Optimal Control Theory.

Malitz: Introduction to Mathematical Logic.

Marsden/Weinstein: Calculus I, II, III. Second edition.

Martin: The Foundations of Geometry and the Non-Euclidean Plane.

Martin: Geometric Constructions.

Martin: Transformation Geometry: An Introduction to Symmetry.

Millman/Parker: Geometry: A Metric Approach with Models. Second edition.

Moschovakis: Notes on Set Theory.

Owen: A First Course in the Mathematical Foundations of Thermodynamics.

Palka: An Introduction to Complex Function Theory.

Pedrick: A First Course in Analysis.

Peressini/Sullivan/Uhl: The Mathematics of Nonlinear Programming.

Prenowitz/Jantosciak: Join Geometries.

Priestley: Calculus: A Liberal Art. Second edition.

Protter/Morrey: A First Course in Real Analysis. Second edition.

Protter/Morrey: Intermediate Calculus. Second edition.

Roman: An Introduction to Coding and Information Theory.

Ross: Elementary Analysis: The Theory of Calculus.

Samuel: Projective Geometry.
Readings in Mathematics.

Scharlau/Opolka: From Fermat to Minkowski.

Schiff: The Laplace Transform: Theory and Applications.

Sethuraman: Rings, Fields, and Vector Spaces: An Approach to Geometric Constructability.

Sigler: Algebra.

Silverman/Tate: Rational Points on Elliptic Curves.

Simmonds: A Brief on Tensor Analysis. Second edition.

Undergraduate Texts in Mathematics

Singer: Geometry: Plane and Fancy.

Singer/Thorpe: Lecture Notes on Elementary Topology and Geometry.

Smith: Linear Algebra. Third edition.

Smith: Primer of Modern Analysis. Second edition.

Stanton/White: Constructive Combinatorics.

Stillwell: Elements of Algebra: Geometry, Numbers, Equations.

Stillwell: Mathematics and Its History.

Stillwell: Numbers and Geometry. *Readings in Mathematics.*

Strayer: Linear Programming and Its Applications.

Toth: Glimpses of Algebra and Geometry. *Readings in Mathematics.*

Troutman: Variational Calculus and Optimal Control. Second edition.

Valenza: Linear Algebra: An Introduction to Abstract Mathematics.

Whyburn/Duda: Dynamic Topology.

Wilson: Much Ado About Calculus.